MINE ACTION - THE RESEARCH EXPERIENCE OF THE ROYAL MILITARY ACADEMY OF BELGIUM

Authored by **Charles Beumier, Damien Closson, Vinciane Lacroix, Nada Milisavljevic** and **Yann Yvinec**

Mine Action - The Research Experience of the Royal Military Academy of Belgium
http://dx.doi.org/10.5772/60107
Authored by Charles Beumier, Damien Closson, Vinciane Lacroix, Nada Milisavljevic and Yann Yvinec

Contributors

Damien Closson, Vinciane Lacroix, Yann Yvinec, Olga Lopera, Nada Milisavljevic, Yvan Baudoin, Georgios Kechagiadakis

Published by InTech
Janeza Trdine 9, 51000 Rijeka, Croatia

Publishing Process Manager Ana Simcic
Technical Editor SPi Global
Cover InTech Design team

First published July, 2017
Printed in Croatia
Legal deposit, Croatia: National and University Library in Zagreb

Additional hard copies can be obtained from orders@intechopen.com

Mine Action - The Research Experience of the Royal Military Academy of Belgium , Authored by Charles Beumier, Damien Closson, Vinciane Lacroix, Nada Milisavljevic and Yann Yvinec
p. cm.
Print ISBN 978-953-51-3303-2
Online ISBN 978-953-51-3304-9

For EU product safety concerns:
IN TECH d.o.o., Prolaz Marije Krucifikse Kozulić 3, 51000 Rijeka, Croatia,
info@intechopen.com or visit our website at intechopen.com.

Meet the authors

Charles Beumier got an electrical engineering diploma from the Free University of Brussels (ULB) in 1987. He joined the Royal Military Academy as a scientist in 1988 and obtained a PhD degree in image processing from the Ecole Nationale Supérieure de Télécommunication (Paris, France) in 2003. He has extensive experience with low-level image processing, computer vision, and graphical user interface development. He has been mainly active in face recognition, 3D acquisition, humanitarian mine detection, remote sensing, and 3D analysis. In humanitarian demining, he participated in a European project concerning the prototype design of a mine detector combining a metal detector, a ground-penetrating radar, and a radiometer.

Damien Closson obtained a Master degree in cartography and remote sensing at the Catholic University of Louvain (1992) and a PhD in Sciences (Geography) at the University of Liege, Belgium (2005). From 2003 to 2016, he worked as a researcher at the Royal Military Academy. He developed his expertise in spaceborne radar remote sensing for the General Intelligence and Security Service and specific competencies in the teaching of radar to the military analysts as well. Since 2014, he is invited lecturer at the European Union Satellite Centre (SatCen) for the teaching of advanced radar remote sensing techniques (SBAS and PS). Author, Vinciane Lacroix has a "Licence en Physique" from ULB (Belgium), a master in Computer and System Engineering from RPI (USA), and a PhD from ENST (France). She worked at the Royal Military Academy of Belgium since 1994, involved in humanitarian demining projects, remote sensing for security applications, and image processing for map updating. She has been a part of the management team of the FP6 GMOSS network of excellence, a European think tank for Earth Observation Technologies in Support of Security Issues, and was the technical coordinator of the FP7 TIRAMISU Humanitarian Demining Toolbox project (2012–2016).

Nada Milisavljevic earned an electrical engineering degree from the University of Novi Sad (1992), a Master of Science degree in optoelectronics and laser technics from the University of Belgrade (1996), and a PhD degree in applied sciences from École Nationale Supérieure des Télécommunications, Paris (2001). From 1992 until 1997 she was a research and teaching assistant at the University of Novi Sad. She worked as a researcher at the Royal Military Academy from 1997 until 2016, involved in humanitarian demining, image processing and data fusion.

Yann Yvinec has been working on technology for mine action and security at the Royal Military Academy of Belgium since 1996. His experience includes the coordination of several international research projects covering various topics relevant to mine action, with a focus on providing tested and validated tools, and the detection of sea-mines for the Belgian Navy. He has chaired a workshop created by the European Committee for Standardization (CEN) to write guidelines in the domain of mine action. He has also participated in the International Test and Evaluation Program for Humanitarian Demining (ITEP).

Contents

Preface IX

Chapter 1 **Research at RMA in the Evolving Context of Mine Action 1**
Vinciane Lacroix, Yann Yvinec and Marc Acheroy

Chapter 2 **Positioning System for a Hand-Held Mine Detector 21**
Charles Beumier and Yann Yvinec

Chapter 3 **Ground-Penetrating Radar for Close-in Mine Detection 51**
Olga Lucia Lopera Tellez and Bart Scheers

Chapter 4 **Data Fusion for Close-Range Detection 83**
Nada Milisavljevic

Chapter 5 **Remote Sensing for Non-Technical Survey 101**
Yann Yvinec, Nada Milisavljevic, Charles Beumier, Idrissa Mahamadou, Dirk Borghys, Michal Shimoni and Vinciane Lacroix

Chapter 6 **InSAR Coherence and Intensity Changes Detection 155**
Damien Closson and Nada Milisavljevic

Chapter 7 **PARADIS: Information Management for Mine Action 177**
Vinciane Lacroix

Chapter 8 **Assessing the Performance of Personal Protective Equipment1 195**
Georgios Kechagiadakis and Marc Pirlot

Chapter 9 **Unmanned Ground and Aerial Robots Supporting Mine Action Activities 219**
Yvan Baudoin, Daniela Doroftei, Geert de Cubber, Jean-Claude Habumuremyi, Haris Balta and Ioan Doroftei

Chapter 10 **Testing and Evaluating Results of Research in Mine Action 233**
Yann Yvinec

Chapter 11 **The Special Case of Sea Mines 251**
Olga Lucia Lopera Tellez, Alexander Borghgraef and Eric Mersch

Preface

Every day, civilians in dozens of countries around the world are injured and killed by landmines and other lethal leftovers of conflict, years after hostilities of war have ended. In 2014 alone, an estimated 3.678 people were killed or maimed by mines and other explosive devices that have been left behind by armed forces. Approximately 80% of casualties in 2014 were civilians, 39% of whom were children. Unfortunately, most of the accidents caused by landmines happen in countries with limited medical assistance. The amputees not only must be fitted with a prosthesis to recover mobility but also they need help to overcome mental distress and regain dignity. The economic consequences are also serious, especially due to reconstruction efforts after an armed conflict. Further, large polluted areas of territory prevent the production of food.

Laos is the most heavily bombed nation in the world on a per capita basis. Between the years 1964 and 1973, the United States flew more than half a million bombing missions, delivering more than 2 million tons of explosive ordnance in Laos as part of the CIA *Rolling Thunder* operation, in an attempt to block the flow of North Vietnamese arms and troops through Laotian territory. The ordnance dropped includes more than 266 million submunitions released from cluster bombs. One-third of the bombs failed to explode on impact and have since claimed an average of 500 victims a year, mainly children and farmers forced to work on their contaminated fields to sustain their families. Despite tens of millions of dollars spent, only 1% of Laos territory has been cleared so far.

It soon became apparent that the only real solution to address the landmine crisis was a complete ban on antipersonnel mines and later on the cluster munitions. No technical changes or changes to the rules on their use could change the fact that antipersonnel mines or bomblets are inherently indiscriminate. Once planted, a mine will never be able to tell the difference between a military and civilian footstep, and a bomblet will continue to attract children and metal dealers.

In order to put an end to the suffering and casualties caused by antipersonnel mines, the *Convention on the Prohibition of the Use, Stockpiling, Production and Transfer of Anti-Personnel Mines and on their Destruction*, also called the Ottawa Convention or Mine Ban Treaty, was adopted in 1997. Further, in order to put an end for all time to the suffering and casualties caused by cluster munitions at the time of their use, when they fail to function as intended or when they are abandoned, the *Convention on the Use, Stockpiling, Production and Transfer of Cluster Munitions,* also known as the Oslo Convention, was adopted in 2008.

As a deminer using conventional tools clears an area of approximately 10 m2; every working day, humanitarian mine clearance operations must be understood and designed correctly, with the conviction that their main goal is to provide efficient aid to innocent people, who may be severely injured by this dreadful pollution. In this context, several questions arise such as: (i) Who is to fund the costs of development and trials? (ii) Who is to decide which technologies should be supported and which should be left on the shelf? The last question is made much harder by the considerable confidence that many inventors appear to have in their demining machine, even if it is of no interest to field-based deminers. Therefore, we need to consider the recommendations made during the ad hoc Standing Committee stating that (i) technologists should avoid building technologies based on assumed needs and work interactively with end users, (ii) appropriate technologies could save human lives and increase mine action efficiency, and (iii) nothing is more important than understanding the working environment. Nevertheless, even if an answer is given to the above mentioned questions, the challenge of bridging the gap between successful laboratory prototypes and commercially available equipment remains. This tends to make mine clearance a very complex problem for which no silver bullet solution exists. Therefore, the first priority of mine action consists in allowing affected people to reach a normal way of life according to local socioeconomic standards. Thus, the concept of a mine-free world has evolved softly toward the concept of a mine impact–free world. This particular vision tends to increase the development and the importance of tools to facilitate prioritization and contribute to a rational and efficient distribution of the resources available. In this context, different information management systems have been designed and are in use. A good example is the Information Management System for Mine Action, developed under the umbrella of the Geneva International Centre for Humanitarian Demining and in use in more than 80% of the mine action programs.

Belgium is far from having remained inactive regarding landmines and cluster munitions. Belgium is indeed the first country to have adopted, in March 1995, a national legislation forbidding antipersonnel landmines. Further, a legislative proposal forbidding cluster munitions was adopted by Belgium in May 2006. Belgium is also the first country to the world to have adopted such a law.

In 1996, the Royal Military Academy (RMA) opted for the implementation of mine action technological projects funded by the Belgian Ministry of Defense and the Belgian State Secretariat for Development Cooperation. It further decided to set up a close collaboration with other Belgian universities (not always an easy task!), which started organizing their own research activities on mine action. Later, other funding sources were granted to RMA by the Belgian Science Policy, the European Commission, and the European Committee for Standardization. At a more politico-administrative level, RMA participates in the States Parties Meetings of the Mine Ban Treaty, and in this context, Prof. Acheroy created an expert group on mine action technologies with representatives of different organizations and countries, aiming at informing the States Parties of the Mine Ban Treaty about the evolution of the mine action technologies. Further, Prof. Y. Baudoin created working groups dedicated to robotics in mine action within international organization.

This book only reports research activities achieved by RMA. The first chapter sheds new light on mine action evolution and perspectives. As far as close-in detection is concerned, metal detectors and ground-penetrating radars, among the most cited sensors, are presented in the second and the third chapters. The fourth chapter addresses the problem of data fusion in both close-in detection and area reduction. While Chapter 5 presents area reduction, dealing with the problem of minefield delineation and nontechnical survey by remote sensing, Chapter 6 addresses change detection methods for humanitarian demining land release. Chapter 7 describes PARADIS, an information management system for mine action, and

Chapter 8 deals with protective equipment issues against mine explosions. Afterward, Chapter 9 tells how robotics can be useful to organize laboratory and field trials but also how unmanned robots can contribute to mine action activities. Chapter 10 explains how to test and evaluate results of research in mine action. Finally, Chapter 11 develops a subject beyond the problem of landmines: the detection of sea mines.

Note that this book does not cover topics concerning mine risk education, victim assistance, socioeconomic reintegration, and stockpile destruction.

Prof. Dr. Ir. Marc Acheroy
Royal Military Academy (RMA),
Brussels, Belgium

Chapter 1

Research at RMA in the Evolving Context of Mine Action

Vinciane Lacroix, Yann Yvinec and Marc Acheroy

Additional information is available at the end of the chapter

http://dx.doi.org/10.5772/66710

The White Rabbit put on his spectacles.

"Where shall I begin, please your Majesty?" he asked.

"Begin at the beginning", the King said, very gravely, "and go on till you come to the end: then stop".

– Alice's Adventures in Wonderland, Lewis Carroll

Abstract

The purpose of this chapter is to put the research of the Royal Military Academy (RMA) in mine action in a historical perspective by providing some background information. The vocabulary used in mine action and the landmine contamination problem are first presented. Formalisation of close-in detection and of area reduction is then proposed. An overview of the research projects, the involved partners and the objectives as well as the list of PhDs performed at RMA are then provided. The chapter ends with an overview of the book and their link with the cited projects.

Keywords: area reduction, close-in detection, landmine contamination, mine action, Royal Military Academy

1. Introduction

RMA has been involved in research for humanitarian demining since 1996. Since these early days, the field of mine action has evolved thanks to the interactions of the different organisations involved: international and national government agencies, Non-Governmental Organisations (NGO), commercial companies, universities and research centres, affected population and individuals involved in that cause. Scope, processes, standards and vocabulary, all defined by the United Nations in the International Mine Action Standards (IMAS) [1], changed along this 20-year period; they are continuously reviewed. This is why having in mind the time frame in which RMA research activities took place is important. In particular, the focus in the early years was put on anti-personnel mines; the problem of Explosive Remnants of War (ERW) was then recognised as an important issue too.

In fact, when drafting a national law to organise mine action, each state may customise the IMAS to fit its environment and context, adapt them to the local threat and produces National Mine Action Standards so that, in practice, the mine action vocabulary does vary not only in time but also in space!

Nevertheless, at the time of this writing in 2016, the mine action community usually agrees on a basic vocabulary; a selection of this vocabulary and the correspondence with the former terminology is provided in the next subsection.

2. Vocabulary in mine action

The terminology provided in this section is defined in Ref. [2]. Terms in italics are defined in the current section.

Table 1 will help the reader linking former and today's terminology.

Current terminology	IMAS	"Old" terminology
General Mine Action Assessment (GMAA)	IMAS 8.10	
Non-Technical Survey (NTS)	IMAS 8.21	Level 1 Survey
Technical Survey	IMAS 8.22	Level 2
Clearance	IMAS 9.10	Level 3

Table 1. Current and old terminology in mine action.

2.1. Mine action

"Activities which aim to reduce the social, economic and environment impact of mines and other *Explosive Remnants of War (ERW)* including cluster munitions". The activities are grouped in five "pillars":

1. Mine/*ERW* risk education.
2. Demining (*survey*, mapping, marking and *clearance*); note that demining is only a part of mine action.
3. Victim assistance.
4. Stockpile destruction.

http://dx.doi.org/10.5772/66710

5. Advocacy against the use of anti-personnel mines and cluster munitions. Note that US do not recognise this as a pillar, since they promote (at least until now) the use of non-persistent mines.

In order to support these five pillars, the United Nations recognises the following necessary activities:

1. Assessment and planning.

2. Mobilisation and prioritisation of resources.

3. Information management.

4. Human skills development and management training.

5. Quality management.

6. Application of effective, appropriate and safe equipment.

2.2. Land release

Land release describes the process of applying all reasonable efforts to identify, define and remove all presence and suspicion of mines/*ERW* through *non-technical survey (NTS)*, *technical survey* (*TS*) and/or *clearance*. The process is detailed in Section 2.11 and in **Figure 1**.

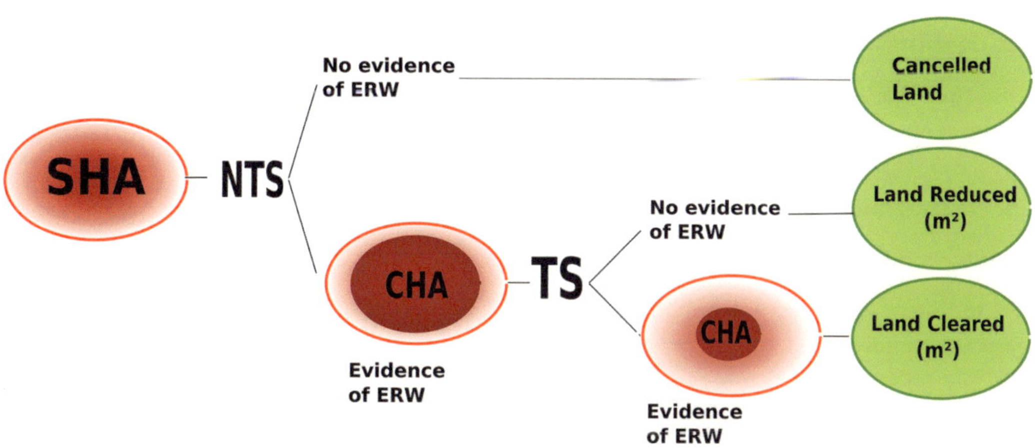

Figure 1. Land release process (adapted from Ref. [5]).

2.3. Non-technical survey (NTS)

Non-technical survey includes the collection and analysis of data, without the use of technical interventions, about the presence, type, distribution and surrounding environment of mine/*ERW* contamination, in order to define better where mine/ERW contamination is present, and where it is not, and to support land-release prioritisation and decision-making processes through the provision of evidence.

2.4. Technical survey (TS)

Technical survey includes the collection and analysis of data, using appropriate technical interventions, about the presence, type, distribution and surrounding environment of mine/*ERW* contamination, in order to define better where mine/*ERW* contamination is present and where it is not, and to support land-release prioritisation and decision-making processes through the provision of evidence.

2.5. Clearance

Clearance includes activity to insure the removal and/or the destruction of all mine and *ERW* hazards from a specified area to a specified depth.

2.6. Explosive remnants of war (ERW)

Explosive remnants of war refer to ordnance left behind after a conflict. They are made of unexploded ordnance (UXO) and abandoned ordnance (AXO).

2.7. Suspected hazardous area (SHA)

Suspected hazardous area (SHA) is an area where there is reasonable suspicion of mine/*ERW* contamination on the basis of indirect evidence of the presence of mines/*ERW*.

2.8. Confirmed hazardous Area (CHA)

Confirmed hazardous area (CHA) is an area where the presence of mine/*ERW* contamination has been confirmed on the basis of direct evidence of the presence of mines/*ERW*.

2.9. General mine action assessment (GMAA)

This is a continuous process, which aims

- to assess the scale and impact of the landmine problem on the country and individual communities;
- to investigate all reported and/or suspected areas of mine or UXO contamination, quantities and types of explosive hazards and
- to collect general information such as the security situation, terrain, soil characteristics, climate, routes, infrastructure and local support facilities, and to assist the planning of future mine action projects.

The information gathered during the general mine action assessment (GMAA) process should be sufficient to enable priorities to be established or updated and plans to be developed.

Impact survey, technical survey and post-clearance completion activities are still functional areas of the overall GMAA and new and existing programmes are still requested to make a GMAA if they follow IMAS.

2.10. What is a mine?

A mine is basically an explosive trap that is victim-activated. Its main danger comes from the fact that it might be activated by civilians many years, if not decades, after its main reasons to have been used in the first place have disappeared.

Mines have usually three main components: the explosive, the casing and the fuzing mechanism. Many sensors used to look for mines that actually detect the casing. Metal detectors detect the metallic parts of a mine that could be in the casing or in the fuzing mechanism. Dogs, on the other hand, detect the explosive.

Mines intended to target people are called anti-personnel mines and might include up to 250 g of explosive. Mines intended to destroy vehicles are called anti-tank or anti-vehicle mines and may include a much more important charge of explosive. Mines can also be designed to destroy ships. These are called sea-mines to distinguish them from the others, which are then called landmines.

Landmines are legally defined in two treaties. The Protocol II to the 1980 CCW Convention as amended on 3 May 1996 [3] defines a mine as "a munition placed under, on or near the ground or other surface area and designed to be exploded by the presence, proximity or contact of a person or vehicle".

The Convention on the Prohibition of the Use, Stockpiling, Production and Transfer of Anti-Personnel Mines and on their Destruction, 18 September 1997 [4], defines a mine in a very subtly different way: "a munition designed to be placed under, on or near the ground or other surface area and to be exploded by the presence, proximity or contact of a person or a vehicle". The International Mine Action Standards (IMAS) [1] use this definition.

More information on anti-personnel mines, fragmentation mines and blast mines is provided in Chapter 8.

2.11. Land-release process

The land-release process and the related vocabulary are defined in Section 2 and described in **Figure 1** (adapted from Ref. [5]). The process aims at bringing back to the population the land which should be if not "Mine-Free" as requested by the Treaty of Ottawa, at least "Mine-Safe" [6]. In this section, the specific terms are put in light by explaining the whole process.

The removal and destruction of landmines and ERW are relatively straightforward once the location has been clearly identified. The main problem is to precisely define this location. In a study on landmine impact survey, it has been shown that roughly half of the areas identified by the survey is not affected by mines [7], leading to a bad connotation to the term "Land Impact Survey", which has been abandoned since. Usually a mine action centre has only imprecise or incomplete information about these locations. This information may be derived from a database of accidents, or military records or is collected from surveys and discussions with the local population. This broad area with unclear boundaries is called "suspected hazardous area" (SHA). Clearing it would take much time, effort and resources. Indeed, clearing an area requires to prepare the land, that is, to cut the vegetation, divide the areas into smaller areas, use detection tools over each of them, excavate and dispose of suspected objects either on the site or in remote areas, which also means to transport them. If the area is not densely contaminated, the cost per square metre cleared can be huge and, paradoxically, the security of deminers less guaranteed: if they work for a long time in mine-free areas, they become less vigilant and accidents may be more frequent. Therefore, the whole SHA is usually not considered for clearance.

Non-technical survey (NTS) aims at providing a more precise border and prioritising the areas. During this process, where there is no evidence of ERW, parts of the SHA are excluded and labelled as "cancelled land". The areas that are still suspicious have more precise borders; they are called "confirmed hazardous areas" (CHA). These areas are then considered for technical survey (TS). The parts of CHA where there is no evidence of contamination are referred as "reduced land", while the remaining parts are cleared.

3. Mine contamination

3.1. Introduction

The magnitude of the mine problem is often expressed in terms of the number of mines that are still buried, the number of victims of mines, the area that is contaminated by mines and the area that is suspected to be contaminated by mines. Various estimates for these figures have been given throughout the years.

The early estimate of 110 million buried mines was shown to be wrong in 1998 [8] by its own authors. Moreover, the number of mines is no longer considered as a relevant indicator of the scope of the mine problem.

The estimate used in 1993–1997 was about 25,000 casualties every year.[1] The more recent estimate, from Landmine Monitor 2015 [9], is about 4000 every year, although this figure is expected to be underestimated. It includes casualties of mines, victim-activated-improvised explosive devices, cluster munition remnants and other explosive remnants of war.

To our knowledge, there is no global estimate currently available about the area suspected to be mined. Land Mine Monitor 2015 writes that it is not even possible to provide such an estimate. It lists nine countries and one other area where the mine contamination is expected to exceed 100 km^2: Afghanistan, Angola, Azerbaijan, Bosnia and Herzegovina, Cambodia, Chad, Croatia, Thailand, Turkey and Western Sahara.

We propose here a summary of the history of the estimation of the mine problem over the years. It shows the difficulty to estimate the scale of the problem and the evolution in both the estimation and the criteria to be estimated to describe the mine contamination.

---

[1]That is, more than 2000 every month, i.e., around one victim every 20 min.

3.2. Montreux symposium on anti-personnel mines, April 1993

The Montreux symposium on anti-personnel mines was organised by the International Committee of the Red Cross (ICRC) to collect information to coordinate action on mine action. In its report of the results of the symposium [10], ICRC estimated that 800 people worldwide die from mines each month. Moreover, it estimated that it would take 4300 years to clear Afghanistan of landmines.

Some people have reported that the Director of the United Nations Development programme stated at this symposium that the number of mines still in the ground was around 100,000 with another 100,000 mines in stockpiles.

3.3. US Department of state's first landmine report, September 1993

According to Hidden Killers 1994: The Global Landmine Crisis ([11] see below), "The Department of State's first landmine report, issued in September 1993, indicated that the total number of uncleared landmines in the world was between 65 and 110 million, scattered through 62 countries".

3.4. Hidden Killers 1994: The Global Landmine Crisis

Hidden Killers 1994: The Global Landmine Crisis [11] is a report to the U.S. Congress on the problem with uncleared landmines and the United States strategy for demining and landmine control prepared by the Office of International Security and Peacekeeping Operations and released on 27 January 1995.

The report states that the global landmine problem cannot be quantified with precision. It nevertheless estimates that there were between 80 and 110 millions of landmines. The report concludes: "The U.S. Government considers the figures in this report to be rough estimates at best".

As for the number of mine victims, the report states: "The three nations with the largest landmine problem are Afghanistan, Angola, and Cambodia. Collectively, they are besieged by an estimated 28 million mines and suffer 22,000 casualties every year (85 percent of the world's total)". This leads to a world's total of some 26,000 casualties every year or 2200 every month. This figure includes deaths and injuries.

3.5. United Nations Mine Clearance and Policy Unit, Department of Humanitarian Affairs, September 1997

In September 1997, the UN Department of Humanitarian Affairs gave the estimation of 110 millions of mines and added that 2000 people were killed or maimed by mine explosions every month. They also added that it would take more than 1100 years to clear the entire world of mines [12].

This website cites the following source: United Nations Mine Clearance and Policy Unit, Department of Humanitarian Affairs, September 1997. It is not clear if the UN cited estimations from Hidden Killers 1994: The Global Landmine Crisis or if they reached their conclusions independently.

3.6. Hidden Killers 1998

Hidden Killers: The Global landmine Crisis 1998 [9] was released in September 1998 by the U. S. Department of State, Bureau of Political-Military Affairs, Office of Humanitarian Demining Programs, Washington, DC.

New estimates of the number of uncleared landmines are in the range of 60–70 million.

3.7. Landmine Monitor Report 1999

Landmine Monitor[2] is an "initiative by the International Campaign to Ban Landmines (ICBL) to monitor implementation of and compliance with the 1997 Mine Ban Treaty, and more generally to assess the efforts of the international community to resolve the landmines crisis".

Landmine Monitor released its first report in 1999 [13]. About the estimates from 1993 to 1997, the report warns that previous estimates were repeated and reprinted so many times that they became 'reality'.

The report states that for mine action the sizes of areas affected by mines are a better indicator of the problem than the number of mines in the ground. But they still reprinted the estimates from the U.S. State Department's 1998 report, Hidden Killers.

3.8. Landmine Monitor Report 2015

Landmine Monitor Report 2015 [9] is currently the latest release of the Landmine Monitor Report. It considers that early estimates were inaccurate due to a lack of data.

The report gives the following information about the mine contamination:

It gives a list by expected magnitude of contamination. Nine countries and one other area have a mine contamination expected to exceed 100 km^2: Afghanistan, Angola, Azerbaijan, Bosnia and Herzegovina, Cambodia, Chad, Croatia, Thailand, Turkey and Western Sahara.

In 2014, more than 3500 casualties were recorded. The incidence rate is 10 casualties per day for 2014.

The report, however, warns: "In many states and areas, numerous casualties go unrecorded, especially in conflict settings; therefore, the true casualty figure is anticipated to be much higher".

---

[2]http://www.icbl.org/.

4. Formalising the problem

The next two subsections have been partly published in Refs. [14, 15] (©2007 EAGE Publications, The Netherlands); content from the latter is reproduced with kind permission from EAGE.

4.1. Formalising close-in detection

In the context of humanitarian demining, detection refers to the discovery of the presence of mines or unexploded remnants of war (ERW) [2]. Below, "close-in detection" refers to the discovery at short range of the presence of mines or unexploded ordnances. Typical examples of close-in detectors include manual prodders, metal detectors, ground-penetrating radars, and so on. Most of them emit an alarm close to a suspect object. In order to understand the incurred risks, let us consider the event A: "occurrence of an alarm" in a given position $x = (x,y)$. Clearly, this event depends on the detection system used and especially on the field reality. Further, let us define the following events M: "the presence of mines" and $\overline{M}$ "the absence of mine" which both are contradictory. Then, the alarm occurrence probability $p_A(x)$ in a location $x = (x,y)$ is given by

$$p_A(x) = p_M(x) \cdot p_{A/M}(x) + p_{\overline{M}}(x) \cdot p_{A/\overline{M}}(x) \tag{1}$$

which clearly explains the importance of the parameters describing the mine detection problem, that is, the detection probability and the false alarm probability:

- The *mine occurrence probability* in a given position x of a minefield, $p_M(x) = 1 - p_{\overline{M}}(x)$ expresses the local mine density of that minefield. It is impossible to control this parameter which depends on the field reality. Nevertheless, this parameter is very important for assessing the probability of an alarm in a given location x of the minefield.
- The *detection probability,* $p_{A/M}(x)$ is the probability of having an alarm in a given position x of a minefield for a given sensor system, if there is at least one mine in that position. This probability gives an indirect measure of the non-detection probability of that sensor system as well.
- The *probability of false alarm,* $p_{A/\overline{M}}(x)$ also called false alarm rate, is the probability of having an alarm, for a given sensor system, in a given location x if there is no mine in that location.

The two latter definitions are extremely important to understand the mine action problem and to design mine clearance systems. Note that a more precise expression for Eq. (1) should in particular take into account additional factors such as a sensor sensitivity area around the mine, but this is out of the scope of this summary.

The detection probability $p_{A/M}(x)$ should be as close as possible to one. Evaluating the detection probability also amounts to evaluating the risk $p_{\overline{A}/M}(x)$ of the occurrence of a mine that is not detected, since $p_{\overline{A}/M}(x) = 1 - p_{A/M}(x)$. This risk is linked to *human safety* and is therefore of the utmost importance. It is an absolute requirement that a mine clearance system should decrease the probability of such a risk to the lowest upper bound possible.

As a matter of fact, it is imperative to evaluate the detection probability when optimising the performances of a system. However, the detection probability $p_{A/M}(x)$, as defined above, assumes that a mine is present in the considered position x. During organised trials, the position of the mines is well known, so the condition of the presence of a mine in the given position x, where the performances of a system must be evaluated, is always fulfilled. This is of particular importance because it justifies the organisation of trials and the construction of models, to be validated by trials, in order to evaluate the detection probabilities. It also justified the implementation of test and evaluation organisations for mine action technologies such as the International Test & Evaluation Programme (ITEP).[3] A demining method, which minimises the false alarm rate $p_{A/\overline{M}}(x)$ leads to an acceleration of the demining operations. All other things being equal, a faster demining results indirectly in spending less money and reducing the risk that human being entered a mined area.

Therefore, any demining operation enhancement must result in the highest possible detection probability $p_{A/M}(x)$ (close to one) and in the smallest possible false alarm rate $p_{A/\overline{M}}(x)$, while keeping the price reasonably low. Unfortunately, increasing the detection probability generally results in increasing the false alarm rate. The most efficient way for increasing the detection probability while minimising the false alarm rate consists in using several complementary sensors in parallel and in fusing the information collected by these sensors.

Let us first consider the case of the detection probability. If the event $\overline{A}$: "absence of an alarm", means the event $\overline{A}_1\overline{A}_2\overline{A}_3...\overline{A}_N$: *"absence of an alarm for all sensors and associated signal processing"*. Then, the overall detection probability is described by

$$p_{A/M}(x) = 1 - p_{\overline{A}/M}(x) = 1 - p_{\overline{A}_1/M}(x) \cdot p_{\overline{A}_2/M\overline{A}_1}(x) \cdot p_{\overline{A}_3/M\overline{A}_1\overline{A}_2}(x).... \quad (2)$$

The latter expression shows that the detection probability increases with the number of sensors. This justifies the use of several sensors to increase the detection probability. Moreover, if the events $A_1, A_2, ..., A_N$ are statistically independent,

$$p_{A/M}(x) = 1 - p_{\overline{A}/M}(x) = 1 - p_{\overline{A}_1/M}(x) \cdot p_{\overline{A}_2/M}(x) \cdot p_{\overline{A}_3/M}(x).... \quad (3)$$

In this case, maximising the overall detection probability $p_{A/M}(x)$ of a set of sensors, acting independently, clearly amounts to the same as maximising the detection capabilities of each sensor individually (e.g. $p_{A/M}(x) = 1 - p_{\overline{A}_k/M}(x)$ for the kth sensor), by optimising separately the design of each sensor and its associated signal processing.

---

[3]ITEP's mission was to develop standards, coordinate and perform tests of materials and methods, and spread information about the results to all other interested parties.

Let us analyse the probability of false alarm in more detail. Current sensor systems (e.g. metal detectors) are affected by their high false alarm rate ($p_{A/\overline{M}}(x)$), that is, the probability of having an alarm if there is no mine (e.g. metallic garbage detected with a metal detector), making demining operations slow, tedious and demanding many resources. Unfortunately, the use of a set of different sensors without associated data fusion techniques will dramatically decrease efficiency. Indeed, the probability of false alarm can be expressed as

$$p_{A/\overline{M}}(x) = 1 - p_{\overline{A}/\overline{M}}(x) = 1 - p_{\overline{A}_1/\overline{M}}(x) \cdot p_{\overline{A}_2/\overline{M}\overline{A}_1}(x) \cdot p_{\overline{A}_3/\overline{M}\overline{A}_1\overline{A}_2}(x).... \quad (4)$$

The latter expression shows that the probability of false alarm increases if the number of sensors increases. Further, if the events $A_1, A_2, \ldots, A_N$ are independent, we can write

$$p_{A/\overline{M}}(x) = 1 - p_{\overline{A}/\overline{M}}(x) = 1 - p_{\overline{A}_1/\overline{M}}(x) \cdot p_{\overline{A}_2/\overline{M}}(x) \cdot p_{\overline{A}_3/\overline{M}}(x).... \quad (5)$$

In this case, minimising the overall false alarm rate $p_{A/M}(x)$ of a set of sensors, acting independently, clearly amounts to the same as minimising the false alarm rate of each sensor individually (e.g. $p_{A_k/\overline{M}}(x) = 1 - p_{\overline{A}_k/\overline{M}}(x)$ for the kth sensor), by optimising, separately, the design of each sensor and its associated signal processing.

Since most sensors used for mine detection are, actually, anomaly detectors (detector of metal, of the difference in dielectrical permittivity, etc.), it is very difficult to estimate the risk of false alarm as it is especially difficult to define, in a general way, what is not a mine. The problem of estimating the false alarm risk becomes even more complex when data coming from different sensors are fused, which favours the manual or automatic cancellation of false alarms. In this context, it should be particularly inappropriate that a demining system, whatever it may be, makes a decision instead of the final user whose own safety is involved. Therefore, a well-designed system should help the user in the decision-making, not by replacing him, but by implementing efficient data fusion methods. To make the process more feasible and accurate, methods, which are able to deal with uncertainty by making proposals including the doubt to the user, as a Dempster-Shafer framework does, prove to be promising (see Chapter 4).

4.2. Formalising area reduction

Area reduction is an important challenge consisting of finding where the mines are not. Mine-suspected area reduction, recognised by the mine action community as a mine action activity at least as crucial as close-in detection, enables to reduce mine clearance time and resources.

Mine-suspected area reduction means finding the set of positions x for which $p_M(x)$ equals zero. Under this condition, Eq. (1) yields

$$p_A(x) = p_{A/\overline{M}}(x) \quad (6)$$

and mine-suspected area reduction with classic tools (e.g. metal detectors) is affected by the high false alarm rate ($p_{A/\overline{M}}(x)$) of current sensors, making the corresponding operations slow, tedious and resource demanding. Further, long-term empirical data from the Croatian Mine

Action Centre (CROMAC) show that around 10–15% of the suspected area in Croatia is actually mined. The minefield records alone, beyond the fact that they are not always reliable and complete, do not contain enough information for the proper allocation of limited mine clearance resources to really mined areas. Decision-makers need additional information.

This means that a broader approach is needed, which has to include a priori knowledge. Indeed, if no a priori knowledge is available about context as conflict history, strategies and tactics of the parties, communication networks, terrain configuration, power lines, land use, and so on, the a priori probability of having a mine in a given location is distributed uniformly and the only method to clear mines is the classic close-in detection $p_M(x)$. If on the contrary a priori information is available on the distribution of $p_M(x)$, especially by deducing from the context where the mines are certainly not (e.g. agricultural fields in use) and where the mines are possibly present (e.g. along the confrontation lines, in the vicinity of trenches, on top of hills that are possible artillery positions, etc.), it makes sense to build a risk map ($p_{M/\text{context}}(x)$) of the affected areas. This assumes to define a list of indicators of mine presence ($p_{M/\text{context}}(x)$ is not negligible) and absence ($p_{M/\text{context}}(x)$ is close to zero) as well as a list of tools and methods to detect them. One of the most appropriate methods to build a risk map is associating airborne and satellite data, with context and ground truth data collected during field campaigns (see Chapters 5, 6 and 9). Analysing the collected data with modern remote-sensing tools, such as land use classification, anomaly detection and change detection, considered as experts, and fusing the "opinions" of those experts enable to produce the so-called risk or danger maps. The main advantage of airborne methods rests in the possibility to reduce areas located in regions that cannot be accessed without very costly safe lanes and full safety procedures that are mandatory when entering the minefields. Further, the assessment of areas for reduction and assessment of spatial danger distribution can be performed in short time over large areas.

In the context of mine action, a frequently asked question is the possibility of directly detecting mines using high-resolution airborne sensors. As part of their policy to assist developing countries, the European Commission (the former DG-VIII) and the participating member states have funded in 1997 a "Pilot project on airborne minefield detection in Mozambique" which has clearly shown that it is impossible to find reliably anti-personnel landmines even with a very high-resolution (order of magnitude of a few millimetres) airborne sensors neither using objective signal-processing tools nor using subjective photo-interpretation.

In summary, instead of addressing the mine detection it is suggested to address "Indicators of Mine Presence" and "Indicators of Mine Absence" (see Chapters 5 and 6).

4.3. Formalising current dual sensors

There are now several dual sensors combining metal detectors and ground-penetrating radars: AN/PSS-14 (formerly known as HSTAMIDS), VMR3 and VMR3G (MINEHOUD) and ALIS. Although these systems could theoretically be used to detect plastic mines, the ground-penetrating radar is usually used to confirm, or not, an alarm from the metal detector.

Let $p_{A^{MD}/\overline{M}}(x)$ be the probability that the metal detector gives an alarm at location x if there is no mine and $p_{A^{GPR}/\overline{M}}(x)$ the probability that the ground-penetrating radar gives an alarm at location x if there is no mine.

The probability that a dual sensor as those cited above gives a false alarm at location is therefore

$$p_{A/\overline{M}}(x) = p_{A^{MD}/\overline{M}}(x) \cdot p_{A^{GPR}/\overline{M}}(x) \quad (7)$$

We can see that if the probability of false alarm of the ground-penetrating radar $p_{A^{GPR}/\overline{M}}(x)$ is low then the probability of false alarm of the dual sensor is reduced compared to the usually high probability of false alarm of a metal detector. This is the main advantage of these dual sensors.

This advantage, however, does not come without a drawback. The probability that the dual sensor gives an alarm at location x when there is a mine is

$$p_{A/M}(x) p_{A^{MD}/M}(x) \cdot p_{A^{GPR}/M}(x) \quad (8)$$

We can see that in this condition, the detection rate of the dual sensor cannot be higher that the detection of the metal detector alone. Therefore, the ground-penetrating radar used as a confirmation sensor can only reduce the probability of detection. It is therefore paramount that the radar does not miss any mine detected by the metal detector.

5. Overview of RMA research in mine action

RMA research in mine action has been mainly focused on the "Demining" pillar except for the TIRAMISU project (see www.fp7-tiramisu.eu/and Chapters 5, 6, 8 and 9) coordinated by RMA, where the aim was to develop tools for pillars 1, 2 and 4.

In the late 1990s, the Royal Military Academy coordinated a Belgian research programme, called HUDEM for Humanitarian Demining, where many Belgian universities worked together to improve the detection of landmines by developing many technologies. This led to a project called HOPE co-funded by the European Commission which developed a prototype for a multi-sensor handheld mine detector. The research on detection continued in the late 2000s with the Belgian programme BEMAT for Belgian Mine Action Technologies.

In parallel, work was done to help non-technical survey by using remote sensing. It stated in the late 1990s by a project co-funded by the European Commission to detect mines by airborne surveys. In the 2000s, the focus moved to detect indirect indicators of the presence or absence of mines by remote sensing, which was implemented in the SMART project co-funded by the European Commission.

The investigation of the use of information management system could be carried out with the programme PARADIS, which is described in more detail subsequently.

The Robotics Department of the Royal Military Academy also investigated the use of robots for demining in the CLAWAR and VIEWFINDER programmes, both co-funded by the European Commission.

The Royal Military Academy was also the representative of Belgium to the International *Test and Evaluation* Program for Humanitarian Demining (*ITEP*) during the 10 years of its duration, from 2000 to 2010.

All these research activities later gave birth to the TIRAMISU programme, co-funded by the European Commission, which, together with the SPRINT programme, deepened and extended the experience gather in the last 15 years or so.

During these years, 10 phD thesis have been produced by researchers working at the Royal Military Academy in partnership with European schools or universities, and in particular, the first thesis delivered by the Royal Military Academy in 2006.

Below, the research projects are listed; the funding source and the time frame are put in parenthesis; the other collaborating partners (if any), and the objectives of the research and the references to the chapters are also provided. The list of PhD thesis is also provided afterwards.

5.1. Research projects

Airborne minefield detecion: pilot project

(EC, Goverments of B, G, P, UK, and ITC: 1997–1999) see Chapter 5

Partners: ITC (N), CAE Aviation (L), EOS (UK), VUB(B), Eurosense (B), Geograf (P), SSC (S), ZEO (G), Aerodata (B), Aerosensing (G), IGI (G), Recon Optical (UK), CND (Moz), NPA (N).

Objective: Detecting minefield by airborne survey

BEMAT (MoD: 2006–2009) see Chapters 3 and 4

Belgian Mine Action Technology

Objective: Develop technology for mine action

CLAWAR (EC: 1998–2002, 2002–2005) see Chapter 9

Thematic network on Climbing and Walking Robots including the support technologies for mobile robotic machines

Partners: Consejo Superior de Investigaciones Científicas (ES), H Enwesa Operaciones SA—Madrid (ES), Forschungszentrum Informatik an der Universität Karlsruhe (DE), Robotsysteme Yberle GmbH. (DE), Universität Kaiserslautern (DE), Forward Industries LTD (GB), Gravatom Engineering Systems Ltd. (GB), Secretary of State for Defence—Ministry of Defence (GB), Portsmouth Technology Consultants Ltd. (GB), Transtech Parallel Systems Limited (GB), University of Salford (GB), Helsinki University of Technology (FI) Inox Pneumatic AS (DK), Instituto de Soldadura e Qualidade (PT) Instituto Nacional de Engenharia e Tecnologia Industrial (PT), Kentree Ltd.—Cork (IE) Risø National Laboratory (DK) SGS—Thomson Microelectronics s.r.l. (IT), Università degli Studi di Catania (IT).

Objective: Coordinate WG Robotics in Humanitarian Demining

ITEP (MoD: 2000–2010) see Chapter 1, Section 1.5, and Chapter 10

http://dx.doi.org/10.5772/66710

International Test and Evaluation Program for Humanitarian Demining

Partners: EU (JRC), MoD CAN (CCMAT), NLD (MoD), GER (MoD), SWE (SWEDEC), UK (QinetiQ), US (HDP)

Objective: Develop standards, coordinate and perform tests of materials and methods and disseminating results to interested parties.

HOPE (EU: 1999–2001) see Chapter 2

Handheld multi-sensor operational demining system

Partners: Vallon GmbH, BATS, EC DG-INFSO, ISL, MAG, NPA, ONERA CERT, POLIMI – DEI, RST AG.

Objective: Develop and build an efficient handheld demining tool.

HUDEM (MoD: 1996–2001) see Chapters 3, 4 and 9

Humanitarian Demining

Partners: Belgian Universities (KU Leuven in collaboration with Kings College London, UGent, UCL, ULB, and VUB) and APOPO.

Objective: study tools that could help landmine detection (sensors, robots, rats, etc.)

MRN Studies (MoD: 2005–2015) see Chapter 11

MRN06 (2005–2011), MRN13 (2011–2015) and EDA-DMD (2011–2014), research was conducted under two CMRE NATO-STO Joint Research Programs, MCM-D&C (Mine Countermeasures, Detection and Classification) and MIAMS (Machine Intelligence for Autonomous Mine Search).

Partners: Belgian Navy, EDA (EU), DGA (FR), Thales (FR), DCNS.

Objective: the automatic detection of small threats on the sea surface, such as drifting mines, swimmers, debris and small boats.

MRN10 (2010–2013) and MRN16 (2014–2017)

Under the UMS Project BURMIN

Partners: TNO (NL), CTM (PL), DGA/Technology Naval (FR), ECA (FR), Thales (FR), WTD-71 (GE), IPHT (GE), Atlas Elektronik (GE), as well as the Ministries of Defense of Belgium, France, Germany, Poland and The Netherlands.

Objectives: to eliminate technological gaps in the field of bottom/buried mine detection and neutralisation, and to establish common standards for the future European Unmanned Maritime Systems.

MRN09 (2008–2014) and MRN17 (2015–2018) are conducted under two CMRE NATO-STO Joint Research Programmes, MCM-D&C (Mine Countermeasures, Detection and Classification) and MIAMS (Machine Intelligence for Autonomous Mine Search).

Objective: to develop automatic target recognition algorithms and environmentally adaptive sensing for mine search.

MSMS (JRC: 1999–2003) see Chapters 2–5 and 10

Joint Multi-Sensor Mine Signature Measurement Campaign

Partners: EU/JRC, DLR, FGAN,Kayser-Threde GmbH

Objective: execute a campaign for collecting data of buried landmines with multiple sensors.

PARADIS (MoD and Belspo: 1999–2001; MoD: 2002–2006; APOPO: 2007) see Chapter 7.

A Prototype for Assisting Rational Activities in Humanitarian Demining using Images from Satelites.

Partners: SEDEE-DOVO, IGEAT (ULB)

Objective: Plan and report demining campaigns

SMART (EU: 2001–2004) see Chapter 5

Space and airborne Mine Area Reduction Tools

Partners: DLR, CROMAC, IGEAT (ULB), TRASYS, ENST, Zeppelin, RST, IXL.

Objective: finding parts of mine-suspected areas that are not mined

SPRINT (Belspo: 2013) see Chapter 6

Spaceborne Radar Interferometric Techniques for Humanitarian Demining Land Release

Partner: CROMAC

Objective: Differentiate human-induced effects from natural ones based on the analysis of time series SAR satellite data

TIRAMISU (EU fp7: 2011–2015) see Chapters 5, 8 and 9.

Toolbox Implementation for Removal of Anti-Personnel Mines, Submunitions and UXO

Partners: DLR, ISR, CSIC, Univ. of Catania, Univ. of Genova, PLUS, USTAN, ULB-IGEAT, FGUNIZ, IMM, CTRO, WITI, SPINATOR, PROTIME, SPACETEC, SatCen, DIALOGIS, Brimatec, Vallon, IDS, Cen, NOVELTIS, Pierre, Snail Aid, Tohoku University.

Objective: Build a toolbox for Mine Action

VIEW-FINDER (EU, 2006–2009) see Chapter 9

Partners: Sheffield Hallam University,Eidikos Logariasmos Erevnon Dimokriteiou Panepistimiou Thrakis, Galileo Avionica—S.P.A., Intelligence For Environment & Security Srl—Ies Solutions Srl, Przemyslowy Instytut Automatyki I Pomiarow, South Yorkshire Fire And Rescue Service, Space Applications Services, Universita Degli Studi Di Roma "La Sapienza".

Objective: Robotics assistance for dangerous tasks

http://dx.doi.org/10.5772/66710

5.2. PhD thesis in mine action

Pascal Druyts: Analysis of Environmental Effects on Electromagnetic Induction Sensors, PhD dissertation, Université catholique de Louvain & Royal Military Academy, 2011.

Diederik Borghys: Interpretation and Registration of High-Resolution Polarimetric SAR Images, PhD Dissertation, Ecole Nationale Supérieure des Télécommunications, 2002

Damien Closson: Exploiting SAR coherence in time, PhD dissertation, Université de Liège, 2005.

Eric Colon: COROBA, an open framework for distributed robot control and sensor networks. PhD dissertation, Vrije Universiteit Brussel & Royal Military Academy, 2006.

Jean-Claude Habumuremyi: Adaptative neuro-fuzzy control for a walking robot with six pantograph-based legs, PhD dissertation, Free University of Brussels, 2004. C

Olga Lopera: Development of data fusion techniques to support multi-sensor landmine detection, PhD dissertation, Université catholique de Louvain & Royal Military Academy, 2008.

Nada Milisavljevic: Analysis and fusion using belief function theory of multisensor data for close-range humanitarian mine detection, PhD Dissertation, Ecole Nationale Supérieure des Télécommunications (Paris), 2001.

Bart Scheers: Ultra-Wideband Ground Penetrating Radar, with Application to the detection of Anti Personnel Landmines, PhD dissertation, Université Catholique de Louvain, 2001.

Catherine Steukers: Etude de la faisabilité de la neutralisation de mines antichar enterrées à l'aide de micro-ondes de forte puissance,Thèse de doctorat, Fondation Universitaire Notre Dame de la Paix, Namur, 1998.

Idesbald van Den Bosch: Modeling the ground penetration radar, PhD dissertation, Université catholique de Louvain & Ecole Royale Militaire, 2006.

6. Conclusions

During 20 years, since 1996, RMA has been involved in research for mine action, associating end users such as the SEDEE-DOVO (the Mine Clearance Service of the Belgium Defence also known as the "explosive ordnance disposal (EOD) Battalion") or the CROMAC (the Croatian Mine Action Centre) in order to better answer the mine action needs and to collaborate in the design of field tests and missions. In this sense, RMA answers a critic often addressed to the Mine Action Research Community that would "ignore field reality". Moreover, RMA research addressed several activities of mine action; they are presented in the various chapters of the current book [16]. **Close-In Detection** is addressed in Chapter "Positioning System for a Hand-Held Mine Detector", in Chapter "Ground-penetrating Radar for Close-in Mine Detection" and in Chapter "Data fusion for close-range detection". **Area Reduction** and **Land Release** are addressed in Chapter "Remote Sensing for Non-Technical Survey" and "InSAR Coherence and Intensity Changes Detection" where remote-sensing data and dedicated processing are presented. Information Management, a concern for all mine action activities, is presented in Chapter "PARADIS: Information Management for Mine Action" with a focus on the data needed for the planning of humanitarian demining campaigns. Protective Equipment used for **Clearance** is addressed in Chapter "Assessing the Performance of Personal Protective Equipment". Vehicles presented in Chapter "Unmanned Ground and Aerial robots for Mine Action" are designed for **Non-Technical** and **Technical Survey**. Sea-Mines are addressed in Chapter "The Special Case of Sea-mines". Finally, the Chapter "Testing and evaluating results of research in mine action" summarises the involvement of RMA in **Tests and Evaluation**. This large scope shows that the focus of RMA research has not been limited to the detection of buried mines, another critic often addressed to the Mine Action Research Community [17].

Author details

Vinciane Lacroix*, Yann Yvinec and Marc Acheroy

*Address all correspondence to: vinciane.lacroix@elec.rma.ac.be

Royal Military Academy (RMA), Brussels, Belgium

References

[1] IMAS – International Mine Action Standards [Internet]: www.mineactionstandards.org [Accessed: 2016-06-16].

[2] IMAS 04.10, Second edition, Amendment 7, August 2014, Available from: www.mineactionstandards.org/standards/ [Accessed: 2016-06-16].

[3] Protocol on Prohibitions or Restrictions on the Use of Mines, Booby-Traps and Other Devices as amended on 3 May 1996 (Protocol II to the 1980 CCW Convention as amended on 3 May 1996). [Internet] Available from https://www.un.org/disarmament/geneva/ccw/amended-protocol-ii/ [Accessed:2017-01-04].

[4] Convention on the Prohibition of the Use, Stockpiling, Production and Transfer of Anti-Personnel Mines and on their Destruction, 18 September 1997. [Internet] Available from http://legal.un.org/avl/ha/cpusptam/cpusptam.html [Accessed:2017-01-04].

[5] H. Gray, "Amendments to IMAS Land Release Series", The Journal of ERW and Mine Action, Issue 18.1, 2014.

[6] T. Gabelnick, "We can Only Be 'Mine-Safe' When We Are 'Mine-Free'", The Journal of ERW and Mine Action, Issue 15.2, 2011.

[7] R. Gasser, "The Surprisingly Constant Cost of Landmine Impact Survey", The Journal of ERW and Mine Action, Issue 15.1, 2011.

[8] H. Killers, The Global Landmine Crisis, Report released by the U.S. Department of State, Bureau of Political-Military Affairs, Office of Humanitarian Demining Programs, Washington, DC, September 1998 Available from: http://www.state.gov/www/global/arms/rpt_9809_demine_toc.html [Accessed: 2016-06-16].

[9] Landmine Monitor 2015 [Internet]. Available from: http://www.the-monitor.org/en-gb/reports/2015/landmine-monitor-2015.aspx [Accessed: 2016-06-16].

[10] A. Parlow, "Toward a Global Ban on Landmines", International Review of the Red Cross, No. 307, 1995. Available from: http://www.icrc.org/web/eng/siteeng0.nsf/html/57JMCS [Accessed: 2016-06-16].

[11] H. Killers, The Global Landmine Crisis [Internet] 1994. Available from: http://dosfan.lib.uic.edu/ERC/arms/arms_briefing/950127arms_briefing1.html [Accessed: 2016-06-16].

[12] Landmines Factsheet, New York: United Nations Mine Clearance and Policy Unit, Department of Humanitarian Affairs, September 1997 [Internet] https://sites.google.com/a/neisd.net/isa-global-impact/faq-s [Accessed: 2016-06-17] was http://www.un.org/cyberschoolbus/banmines/facts.asp.

[13] Landmine Monitor 1999 [Internet]. Available from: (http://www.the-monitor.org/en-gb/reports/1999/landmine-monitor-1999.aspx).) [Accessed: 2016-06-21].

[14] M. Acheroy and Y. Yvinec. "Mine-suspected Area Reduction Using Aerial and Satellite Images". In: M.K. Habib, editor. Humanitarian Demining. I-Tech Education and Publishing, 2008. DOI: 10.5772/5409. Available from: http://www.intechopen.com/books/humanitarian_demining/mine-suspected_area_reduction_using_aerial_and_satellite_images [Accessed: 2016-10-31].

[15] M. Acheroy. "Mine Action: Status of Sensor Technology for Close-in and Remote Detection of Anti-personnel Mines" – EAGE – Near Surface Geophysics: Vol 5, No 1, February 2007, pp. 43–55, DOI: 10.3997/1873-0604.2006017. Special Topic: Ground-Penetrating Radar.

[16] "Mine Action: The Research Experience of the Royal Military Academy of Belgium", ISBN 978-953-51-4444-1. C. Beumier, D. Closson, V. Lacroix, N. Milisavljevic, Y. Yvinec, Editors. I-Tech Education and Publishing, 2017. Available from: http://www.intechopen.com/books/

[17] R. Gasser. "Technology Research in Mine Action: Enough is Enough", Journal of Conventional Weapons Destruction: Vol. 20: Issue 1, Article 3, 2016. Available at: http://commons.lib.jmu.edu/cisr-journal/vol20/iss1/3 [Accessed: 2016-06-27].

Chapter 2

Positioning System for a Hand-Held Mine Detector

Charles Beumier and Yann Yvinec

Additional information is available at the end of the chapter

http://dx.doi.org/10.5772/65784

Abstract

Humanitarian mine clearance aims at reducing the nuisance of regions infected by explosive devices. These devices need to be detected with a high rate of success while keeping a low false alarm rate to reduce time losses and personnel's fatigue. This chapter describes a positioning system developed to track hand-held detector movements in the context of close-range mine detection. With such a system, the signals captured by the detector over time can be used to build two- or three-dimensional data. The objects possibly present in the data can then be visually appreciated by an operator to detect specific features such as shape or size or known signatures. The positioning system developed in the framework of the HOPE European project requires only a camera and an extra bar. It adds few constraints to current mine clearance procedures and requires limited additional hardware. The software developed for calibration and continuous acquisition of the position is described, and evaluation results are presented.

Keywords: mine detection, hand-held detector, positioning system, camera calibration

1. Introduction

For several decades humanitarian mine clearance has received much attention to reduce the nuisance of infected regions. Although the recent trend for demining consists in reducing the search area thanks to heterogeneous sources of information [1], the actual detection of explosive devices still needs to be addressed. It requires high detection rates to maximise the chance of detection while keeping a low rate of false alarms to avoid time and energy losses and reduce fatigue of the personnel.

Metal detectors are possibly the most popular equipment to detect mines. The fact that mines may contain less and less metal has forced manufacturers to produce more and more sensitive metal detectors. As a result metal detectors frequently detect small harmless pieces of metal, which requires time to be investigated by the clearance teams. One possible solution consists in showing the operators a display of what is below the ground surface. Another one is to consider other mine detectors such as a ground-penetrating radar (GPR), with the possibility to combine both options.

In the European ESPRIT project *h*and-held *ope*rational demining system (HOPE [2], 1999–2001), higher detection rates and lower false alarm rates were expected from the combination of a metal detector (MD), a GPR (sensitive to metallic and dielectric objects) and a microwave radiometer (MWR, indicating flushed objects). Tracking the position of the sensors during scanning enabled the reconstruction of registered images, allowing for the fusion of the sensor signals and for the discrimination of objects from their shape and size. The advantage of image analysis had already been shown in [3, 4]. However, since a hand-held detector was considered as opposed to a robotic one [5], sensor head positioning was not trivial.

In the late 1990s, local positioning was commonly based on triangulation, scene analysis or proximity, possibly using time-of-flight, infrared or ultrasound emitters/receptors or magnetic sensors [6]. In the HOPE project, we opted for a camera attached to the hand-held detector to report positions relatively to a marker bar laid on the ground close to the area to be scanned.

This chapter describes the optical positioning system developed in the framework of the HOPE project. The different parts of the system are described, such as the camera attached to the detector, the marker bar used for reference and the trigger box aimed at synchronising sensor data and positioning. The mathematical solution for positioning the camera and finally the detector relatively to the bar is explained. A specific calibration procedure was developed to deliver positions of the detector relatively to the bar. The image processing tools for positioning and for calibration are detailed. Finally, the evaluation of the system is given in terms of positioning and orientation precision.

2. Methodology

2.1. Necessity for a positioning system

The major role of the positioning system for a hand-held detector is the ability to reconstruct an image of the detector response from a series of individual measurements obtained over time when scanning the area. This image may serve several purposes. First, it represents a map of detector responses providing a reference for a later device removal. Secondly, it may give a better view and description of the scene such as the object size and shape. Thirdly, the map indicates the scan progress and the areas still to be scanned before moving further.

Besides these direct products of the reconstructed image, the registration of the image pixels allows for the fusion of several captures, possibly with different detectors. The image(s) can be processed by traditional or new image processing tools. For instance, noise reduction with low-pass filtering is efficient and trivial on the image but quite imprecise on a sequence of measures with no localisation.

In the specific case of hand-held detectors for buried mine detection, the positioning system should be able to capture the sensor orientation. A non-vertical orientation of the detector may lead to a shifted localisation of an object in the map. Recording the angles of the detector is useful to compensate for this complication.

2.2. Requirements

In the case of hand-held detectors, we cannot rely on mechanical systems to have position. The freedom of movement makes position estimation more difficult.

The required precision is high since the localisation of suspect objects is of prime importance for their removal. The a priori precision was set to 2 cm for positions and 1° for angles.

The solution must be resilient to practical conditions such as a changing weather, varying light conditions and uncontrolled movements of the operator. In particular, should the positioning system fail for some measurements, it should be able to restart delivering positions without the need of a lengthy starting procedure, if any.

Finally, the developed solution should be consistent with the demining procedure so that it does not disturb the operator and does not ruin the chances for adoption by the demining team.

2.3. Discussion of possible methods

By the time of the HOPE project, there was no existing system for position estimation of a hand-held mine detector.

One classic way to localise an object is to attach an emitter and use triangulation from a few receivers. The setup requires some arrangement to correctly position the receivers, with good reception and appropriate distance and angle for precision. This is not trivial in real conditions, especially if the terrain is unfriendly. In the context of a team demining lanes in parallel, the system installation may become complex, especially if interferences exist.

With the availability of cost-effective, light and small image sensors, positioning has also been dealt with image matching. One possible solution consists in matching the ground part seen by a camera attached to the hand-held detector [7]. A global picture of the area must be first acquired before starting the scan. Or the localisation can be made relative, picture after picture during the scan, but the drift has to be compensated for at the end of the procedure, and the lack of matching during the scan may result in abandoning and restarting the scan. The problem with the image matching approach is that it completely relies on the existence of image features to be matched. These are typically colour, corners, lines or areas that have to be present and properly detected to bring images into correspondence. The sensitivity of the matching relatively to image features is always high since the detectability of features highly depends on image quality. The presence of shadow (induced by the operator or by the detector) and the change in illumination (due to reflections or clouds) are practical conditions which hamper image systems from a perfect functioning.

2.4. Selected method: the optical position monitoring system (OPMS)

In a common demining procedure, the suspected area to be cleaned from explosive devices is divided into lanes of about 1 m width. A bar is laid on the ground in front of the deminer to materialise the limit up to which the lane is considered safe since it was so far inspected and possibly cleaned from mines. In the context of the HOPE project, it has been found that the bar, used during most manual demining procedures, can appropriately serve the purpose of spatial reference.

We decided to realise positioning by attaching a camera to the arm of the hand-held detector and localising this camera relatively to a marker bar laid on the ground. In a feasibility study, we considered the addition of an accelerometer to get the angular deviation from the vertical direction [8, 9]. Due to the important noise in angular measures and the drift associated with acceleration measurements, we abandoned the accelerometer and turned towards a higher quality camera to grab the orientation from the capture of squares with known size printed on the marker bar.

3. A positioning system for hand-held detectors

3.1. Context

By the time of the HOPE project, building a camera-based positioning system was quite a challenge. Special attention had to be paid to design a precise and near real-time system given the camera speed and resolution, the available standards for image transfer and the limited computational power.

3.2. Principle

In order to get geometrical data and to make correspondences between different sensors for fusion, a position monitoring system is attached to the detector to assign position to the data acquired during scanning. It consists of a camera acquiring images of a marker bar laid on the ground being scanned. Reference points of the marker bar are tracked by the camera, which allows for the 3D camera positioning relatively to the bar. The bar must not be moved during scanning. It was decided that the marker bar should be laid at the end of the search area (see **Figure 1**).

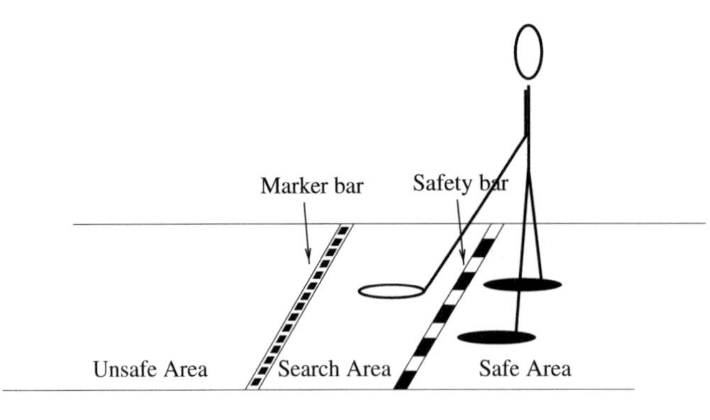

Figure 1. Typical mine clearance setting.

http://dx.doi.org/10.5772/65784

The safety bar used by some EOD teams is not needed by the optical position monitoring system (OPMS). Requesting the marker bar to be placed at the end of the search area is in accordance with some typical mine clearance procedures (e.g. Belgian EOD) which rely on two bars (one before and one after the search area). The new marker bar position reduces the risk of accidental displacement, either by the operator's feet or when scanning the area with the detector. Moreover, the camera can look forward, reducing the occlusion of the bar by the detector and allowing for the adjustment of the carrying bar ('ang' in **Figure 2**) without moving the camera relatively to the detector (**Figure 3**).

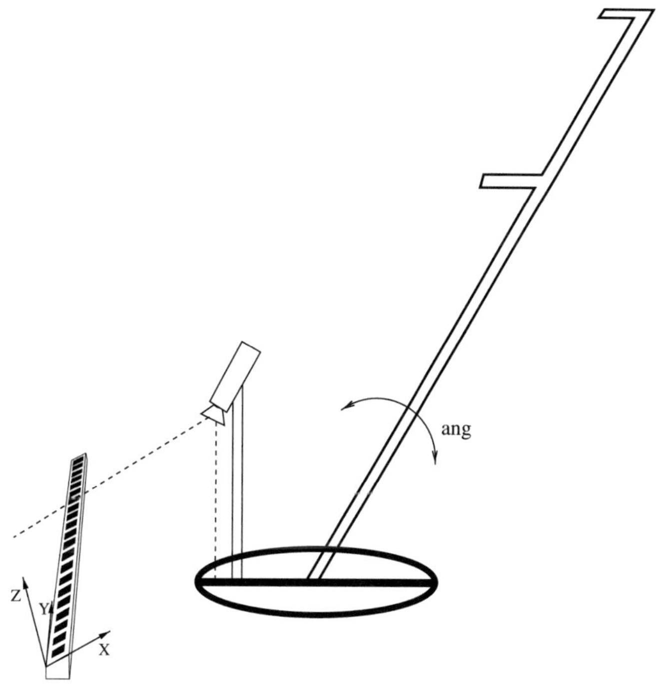

Figure 2. Setting of the camera on the detector.

3.3. System description

The proposed positioning system consists of five components which are detailed in this section. A **marker bar** with a specific pattern, positioned at the end of the search area, serves the purpose of geometrical reference. A **camera**, tracking the marker bar, delivers the camera position in a reference axis system linked to the marker bar. The estimation of the OPMS parameters necessary for the correct estimation of the three-dimensional position is done thanks to a **calibration grid**. A **trigger box** sending trigger signals to the camera helps synchronising the capture of position with other sensor data (GPR, MD, MWR). Finally, a **portable computer** gets camera images in real time, assigns timestamps for synchronisation with other sensor data and derives calibration parameters and XYZ positions.

Figure 3. System at work.

In the context of HOPE, the portable computer is connected to a controlling computer (the Master PC) which sends acquisition requests, displays position data and manages the database for data collection.

3.3.1. The marker bar

The marker bar is 1 m long and has a width of 3 cm. A black pattern consisting of 33 squares (2 by 2 cm) surrounded by small rectangles for square identity coding is printed on a white background (see **Figure 4**). The squares are regularly spaced on a planar and rigid bar. With this assumption, the noise present in image localisation is partially compensated from a least mean square calculation.

Figure 4. Part of the marker bar.

3.3.2. The camera

We considered a digital camera with at least 1000 pixels in one direction to offer maximal precision in the small dimension of the bar which is only 3 cm wide. This direction was used perpendicularly to the bar to get more precision about the size of the squares. Compared to analogue cameras still very common in 2000, a digital camera appeared to be adequate for a better subpixel estimation of the square edges.

With the additional constraint of real-time capture to get enough positions along the trajectory, we selected a SONY camera, model XCD-X700, delivering up to 15 images per second at 1024 × 768.

This model was equipped with FireWire for connection to a portable computer through a PCMCIA card. A repeater was used for electrical power, allowing for a total cable length of 20 m. This camera had an electronic shutter control and an external trigger that allowed for capture synchronisation.

The optics was chosen for its large field of view with yet acceptable distortion. Since the image sensor of the XCD-X700 was large (1 inch), a 6 mm focal length lens was allowed for about 50 × 70 cm of capture at a distance of 40 cm.

3.3.3. The calibration grid

Two calibration procedures are requested by the system. The first one concerns the estimation of the camera parameters to make correct 3D measurements ("intrinsic calibration"). The second procedure ("external calibration") estimates the transform giving the position of the sensor knowing the estimated camera position. **Figure 5** shows the grid used for calibration. The line crossings of the grid are accurate reference points which allow for "intrinsic calibration" if several images are captured from different points of view. For external calibration, the camera is placed at a precise position on the grid (represented by the footprint of the sensor).

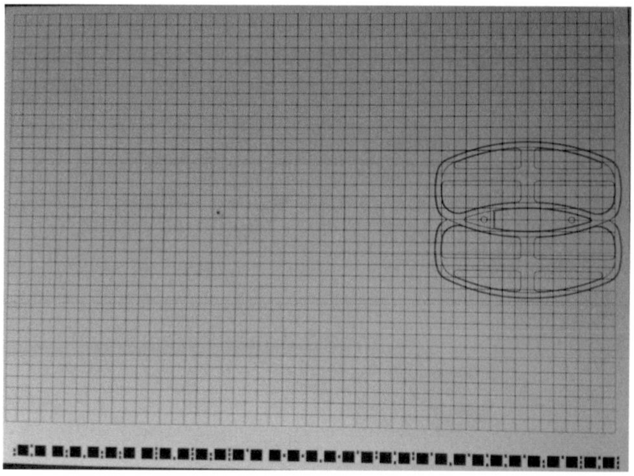

Figure 5. The calibration grid.

3.3.4. The trigger box

The aim of the positioning system is to track the position of the sensors during scanning. For precise estimation, it is necessary to synchronise position gathering with sensor data acquisition.

A common TTL-rising signal is sent to the sensors (GPR, MD, MWR and trigger box of the OPMS) at the start of acquisition. From this TTL signal, the trigger box issues a clock signal with a period of 100 ms to the camera trigger. The camera was operated at a rate of ten images per second, inferior to the limit of 15, to allow for the desired range of shutter speeds.

3.3.5. The OPMS computer

The OPMS computer hosts the image acquisition interface and the procedures for calibration and position estimation involving image processing. In the context of HOPE, considering a system with several sensors, a Master PC centralises data acquisition requests, data display and database management. Acquisition controls like the mode of operation (among "image capture", "calibration" and "position estimation"), and the shutter or the gain setting of the camera is modified through a graphical user interface currently running on the OPMS computer. Images and calibration or position data are locally saved on the OPMS computer on request.

3.4. Functional description

We describe here the interaction between the different components and the algorithms which allow to achieve position estimation. This functional description focuses on the OPMS but describes briefly the role of the **Master PC**, presented in Section 3.4.1. The integration and the role of the OPMS in the detector (consisting of several sensors) need to solve **data synchronisation**, topic of Section 3.4.2. The third subsection presents the **user interface** that allows interacting with the functionality and options of the OPMS. The three main operational modes of the OPMS, namely, **image acquisition**, **calibration** and **position estimation**, are described in the last three subsections.

3.4.1. The Master PC

The OPMS can be driven by a computer, here called the Master PC. In the HOPE context, this PC hosts the database, a graphical man-machine interface (MMI) for data display and a communication protocol (through Ethernet) between the PCs driving the sensors.

In the mode of remote acquisition of the OPMS, when a scan is requested through the MMI of the Master PC, this latter issues a "software start" asking the sensors to get ready and provides a filename under which the data will be stored. A "hardware start" is then sent to the sensors by a rising TTL signal. This signal is used as reference time for synchronisation of the sensors. It fires the clock of the "trigger box" which starts to send periodically (each 100 ms) triggers to the camera. The acquired images are captured by the OPMS application which either saves images, extracts reference points for position estimation or grids nodes for calibration, according to the selected mode of the OPMS. The extracted data is saved locally on the OPMS PC according to the filename assigned by the Master PC database and containing the date and time for easy later reference. The position data is also sent to the MMI of the Master PC for real-time display and direct evaluation and control by the operator. The operator interacts with the MMI of the Master PC to send a "software stop" to the sensors to stop acquisition of the signals (**Figure 6**).

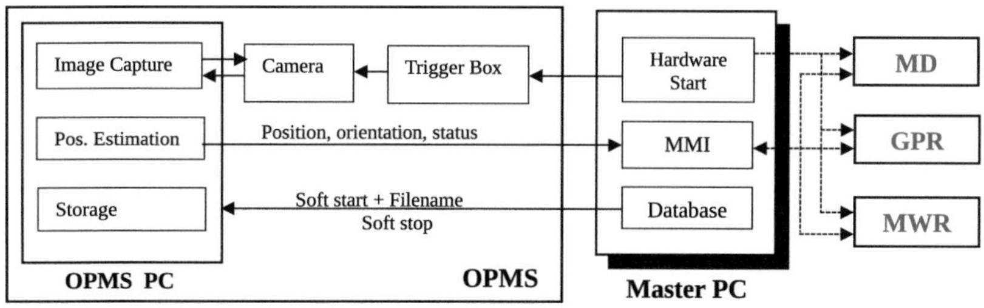

Figure 6. The OPMS synoptic.

3.4.2. *Synchronisation*

Because the signals from the different sensors (MD, GPR, and MWR) must be synchronised for fusion and image reconstruction, a consistent timestamping procedure is required.

Since the precision of time measurements delivered by the PC cannot be guaranteed due to possible operating system latency, and because the transfer of the image from the camera to the PC takes an undetermined time, an external synchronisation mechanism has been developed thanks to the use of a TTL signal ("Hhardware tart") sent by the Master PC at the start of acquisition to all the sensors. When the OPMS trigger box receives the TTL-rising signal, it fires a clock signal that sends trigger signals to the camera. The TTL-rising signal is used as starting reference, and the image acquisition timestamps are derived from the periodicity of the clock issued by the trigger box.

3.4.3. *User interface*

The user interface described below presents the functionality of the OPMS. The minimum functionality (start and stop acquisition, real-time display of acquired data, filename assignment for storage) was located on the Master PC, a few metres behind the HOPE detector. The OPMS computer, hosting data acquisition, camera settings control, position and calibration algorithms, was controlled remotely by the Master PC thanks to a network application (through Ethernet connection) which allows the OPMS graphical interface to be accessible on the Master PC's screen. In real operation, the OPMS computer is intended to be a computation unit without screen in the backpack.

Three main operational modes are available and detailed in the following subsections: "Image acquisition" which acquires one or several images from the camera, "Calibration" which acquires images, extracts reference points from the calibration grid and applies the calibration algorithm, and "Position estimation" which acquires images, extracts reference points from the marker bar and estimates the sensor position.

Image capture can be either single (one shot) or continuous. It can be controlled by an external trigger. The image quality can be enhanced by the proper setting of the shutter (choice of the integration time) to solve the compromise between a low motion blur and enough contrast. The camera gain can also be adapted by modifying the image contrast and lightning level to solve the compromise between good contrast and low noise level without saturation.

Several graphical outputs are available. Acquired images are displayed in real time, with overlaid detected reference points. Position data are displayed in real time on a 2D map with an additional elevation chart for the Z (height) dimension. The orientation is printed in text boxes and shown graphically thanks to a symbolic detector with proper position and orientation. Another representation for orientation consists of two quadrants, one displaying the horizontal orientation and the other one showing the angular deviation from the vertical.

A last option allows for the storage of the acquired data (such as images, reference points and position values) for debugging or logging.

3.4.4. Image acquisition

In order to set up a portable system that can be deployed in the field of operation, a portable PC was selected as the OPMS platform. A FireWire connection was retained as fast camera link with the PC. The SONY camera with its FireWire connection is interfaced thanks to a 'Microsoft Filter' that delivers acquired images. These images receive a serial number and a timestamp. They are bufferised in the application for delayed processing in another thread.

The graphical interface allows to control the camera settings (the shutter for image exposure and the contrast by gain adjustment). On the one side, a small exposure time results in little motion blur in images at the expense of little contrast. On the other side, a large gain for good contrast is limited by the acceptable noise level. A shutter of 10 ms resulting in a motion blur of a few mm seems adequate for correct noise level in normal illumination conditions.

The closing/opening of the camera lens diaphragm also helps solving the previous compromise of good contrast versus sharp image, but here sharpness relates to focusing and not blurring due to movement. In our application, since the distance between the bar reference points and the camera varies in large proportions, a large depth of field was necessary. We thus opted for a rather closed position of the lens and compensated the lower light levels on the CCD by a higher gain of the camera. The electronic noise due to amplification did not disturb localisation of the bar reference points.

The remaining difficult problem is the large illumination variation. Weather conditions drastically change the illumination level, sometimes in a few seconds due to clouds. Shutter and gain might be adapted during image capture. The idea is to optimise the detection of the marker bar by counting the number of detected squares. A grey-level analysis adapts the gain of the camera to avoid low contrast and saturation. Unfortunately, the control of the camera shutter and gain was too slow to allow for a continuous automatic adaptation. An automatic gain control built in the camera would be a better solution.

3.4.5. Calibration

3.4.5.1. Principle

The calibration principle consists in determining the model parameters thanks to grid crosses ('nodes') extracted in images and the corresponding 3D coordinates of these points on the grid. *N* images of the grid from different points of view are captured to avoid the necessity of a 3D calibration object. These different images are taken during a manual vertical movement of the sensor from its original position on the footprint of the calibration grid. The movement must be done in such a way that the images are taken in various positions and orientations.

3.4.5.2. Mathematical formulation

The axis systems used in the presentation are depicted in **Figure 7**.

With (O, X, Y, Z) the world coordinate axis linked to the bar;

(C, X_c, Y_c, Z_c) the camera axis system, C being the optical centre;

(L, X_i, Y_i) the image coordinates, L being the upper left corner.

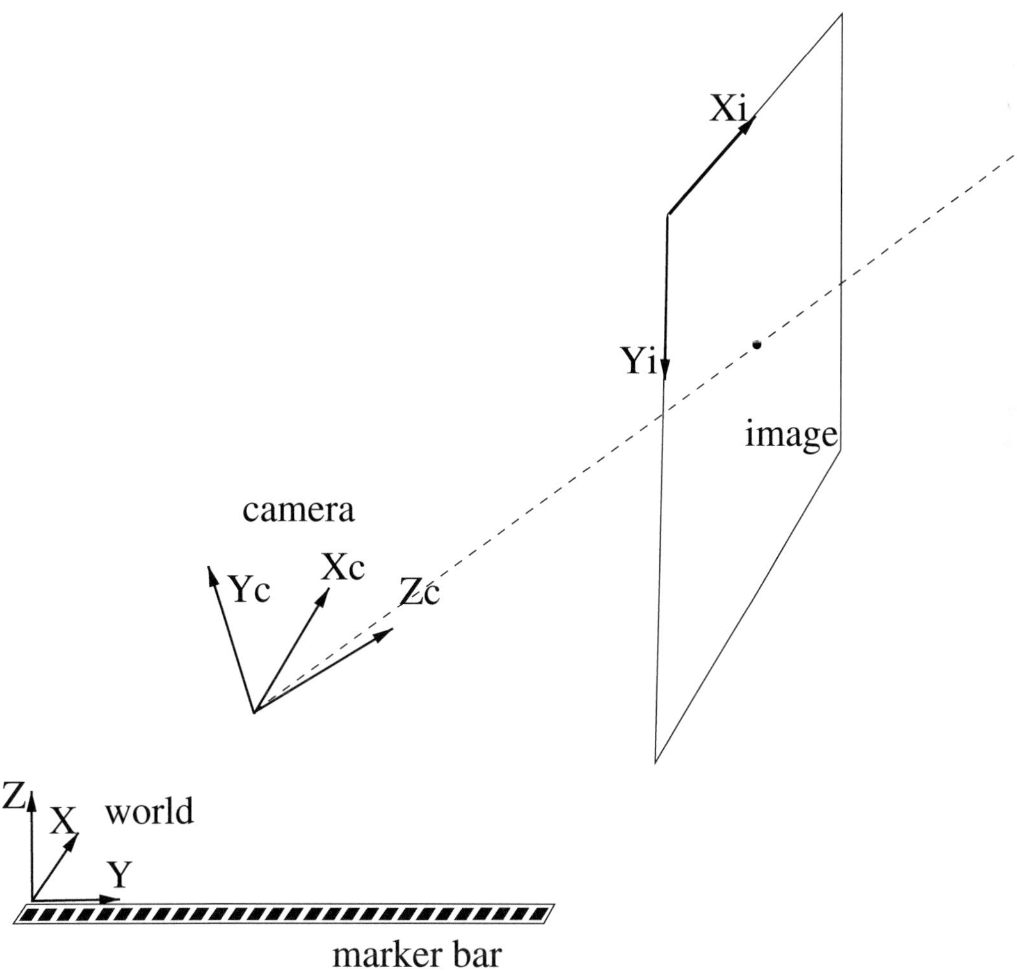

Figure 7. Coordinate systems.

The camera calibration parameters intervene in the transform of the world coordinates of a point in the field of view of the camera into the position in the image where this point can be seen.

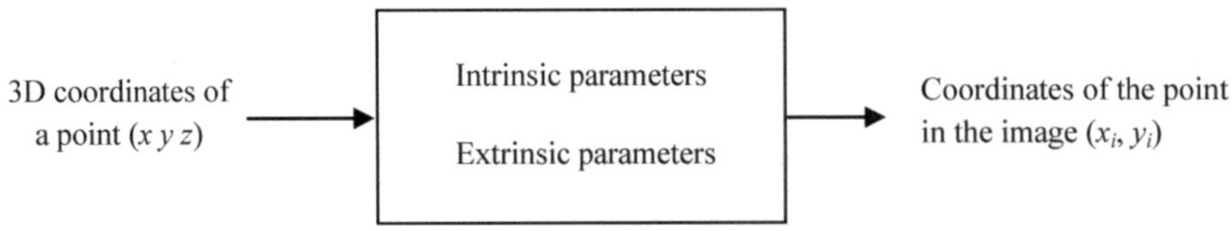

The transformation of the 3D coordinates of a point (x, y, z) into its position in the image (x_i, y_i) can be divided in four steps:

1. Change of coordinates from the world axis system to the camera axis system.

$$\begin{bmatrix} x_c \\ y_c \\ z_c \end{bmatrix} = R(\theta, \varphi\, \psi) \begin{bmatrix} x \\ y \\ z \end{bmatrix} + T(X_0, Y_0, Z_0) \tag{1}$$

where (x, y, z) are world coordinates; (x_c, y_c, z_c) are coordinates in the camera axis system; R is the rotation matrix defined by the angles of the absolute axis system in the camera axis system and T is the translation vector of the absolute axis system relative to the camera axis system.

2. Projection to the image plane.

We used the pinhole model for the camera, with the optical centre at the camera axis system origin. In the following equations linking the coordinates in the camera axis system with the image coordinates x_p and y_p, we used the *model* focal length f, modelling a whole camera system with optics and CCD:

$$\begin{cases} x_p = f\dfrac{x_c}{z_c} \\ y_p = f\dfrac{y_c}{z_c} \end{cases} \tag{2}$$

3. Distortion.

To model the distortion due to the lens, we used the four-parameter Seidel model with two radial distortion parameters R_1 and R_2 and two tangential distortion parameters T_1 and T_2:

$$\begin{cases} x_n = x_p(1+\delta) + 2\,T_1\, x_p y_p + T_2\left(r^2 + 2\,x_p^2\right) \\ y_n = y_p(1+\delta) + T_1\left(r^2 + 2\,y_p^2\right) + 2\,T_2\, x_p y_p \end{cases} \tag{3}$$

with

$$r^2 = x_p^2 + y_p^2 \quad \text{and} \quad \delta = R_1 r^2 + R_2 r^4 \tag{4}$$

4. Translation from camera axis system (mm) into image axis system (pixel).

The position of the principal point P (orthogonal projection of the camera axis system origin on the image plane) gives two more parameters P_x and P_y. If N_x and N_y are the number of pixels in both directions and s_x and s_y are the dimension of the CCD, we have

$$\begin{cases} x_i = \alpha \frac{N_y}{s_x} x_n + P_x \\ y_i = -\frac{N_y}{s_y} y_n + P_y \end{cases} \quad (5)$$

with α the ratio N_x/N_y.

To sum up, we have six extrinsic parameters describing the location (X_0, Y_0 and Z_0) and orientation (θ, φ and ψ) of the camera and eight intrinsic parameters describing the camera itself (the model focal length f; the Seidel parameters for the distortion R_1, R_2, T_1 and T_2; the principal point P_x and P_y and the ratio of the pixel dimension α).

This model takes as input points in the field of view of the camera and provides as output the position in the image of these points (their projections in the image plane). To estimate the parameters, we need a well-chosen set of points with their coordinates in the world axis system and their pixel coordinates in the image.

These points must not lie in a plane. That is why a calibration grid offering reference points on two orthogonal planes is often used. We preferred to use several views of a planar grid since a 2D object is much easier to manufacture than a 3D one. The detector is first positioned at the footprint reference on the calibration grid for extrinsic calibration and then lifts up about 20 cm to deliver several views (about 20 seems enough). To get a unique solution, the displacement must not be parallel to the optical axis of the camera, which is satisfied in our setup when the detector is lifted vertically since the camera is not vertically oriented.

3.4.5.3. Image points: node extraction

The localisation of the nodes, intersections of the lines of the calibration grid, is carried out by a three-step procedure.

First, the centres of the white squares surrounded by grid lines are looked for, starting from the middle of the image. From an initial position, vertical and horizontal crossings with grid lines are searched for, refining the centre point (**Figure 8**). Each centre point is used as a "seed centre" to give rise to four new centres, neighbours in the cardinal directions, with initial position guessed from the seed centre and the size of the square (**Figure 9**). These four new centres are then refined in the same way and will provide new neighbours. This loop goes until no new centres are found in the image.

Secondly, each pair of horizontal or vertical neighbouring centres allows for the detection of a grid point (**Figure 10**), intersection of the line segment joining the two centres with a grid line. The grid point is localised as the minimum of the grey level along this segment. Subpixel accuracy is obtained by interpolation between the two pixels around the sign inversion of the first derivative of grey-level values.

Thirdly, the four grid points around each node are used to localise the node precisely. The left and right grid points and the top and bottom grid points define two line segments whose intersection gives a subpixel accurate node position (**Figure 10**).

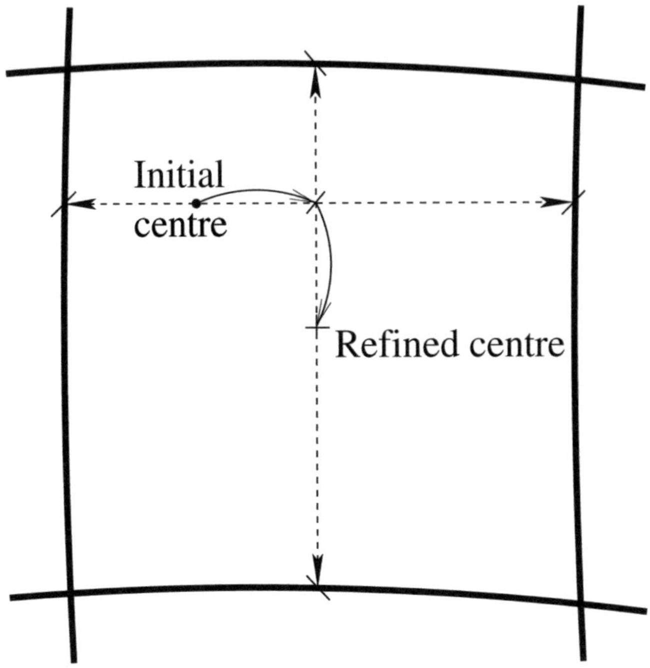

Figure 8. Centre refinement.

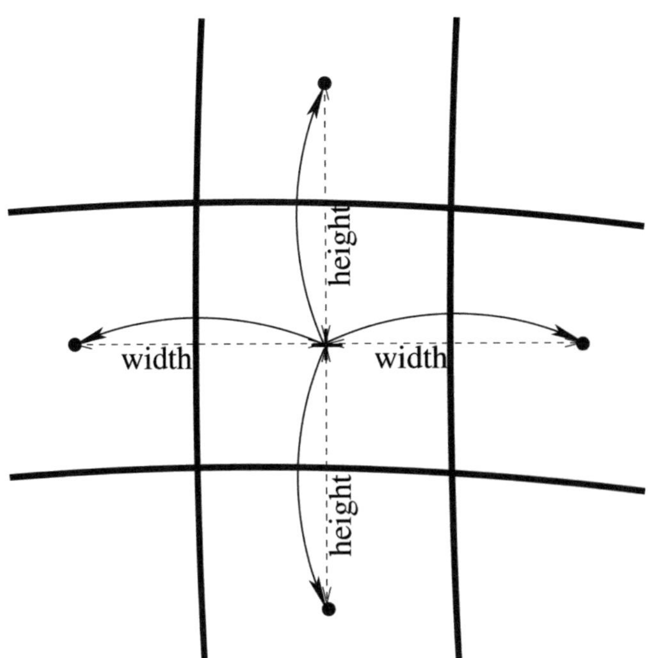

Figure 9. Centre propagation.

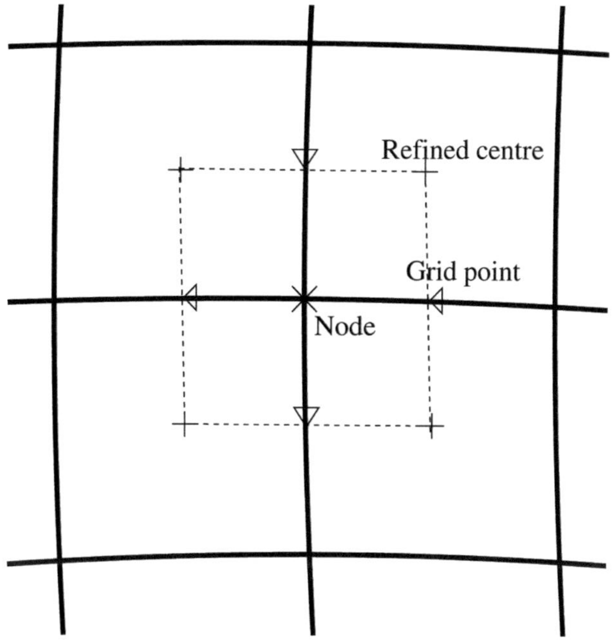

Figure 10. Node localisation.

Once all the nodes have been found, the marker (a small dark square close to the node at the centre of the calibration grid) is looked for to find the reference node. This node is labelled (50,50) in the symbolic coordinate system. It is very important that the marker is correctly detected in the different images for proper calibration.

Figure 11 presents the top left part of node localisation applied to an image of the grid. The superimposed white dots are initial centres, the black '+' are the refined centres and the white '×' are the localised nodes.

The whole image processing procedure was optimised and ran in less than 100 ms on a Pentium II 500 MHz, allowing for fast calibration, although this was not mandatory.

3.4.5.4. Parameter estimation

Each calibration image provides two equations per detected grid node in the images, linking (x_i, y_i) to (x, y, z) thanks to the model. In the specific case of a planar grid, we chose $z = 0$ for all the points.

The total number of parameters to estimate is $8 + 6 \times N$. The eight intrinsic parameters depend only on the camera, while the six extrinsic parameters depend on the position of the camera, different for each image, and a priori unknown.

The error function to minimise is the sum of the image distances between the image coordinates of the extracted grid nodes and the projection of the corresponding 3D points by the model. We used the calibration software from Heikkilä [10], written in MATLAB. All the points of all considered images are processed in the same minimisation procedure.

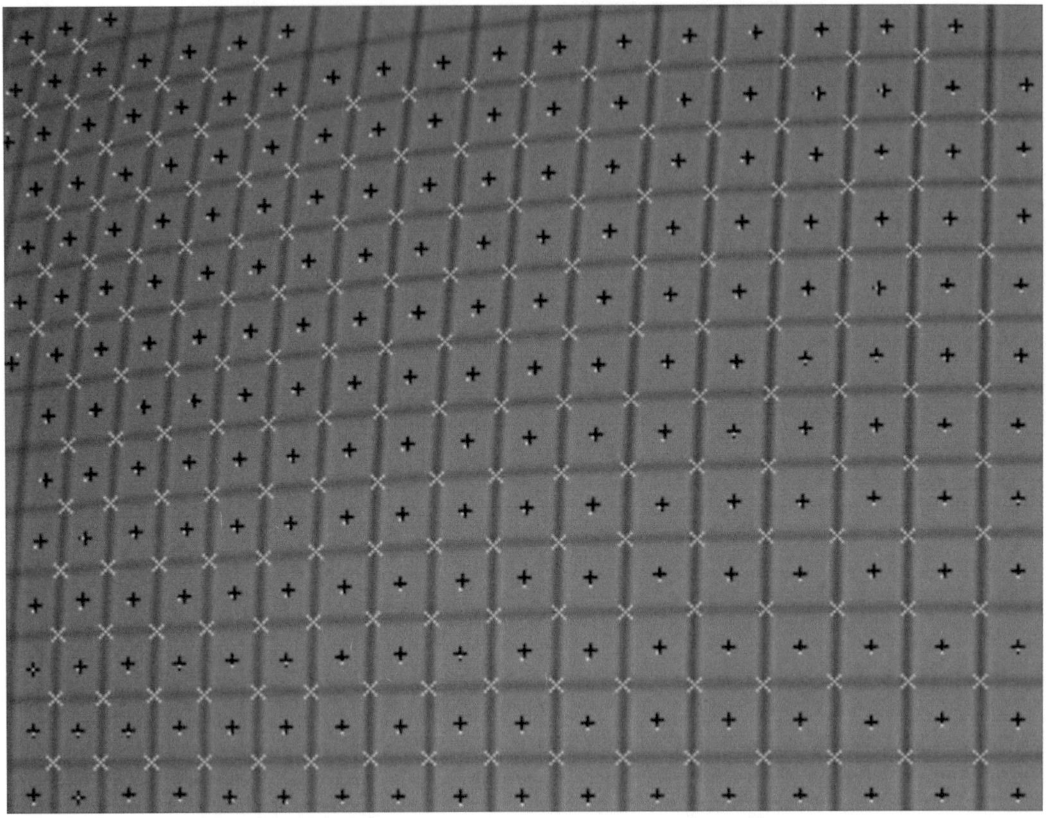

Figure 11. Results of node localisation on the grid.

The camera-to-detector transform, necessary to report position and orientation of the detector, is easily derived from the solution of the full problem. As the first image of the sequence considers the detector at a known position (footprint) on the calibration grid, and as the six extrinsic parameters, position and orientation of the camera, are eventually available for each image relatively to the grid, the camera-to-detector transform is obtained from the six extrinsic parameters of the first image. To derive the position and orientation of the detector, the position and orientation of the camera are combined with this camera-to-detector transform.

Only a dozen of evenly spaced images captured during calibration are considered in order to reduce memory needs, computation time and information correlation between images.

3.4.6. Position estimation

3.4.6.1. Principle

The position estimation of the detector is obtained from the position estimation of the attached camera and the camera-to-detector transform estimated during calibration. Position and orientation are reported relatively to the marker bar upon which the world coordinate system is built (see **Figure 7**). They correspond to the six extrinsic parameters (X_0, Y_0, Z_0, θ, ϕ, Ψ) introduced in the calibration procedure; the eight intrinsic parameters remain constant as long as the camera sensor or the lens is not modified.

3.4.6.2. Problem solving

The position and orientation of the camera in the world axis system are obtained from the equations presented in the previous subsection. The world coordinates of the reference points of the marker bar and their projection in the image offer $N \times 4$ equations with six unknowns (X_0, Y_0, Z_0, θ, ϕ, Ψ), where N is the number of detected squares (each providing two reference points).

The world coordinates of the reference points are known from the six-bit code (see **Figure 12**) surrounding each square to specify its identity. Should the determination of this code fail, an invalid code label is assigned to the square (see 'Reference points extraction').

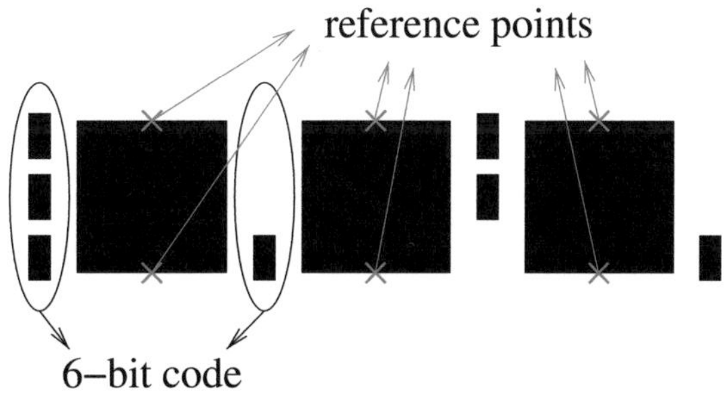

Figure 12. Close-up of the square pattern.

In order to deal with some errors in the data acquisition or point extraction, a first least mean square procedure is applied to one of the two rows of reference points (**Figure 12**) in order to filter and complete the information extracted from the image. The set of reference point coordinates (x_i, y_i) with identified label is reduced to the ones matching the least mean square fit with a tolerance distance of 1 pixel. A consistent labelling of the reference points—necessary to know the corresponding absolute coordinates—is deduced from the labels of the reduced set of reference points and applied to all the detected reference points with an invalid code label. As this least mean square procedure concerns a one-dimensional row of points along Y (the bar), only one row of the matrix R intervenes, leading to linear equations in the unknowns (X_0, Y_0, Z_0, R_{21}, R_{22}, R_{23}). Since the solution is up to a multiplicative factor, the unknowns appearing in their simple form on both sides of the equations, one unknown can be set to 1 (Z_0). Another constraint is the normality of the vector (R_{21}, R_{22}, R_{23}). This partial solution aims at extracting a set of correct reference points with coherent labelling. It also provides initial values to five of the parameters for the whole system solving described below. It cannot estimate the orientation around the bar (φ) since it relies on points aligned on axis Y.

A second least mean square procedure considers the two reference points of the squares with label completion and outlier elimination obtained in the first step. This two-dimensional set of points allows for the determination of the whole rotation matrix. Instead of solving this non-linear overdetermined system in the unknowns (X_0, Y_0, Z_0, θ, φ, ψ), we preferred to consider the linear system in the 12 redundant unknowns contained in the rotation matrix R and translation vector T, as presented in the calibration subsection and add the constraints on the vectors (perpendicularity and normality). Since the pinhole model is projective, multiplying all the unknowns by a constant does not change anything so that we can impose Z_0 to 1, reducing the number of independent unknowns to 11.

The orientation around the bar (φ), and incidentally the precision on Z, was clearly the sensitive point of the design. Since this estimation is based implicitly on the distance between the two rows of reference points, the precision on φ is larger where the derivative of that distance relative to φ is maximal. It is advantageous to tilt the bar so that the squares are less perpendicular to the camera while avoiding a too grazing point of view to ensure visibility of the squares. The value of 45° is a good compromise that implies for our camera setup (25° with the vertical) a tilt of about 20° of the bar.

3.4.6.3. Reference point extraction

The objective of this image processing algorithm is to detect the squares, localise their two reference points and identify the square label coded in small surrounding rectangles (see **Figure 12**).

The first square is looked for by visiting pixels in the image sparsely (one line out of 16), searching for large horizontal grey-level transitions. From each such point, the contour of a hypothetical square is followed until the contrast is below a threshold or the number of visited points is too high. A candidate contour is accepted if the corresponding curve is closed and if its size is in a proper range. Then, the four more prominent points of the retained contour are labelled as the corners. Two of them are detected as the opposite points on the contour with the largest distance (points 1 and 2 on **Figure 13**). The other two corners ('3', '4') are roughly localised as the midpoints on the contour between the first two corners and then refined by picking the points with maximal distance to the segment 1–2. Finally, points '1' and '2' are refined as points with maximal distance to the segment 3–4.

Once the first square has been found, more squares are looked for in the directions of the sides of the first detected square. The corner positions and the linearity at short distances in the image, even with distortion, give hints about the direction and distance about where to find neighbouring squares.

The detected squares are then analysed to offer reference points. The direction of the bar is guessed from the distribution of the position of the squares. Each square provides two reference points: the midpoints of the sides parallel to the direction of the bar (**Figure 12**). They are roughly localised as the midpoint of the two neighbouring corners and refined perpendicularly to the side by a subpixel grey-level analysis (zero crossing of the second derivative of the grey-level profile perpendicular to the side (**Figure 14**)). As a matter of fact, the position along the bar can be well estimated from the distribution of the whole set of reference points, applying a least mean square procedure (see 'Problem Solving' above). On the contrary, the measures across the bar implicitly relate to small distances (2 cm) which gives a high sensitivity to the estimation of the rotation angle φ. Subpixel localisation is thus welcome.

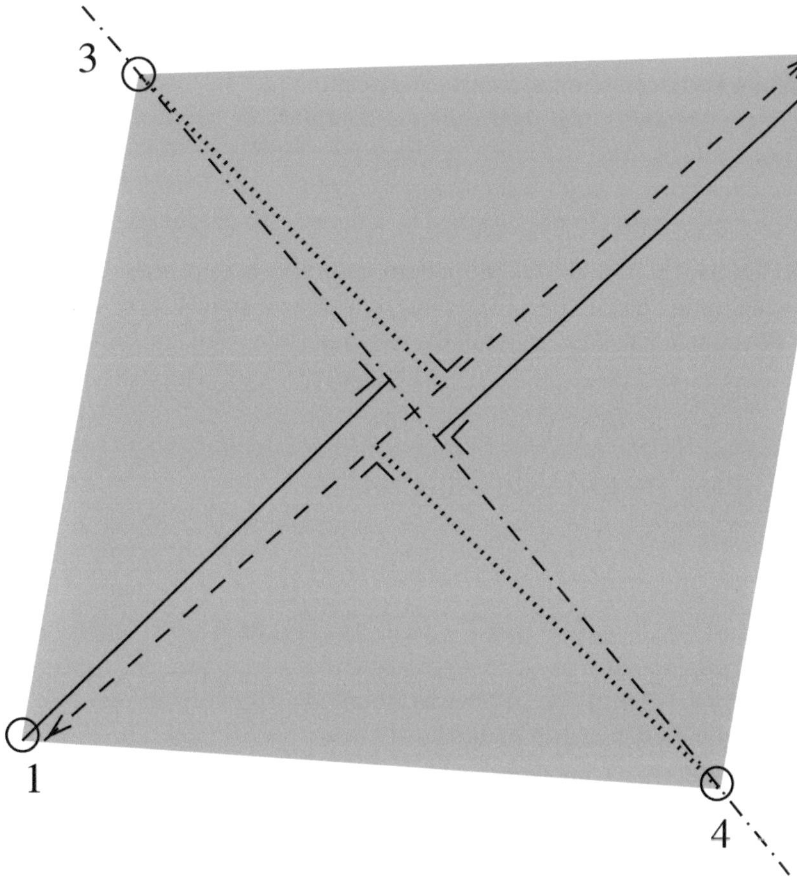

Figure 13. Corner detection.

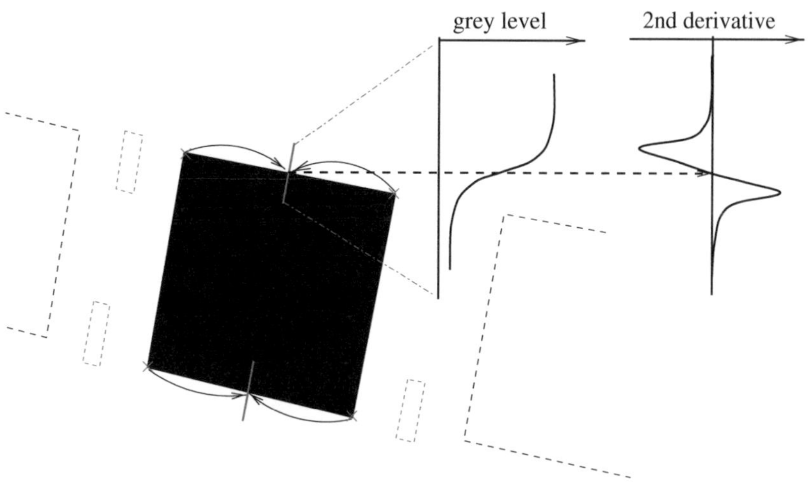

Figure 14. Subpixel localisation of the reference points.

The squares are finally labelled to identify their absolute position on the bar leading to the world coordinates of the reference points. For this, the six-bit code represented by small rectangles surrounding each square is extracted. The position of the small rectangles is known relatively to the square corners, and their colour, either black or white, determines the code. The code always contains at least one black rectangle for visibility checking. The code of a few neighbouring squares bring enough redundancy to enable error detection and correction in labelling. However, an invalid label is assigned to squares with unidentified code.

The design and implementation of the bar pattern and related algorithms have been developed to enable real-time detection and labelling of reference points. Less than 20 ms were necessary on a Pentium II 500 MHz to get all the required processing for camera and detector positioning.

4. Evaluation of the HOPE positioning system

4.1. Principle

The evaluation of the OPMS concerns here the accuracy of position and orientation estimation of the system in a typical range of use (1 m displacement along the bar, 40 cm perpendicularly to the bar and 11.5 cm in height). The OPMS was mounted on the carrying bar (see **Figure 15**), and the detector was placed on a grid manually at known coordinates. Three pieces of wood were added one by one under the detector to change the height to 0, 3.8, 7.7 and 11.5 cm. Other evaluations like the best shutter, gain and lens aperture of the camera were carried out empirically during development and a measure campaign in Bosnia. Although the tests concerned here considered position estimation with little or no motion, the precision estimation should be valid for dynamic estimation thanks to the short shutter setting (around 12 ms).

The evaluation consisted of two phases: calibration and position measurements. The calibration phase, explained in Section 3.4.5, was carried out to ensure that the parameters are adapted to the selected lens aperture and camera position on the detector. Position measurements are made relatively to a marker bar printed with a calibration grid to form the so called "evaluation grid" described in Section 4.2. To complete the tests, an independent marker bar was laid tilted on the printed marker bar to analyse the influence of the bar inclination on the position and orientation accuracy.

4.2. Evaluation grid

The evaluation grid consists of the calibration grid and the marker bar, both presented in Section 3.3 about the system description and printed on an A0 sheet of paper. We are aware that paper cockles easily with humidity so we taped the paper on a metal sheet.

The square grid first gives the reference points for internal and external calibration. It is then used to position the system at known coordinates and extract positions relatively to the printed marker bar pattern. The residual errors after parameter tuning for the internal and then external calibrations are given below and determine the quality of the calibration procedure and of the

http://dx.doi.org/10.5772/65784

proposed model. The estimated positions from the marker bar are compared to the known positions on the grid to reveal error statistics in the area of tests.

Figure 15. Evaluation grid and placement of the detector.

4.3. Calibration

4.3.1. Results

Results of a typical calibration experiment are presented in **Figure 16**. This diagram shows the reference points ($z = 0$) of the grid that were detected in at least one calibration image. Only a dozen of evenly spaced images captured during calibration are used in order to reduce memory needs, computation time and information correlation between images. For each image, the camera axis system resulting from the estimation in position and orientation (extrinsic parameters) is plotted. The positions in the diagram show the vertical trajectory of the detector during calibration acquisition. The orientation reveals the inclination of the forward-looking camera. The corresponding errors for this experiment, expressed in pixel distance in the image, are given in **Table 1**.

4.3.2. Discussion

The residual error after calibration can have three origins. First, the model is not perfect and has, for instance, not considered special lens artefacts. Secondly, the calibration plate for the reported tests was printed on a sheet of hard paper that clearly cockled with humidity so that it lost its planarity. Finally, the image localisation is not perfectly accurate, due to image noise and the hypothesis of segment linearity in the vicinity of the crosses.

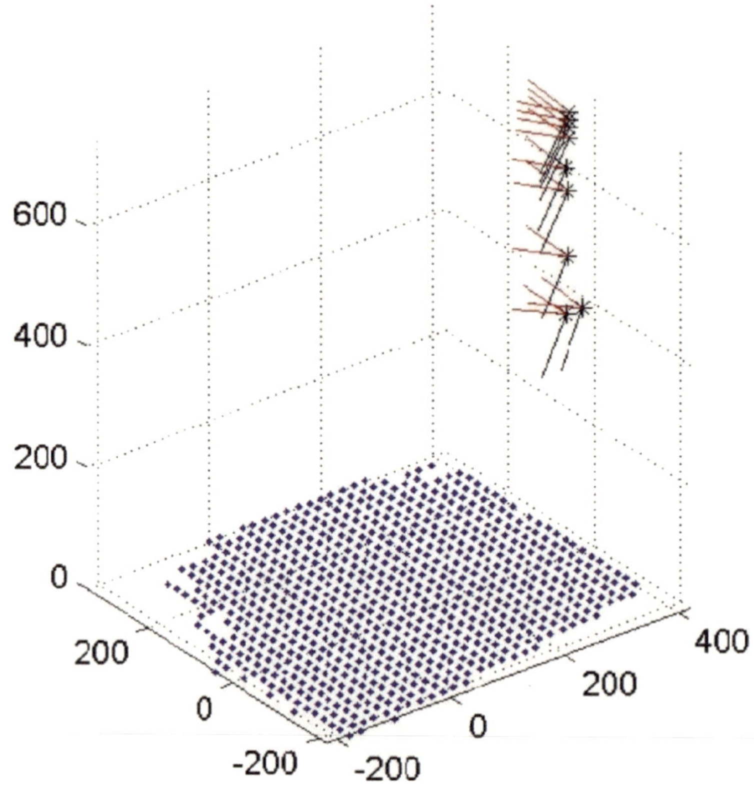

Figure 16. Estimation of the camera position and orientation during calibration. All units in mm.

Standard error	0.259 pixel
Mean error	0.185 pixel
Maximal error	1.56 pixel

Table 1. On-field calibration results.

The error values given here are not to be compared with the results obtained during laboratory calibration of camera with little distortion. The results derive from fast practical tests in a building close to the field, with no special care. The most critical feature of the system is its sensitivity to illumination that largely varies outdoors. This mainly affects position estimation as the calibration can be performed indoors. The fact that this OPMS has no drift is a valuable feature which limits to the minimum the frequency of calibration.

The influence of the calibration imprecision on the overall position accuracy appears to be small in comparison with problems affecting the quality of reconstructed images (the ultimate

http://dx.doi.org/10.5772/65784

aim of position estimation in our application), such as the sensor resolution, the difficulty of sensor capture synchronisation and the influence of the manual scanning (area coverage and height of scanning). No major calibration development was necessary apart from printing the final calibration plate on a plastic substrate, bringing a better stability over time.

The typical processing time amounts to 1 minute on a Pentium II 500 MHz. Further optimisation could be envisaged like a progressive estimation, starting from a reduced set of points, or the conversion of the programme into a compiled language. However, the calibration procedure is an offline task, to be performed from time to time for verification or when a critical parameter is modified.

4.4. Position estimation

4.4.1. Results

As mentioned above, the results concern the position and orientation accuracy of the OPMS when used on the evaluation grid, at several heights and with a flat (the one printed on the calibration grid) or tilted marker bar (tilt angle: 19.3°).

The visualisation of the different coordinates (X', Y' and Z') estimated by the OPMS at different positions on the evaluation grid (X, Y) and at three different heights (Z = 3.8, 7.7 and 11.5 cm) clearly shows on **Figure 17** the qualitative correctness of the system, except in Z for the flat marker bar. This is due to the less precise orientation estimation around the bar. To a smaller extent, we can notice a systematic deviation in the Y' position which should normally look like horizontal rows of crosses on the chart. This is attributed to the distortion compensation which was imperfectly handled at the border of the images.

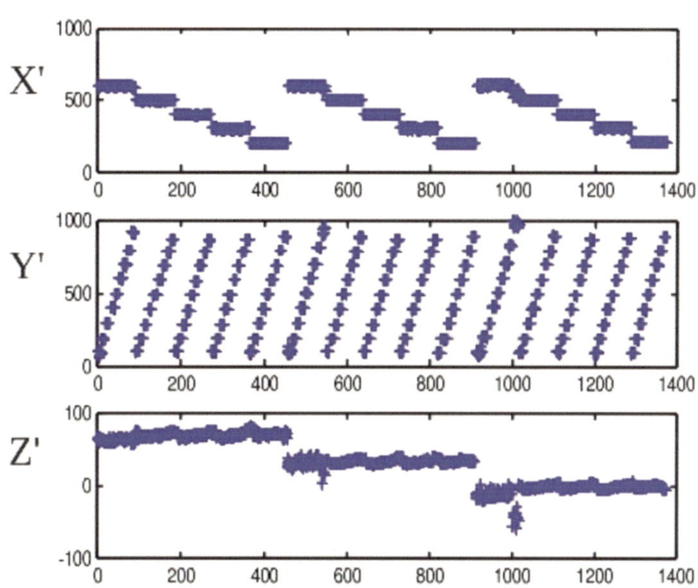

Figure 17. X', Y' and Z' charts (in mm) for all the positions collected during the evaluation at three different heights for a tilted bar.

The histograms of the Euclidean distance (in mm) between the estimated and real 3D positions (**Figure 18**) show again the advantage of the tilted bar, as confirmed by the error levels in **Table 2**. The RMS distance error is the root mean square of 3D distances between the estimated and real positions.

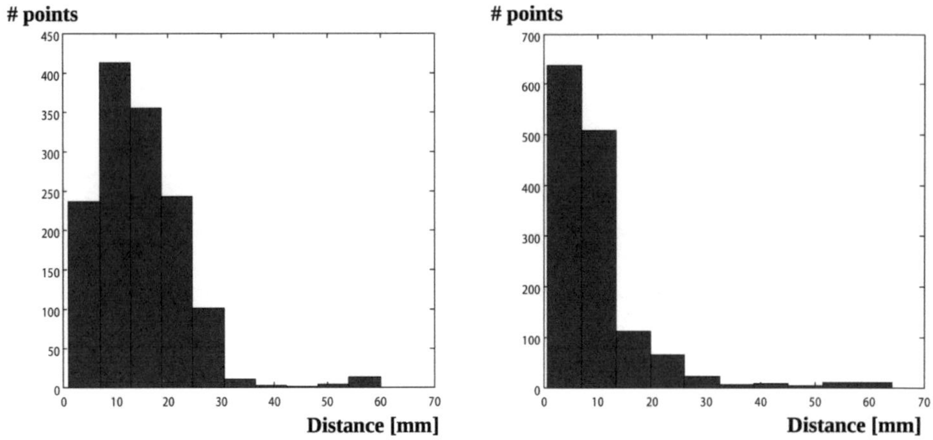

Figure 18. Left: histogram of 3D errors (in mm) for the positions at three different heights with a horizontal marker bar. Right: similar evaluation with a tilted marker bar.

	Flat marker bar	Tilted marker bar
RMS distance error (mm)	16.8	12.93
Std angular error *X* (degree)	1.58	1.16
Std angular error *Y* (degree)	1.51	2.0
Std angular error *Z* (degree)	1.47	1.51

Table 2. Table of position and orientation errors.

The remainder of the chapter will concentrate on results for a tilted marker bar. As they appeared better, we selected this mode of operation.

Figure 19 compares the real and estimated positions in the XY plane. We see here again the problem at the border of the area. As this problem mainly originates from an algorithmic misbehaviour in distortion compensation, we analyse the error level for a central square region with increasing size (see **Figure 20**). Mention that with the constraint of an error level of 5 mm in position and 1 degree in orientation, the acceptable area covers nearly the entire zone of 100 × 50 cm.

Let us finally present in **Figure 21** the matrices of error, representing the scan area 100 × 50 cm in slots of 10 × 10 cm squares, for the scan at *Z* = 11.5 cm. **Figure 21** gives the mean error (Errlm), standard deviation (Stdlm) and maximal value (MaxErrlm) in *X*, *Y* and *Z* and in 3D

http://dx.doi.org/10.5772/65784

with the amplitude coded by a colour corresponding to an error level (in mm) as specified in the legend bar.

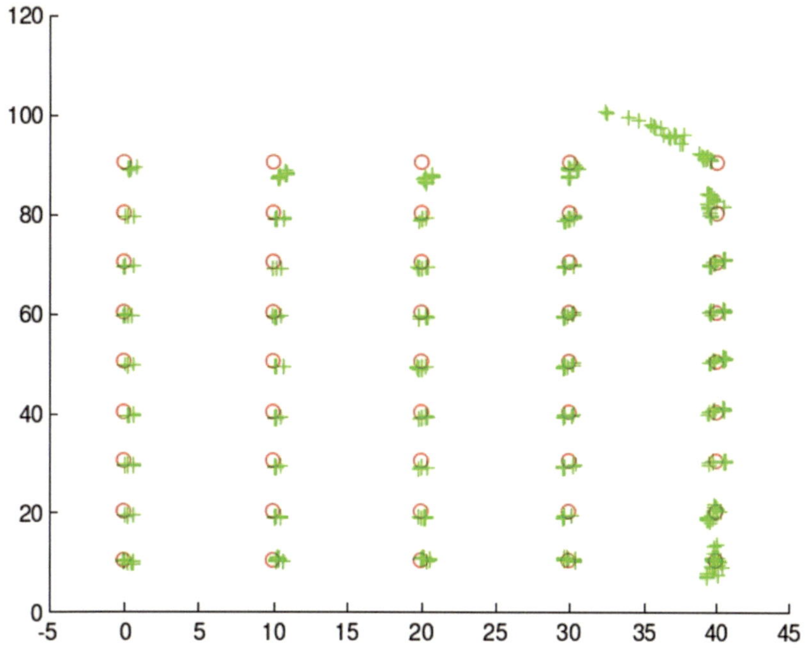

Figure 19. Real (circle) and estimated (cross) positions in the XY plane.

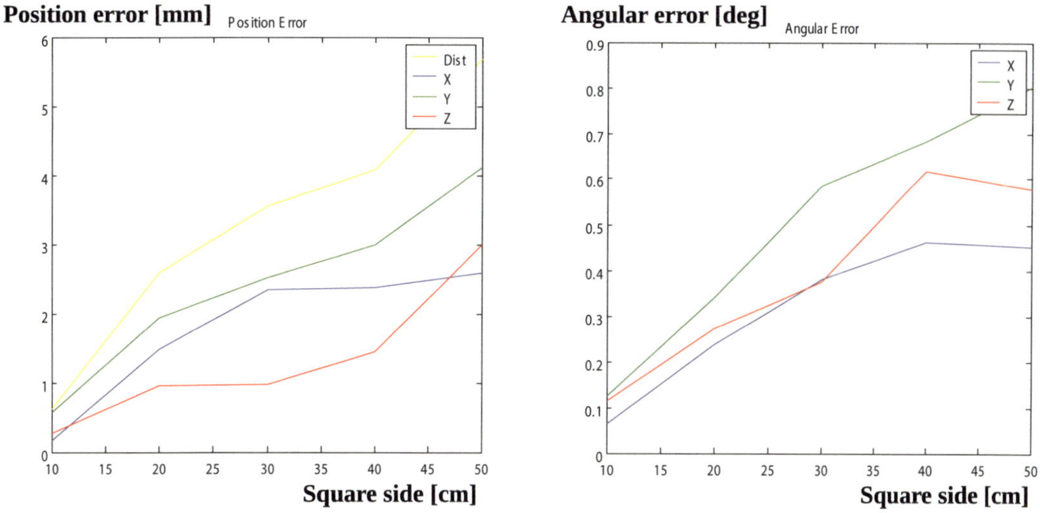

Figure 20. Position error (mm) and orientation error (degree) versus square size (cm) of interest centred in the middle of the 100 × 50 scan area, height = 11.5 cm.

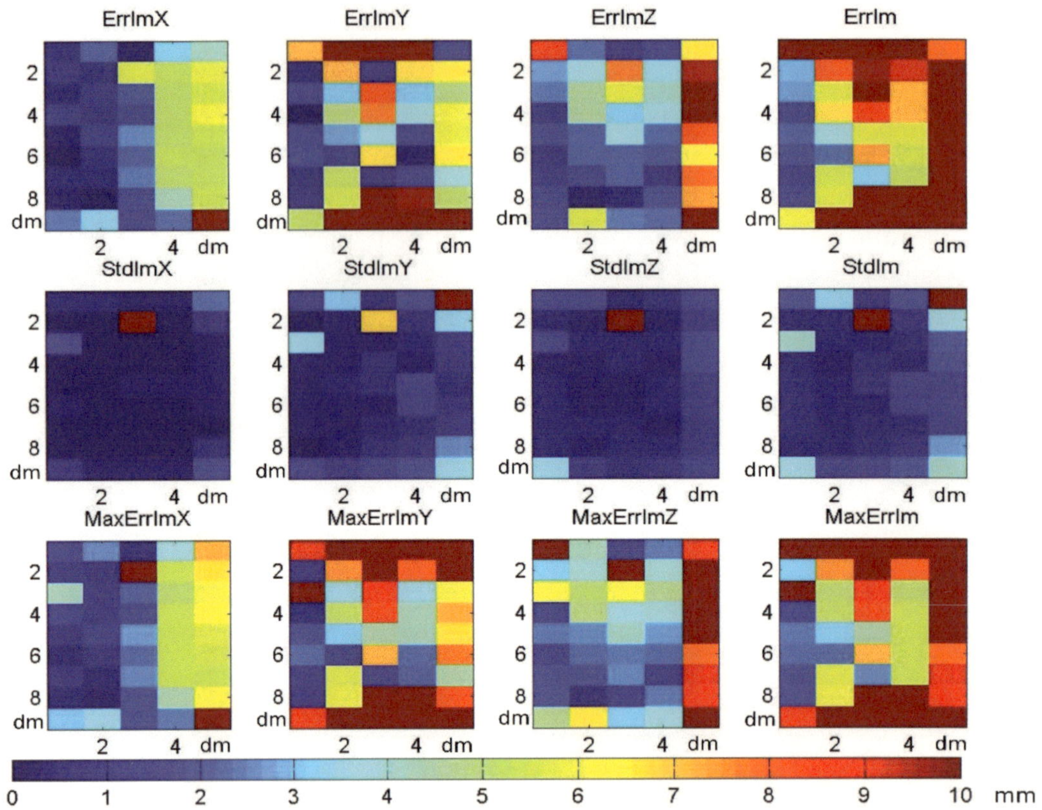

Figure 21. Matrices of error for the scan 100 × 50 cm at $Z = 11.5$, for position (top) and orientation (bottom).

4.4.2. Discussion

Several problems affected the precision of the estimation in position and orientation.

First, the accuracy of the point localisation in the images depends on the algorithm and on the illumination and camera settings (lens aperture, shutter and gain). For instance, the localisation algorithm was optimised with a subpixel accuracy in the direction perpendicular to the bar but simply used the midpoint of two corners, affecting the precision along the bar (Y axis). The results show that the least mean square position of reference points along the bar does not completely compensate for the localisation imprecision in Y. The illumination saturation influences the edge localisation due to the non-linearity of grey levels.

Secondly, a larger number of detected squares contribute to a better estimation of the least mean square procedure. The left and right sides of the scan area are the worst zones for positioning since only a part of the bar is visible from the camera. A low contrast reducing the detectability of the squares also penalises precision.

Thirdly, the parameter estimation is more precise when considering reference points separated by larger distance, due to the general principle of triangulation. Those distances are however limited by the camera field of view which has to stay reasonable to limit distortion.

Finally, the model uses intrinsic parameters which must be determined during a calibration phase. This determination is not error-free and directly influences the position and orientation precision.

4.5. Evaluation conclusions

All the steps required by calibration and position estimation were designed to fulfil the constraints of the application and to realise a prototype that can work in the field. The complete solution is lightweight and cheap. The calibration phase requires the calibration plate and is carried out in a few minutes. As the position estimation of the OPMS has no drift over time, the calibration remains valid until the camera is dismounted from the detector or the lens aperture is modified. Position estimation is performed in real time with about ten measurements per second, which captures the dynamics of the detector movement.

The principal matters of concern in the current implementation are the illumination and synchronisation with the sensors. The quantity of light outdoors varies considerably during the day, and image acquisition has not yet been provided with automatic light adaptation to ensure contrast and avoid saturation. The synchronisation precision of the collected data also conditions the overall accuracy as signals should be captured at the same instants.

In the current implementation, the average position estimation error in the considered area (100 × 50 × 11 cm), evaluated at 1.3 cm, is above the initial requirement of 0.5 cm in each direction (about 0.85 cm of Euclidean distance). The average error in orientation is 2°, also above the initial requirement of 1° in each direction. This however does not affect the quality of image reconstruction for the metal detector and the microwave radiometer. According to **Figure 20**, we are confident that we can reduce these error levels to the requirements thanks to the correction of distortion compensation at the border of the images.

5. Conclusion

Hand-held detectors may benefit from continuous position estimation to localise measurements, to reconstruct 2D (or 3D) measurement maps suited for interpretation and to control if the area is completely scanned.

The OPMS system developed in the HOPE project answers practical constraints of an operational solution. It is weight-, size- and cost-effective since only one camera is necessary. It relies on a reference bar, consistent with many demining protocols, which is typically made of a rigid, straight beam with a specific printed pattern. The additional equipment is easily adapted to an existing detector as long as the reference bar can be kept in the field of view.

The limitations of the approach concern the reference bar and the camera which should not be moved during measurements. The bar is indeed the geometrical reference. Delivered positions are valid since the last calibration as long as the camera is not moved relatively to the detector and its lens is not modified.

The presented solution for positioning requires a camera, a bar with pattern and a calibration procedure. Nowadays, GPS technology offers high precision in very small casing. When differential GPS (DGPS) or Real-Time Kinematic (RTK) can be used, the precision may be high enough to represent a valuable solution. The cost and the needs for satellite visibility may represent an obstacle for these solutions in some circumstances.

More recently a new solution, based on global positioning systems, was designed in the TIRAMISU project. Called TCP box this system uses satellite positioning systems to acquire very precise positions. Mounted in a sensor, it can add the time and location to the sensor's data and transfer the information to a remote ftp server. The TCP boxes can create a local mesh network even when there is no Internet connection.

Acknowledgements

This chapter describes work performed by Yann Yvinec, Pascal Druyts and Charles Beaumier. The authors would like to thank the Belgian Ministry of Defence and the European Commission for the financial support (Esprit Project 29870, DG III).

Author details

Charles Beumier* and Yann Yvinec

*Address all correspondence to: beumier@elec.rma.ac.be

Royal Military Academy, Signal and Image Centre, Brussels, Belgium

References

[1] IMAS 07.11, "Land Release", First Edition 10 June 2009, Amendment 2, March, 2013, Geneva International: Centre for Humanitarian Demining.

[2] HOPE, Hand-held OPErational demining system, Esprit Project 29870, DG III., http://cordis.europa.eu/project/rcn/48853_en.html, accessed on 8 February 2017.

[3] Milisavljevic, N, B. Scheers, O. Thonnard, M. Acheroy, "3D visualization of data acquired by laboratory UWB GPR in the scope of mine detection", In Proceedings of Euro Conference on: Sensor Systems and Signal Processing Techniques Applied to the Detection of Mines and Unexploded Ordnance (Mine '99), Florence, Italy, 1999.

[4] Thonnard, O, N. Milisavljevic, "Metallic shape detection and recognition with a metal detector", In Proceedings of the European Workshp on Photonics applied to Mechanics and Environmental Testing Engineering (PHOTOMEC '99 – ETE '99), Liège, Belgium, 1999.

[5] Havlik, S, P. Licko, "Humanitarian demining: the challenge for robotic research", Journal of Humanitarian Deminig, (2.2), June 1998., http://www.jmu.edu/cisr/journal/2.2/features/havlik.htm, accessed on 8 February 2017.

[6] Hightower, J, G. Borriello, A Survey and Taxonomy of Location Sensing Systems for Ubiquitous Computing, Technical report UW CSE 01-08-03, University of Washington, Dpt of Computer Science and Engineering, Seattle, WA, August 2001.

[7] Sato, M, J. Fujiwara, K. Takahashi, ALIS Evaluation Tests in Croatia, SPIE, Detection and Remediation Technologies for Mines and Minelike Targets XIV, Florida, USA, 2009.

[8] Beumier, C, P. Druyts, Y. Yvinec, M. Acheroy, "Motion estimation of a hand-held mine detector", Signal Processing Symposium, Hilvarenbeek, The Netherlands, 23rd–24th March, 2000.

[9] Beumier, C., P. Druyts, Y. Yvinec, M. Acheroy, "Real-time optical position monitoring using a reference bar", Signal Processing and Communications (SPC2000), IASTED International Conference, Marbella, Spain, September 19-22, 2000, pp. 468-473.

[10] Heikkilä, J, "Geometric camera calibration using circular control points", IEEE Transactions on Pattern Analysis and Machine Intelligence; 22(10), 1066-1077, October 2000.

Chapter 3

Ground-Penetrating Radar for Close-in Mine Detection

Olga Lucia Lopera Tellez and Bart Scheers

Additional information is available at the end of the chapter

http://dx.doi.org/10.5772/67007

Abstract

In this chapter, two of the major challenges in the application of ground-penetrating radar in humanitarian demining operations are addressed: (i) development and testing of affordable and practical ground penetrating radar (GPR)-based systems, which can be used off-ground and (ii) development of robust signal processing techniques for landmines detection and identification. Different approaches developed at the Royal Military Academy in order to demonstrate the possibility of enhancing close-range landmine detection and identification using ground-penetrating radar under laboratory and outdoor conditions are summarized here. Data acquired using different affordable and practical GPR-based systems are used to validate a number of promising developments in signal processing techniques for target detection and identification. The proposed approaches have been validated with success in laboratory and outdoor conditions and for different scenarios, including antipersonnel, low-metal content landmines, improvised explosive devices and real mine-affected soils.

Keywords: ground-penetrating radar, clutter reduction, object detection, landmines

1. Introduction

To make reliable, easily interpreted and less time-consuming operational systems for landmine detection is a real challenge [1, 2]. Nowadays, demining is performed by using different kinds of demining systems, e.g., mechanical excavation, trained dogs/rodents, and metal detectors (**Figure 1**). Metal detector (MD) is one of the most used close-range detection systems for demining. However, antipersonnel (AP) landmines are not made any more with significant amounts of metal but with plastic and other nonmetallic elements. Metal detection-based systems available today do not efficiently detect plastic landmines with minimum metal content in a metal debris contaminated area. In order to compensate for small metal content

Figure 1. An example of classical demining operations in Croatia: deminers scan with metal detectors and trained dogs after the mechanical excavation [4].

in modern landmines, some sensors offer the possibility to the operator to increase their sensibility. However, the number of false alarms rises. Cambodian deminers are confronted with this problem daily. For each detected AP landmine, more than 500 inoffensive metallic debris such as grenade fragments and cartridges are located (some results collected after visiting a Cambodian mine-field, **Figure 2**). False alarm rate as well as misdetection of low and nonmetal content of AP landmines have made mine clearance operations dangerous, time consuming, and expensive.

During the 1990s, several research groups started contributing in solving this problem by developing hardware and software for demining applications [1–3]. However, only a few are currently employed in real mine-affected areas. One of these relatively new technologies is ground penetrating radar (GPR), an attractive choice for landmine detection due to their advantages over other sensors. GPR can detect both metallic and plastic mines in a variety of

Figure 2. Some of the objects found after demining operations with metal detectors in the M7753 minefield, Province of Siem Reap, Cambodia [4].

soils by noninvasive subsurface sensing [3]. GPR sends a series of microwave pulses ranging from about 1 to 4 GHz into the ground. It then looks for anomalies in the reflected signal, which could indicate the presence of a landmine. In terms of buried target detection, the strength of radar echoes is usually associated with contrasts of electromagnetic characteristics between targets and their surrounding soils. For antipersonnel mines, GPR is usually used in combination with a metal detector [3, 4]. The metal detector would detect all metal contents in the soil, and the GPR is used to discriminate on the size of the objects: smaller metal objects are discarded and larger metal objects are confirmed as dangerous. Note that the GPR has the capacity to detect nonmetal mines. But when used alone its possible large false alarm rate makes it more suited to look for antitank mines. Moreover, its weight can be light, so that it can be installed in a hand-held system or in a vehicle-mounted system in the form of an array of multiple antenna elements [3].

This chapter addresses two of the major challenges in the application of GPR in humanitarian demining operations: (i) development of affordable and practical GPR-based systems and (ii) development of robust GPR signal processing techniques for landmine detection and identification. This chapter also reviews research carried out at the Royal Military Academy in these topics.

2. GPR systems

2.1. GPR principles

Electrical properties of materials are determined by electrical conductivity, permittivity and permeability, which are function of frequency. The relative permittivity (or dielectric constant) of a medium impacts the electric field propagation and is the most important parameter for GPR. The relative permeability affects the magnetic field propagation. The electromagnetic wave attenuation in the subsurface is strongly dependent on the electrical conductivity of the medium and its variation. The latter is normally controlled by water [4–8]. For a conductive material, the electromagnetic field is diffusive and cannot propagate as an electromagnetic wave. When it is resistive, or dielectric, an electromagnetic field can propagate as an electromagnetic wave. When an electromagnetic wave is send into the ground, GPR measures the reflected echoes from any electrical property discontinuity in the subsurface structure. **Figure 3** shows a block diagram of a generic GPR system [5].

The velocity and reflectivity of the electromagnetic wave in soil are characterized by the dielectric constant (relative permittivity) of the soil. When the dielectric constant of the soil is ε_r, the propagation velocity of electromagnetic waves in this medium is given by

$$v = \frac{c}{\sqrt{\varepsilon_r}} \tag{1}$$

where $c = 3 \times 10^8$ m/s. Note then that in air ($\varepsilon_r = 1$), the propagation velocity is then $v = c$. Dielectric constants measured at 100 MHz for different materials [3] are listed in **Table 1**.

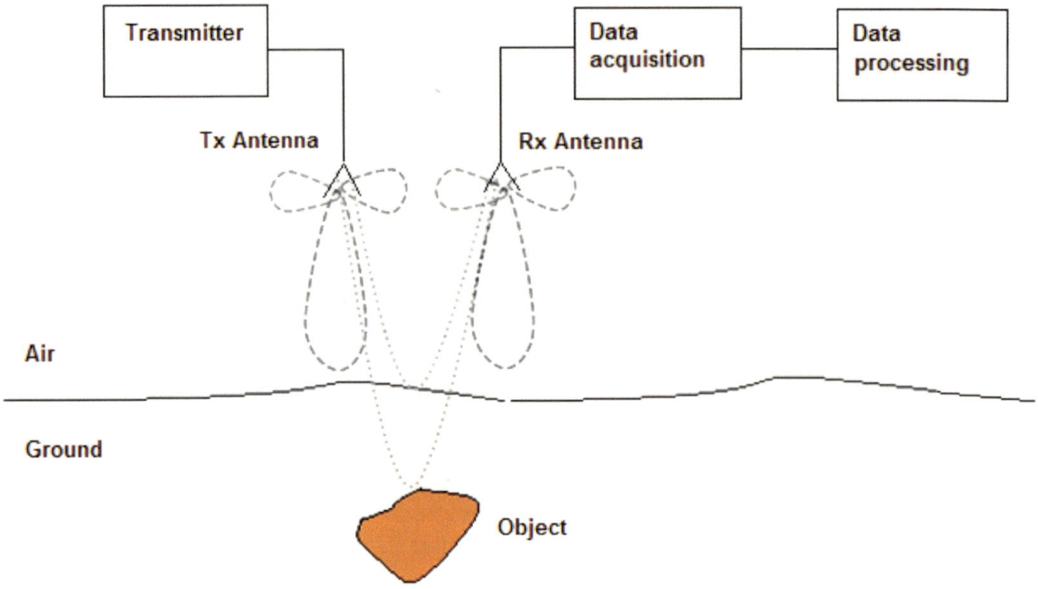

Figure 3. Block diagram of a generic GPR system [5].

Material	Attenuation (dB/m)	Relative permittivity ε_r
Air	0	1
Clay	10–100	2–40
Concrete: dry	2–12	4–10
Concrete: wet	10–25	10–20
Fresh water	0.1	80
Sand: dry	0.01–1	4–6
Sand: saturated	0.03–0.3	10–30
Sandstone: dry	2–10	2–3
Sandstone: wet	10–20	5–10
Seawater	1000	81
Soil: firm	0.1–2	8–12
Soil: sandy dry	0.1–2	4–6
Soil: sandy wet	1–5	15–30
Soil: loamy dry	0.5–3	4–6
Soil: loamy wet	1–6	10–20
Soil: clayey dry	0.3–3	4–6
Soil: clayey wet	5–30	10–15
TNT	–	3
Plastic	–	2–4

Table 1. Attenuation and relative permittivity of subsurface materials measured at 100 MHz [3].

When GPR transmits electromagnetic waves from a transmitting antenna located off-the-ground, signals travel in the air layer and when the electromagnetic wave encounters any dielectrical discontinuity, a reflection occurs. The latter is received by a receiving antenna, located off-the-ground, and it is referred to as an A-scan, e.g., a single waveform recorded by GPR, with the antennas at a given position (x, y). In this data set, the time t is the only variable, related to the depth z by the propagation velocity of the EM waves in the medium. Its representation in the time domain can be seen in **Figure 4** [5]. The time axis or the related depth axis is usually pointed downward.

When moving the GPR antennas on a line along the x-axis, a set of A-scans can be gathered, which form a two-dimensional (2D) data set called a B-scan (**Figure 5**). When the amplitude of

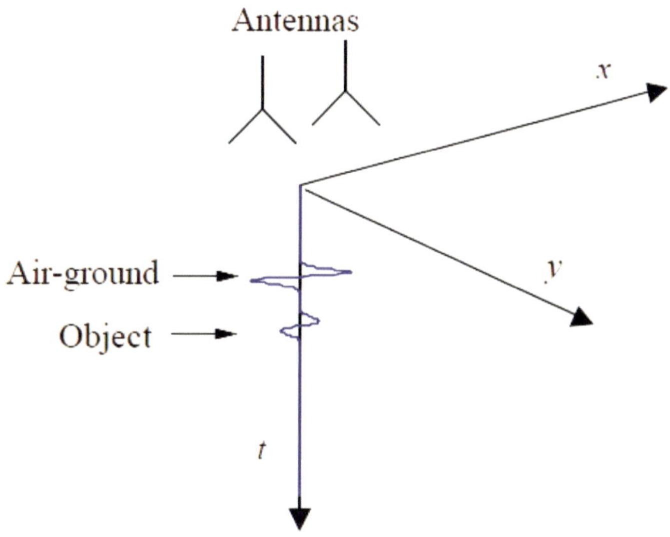

Figure 4. Representation of an A-scan [5].

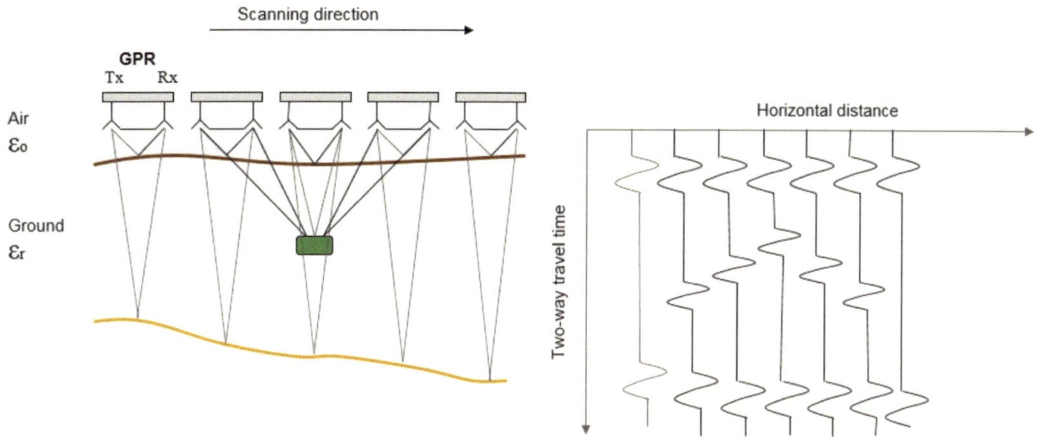

Figure 5. Configuration and representation of a B-scan.

the received signal is represented by a color scale (e.g., gray-scale), a 2D image is obtained and is shown in **Figure 6**. The 2D image represents a vertical slice in the ground. Reflections on a point scatter located below the surface appear, due to the beamwidth of the transmitting and the receiving antenna, as hyperbolic structures in a B-scan. Finally, when moving the antenna over a (regular) grid in the xy-plane, a three-dimensional (3D) data set can be recorded, called a C-scan (**Figure 6**). Usually a C-scan is represented as a two-dimensional image by plotting the amplitudes of the recorded data at a given time t_i. The resulting image represents then a horizontal slice at a certain depth, parallel to the recording plane (**Figure 7**). Nowadays, many user-software packages (e.g., GPRMax2D and 3D [7]) have integrated functions to plot directly two- or three-dimensional representations of the recorded C-scans. Examples of these plots are: (i) an arbitrary cut in the 3D volume and (ii) an isosurface (surface with the same amplitude), both represented in **Figure 7** [5].

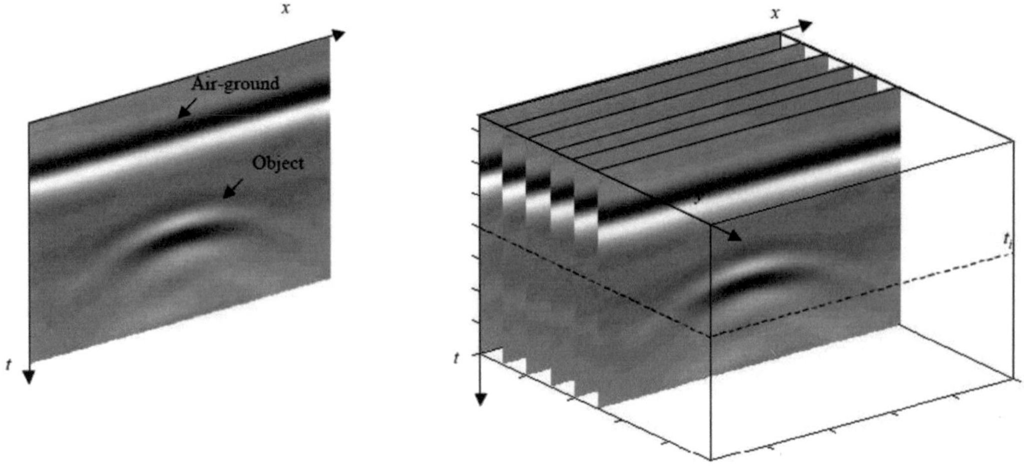

Figure 6. A gray-scale illustration of a B-scan (left) and a series of B-scans forming a C-scan (right).

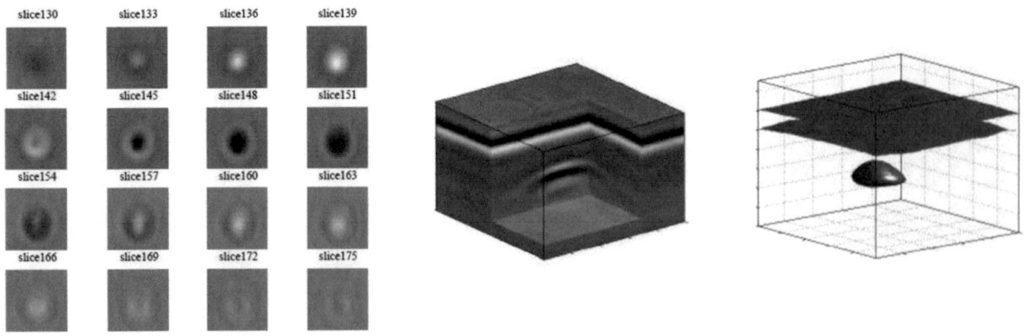

Figure 7. Representation of a C-scan by horizontal slices at different depths (left), arbitrary cut in the 3D volume (center), and isosurface (right) [5].

http://dx.doi.org/10.5772/67007

2.2. GPR design

Table 2 gives a schematic overview of the various possible types of GPR systems that exist today. GPR systems can be classified by the domain in which they work and by the type of modulation. GPR systems operate either in time domain or in frequency domain. In the time domain GPR there are two major categories: the amplitude modulated and the carrier-free GPR. The first one sends a pulse with a carrier frequency. This carrier frequency is modulated by a (square) envelope. Good depth resolution is achieved by reducing the duration of the pulse as short as possible. Most of the commercially available GPRs belong to this group [5].

The need for larger bandwidth has led to the development of a second category of time domain GPR: the carrier-free GPR. The pulse sent by the GPR has no carrier. The shape of the pulse can vary, but typically a Gaussian pulse is used. The carrier-free radar is also called an ultra-wide band (UWB) GPR because of the large bandwidth. Before, GPR systems were developed based on time domain waveform. Nowadays GPRs are also developed in the frequency domain. In the latter, systems can have two possible modulation types: either the frequency modulated (FM) continuous wave (CW) or stepped frequency (SF) GPR [5].

FM systems transmit a carrier frequency, which changes continuously by using a voltage-controlled oscillator over a certain frequency range. The frequency sweeps according to a function within a certain time. After reception, the reflected wave is mixed with the emitted one, and the target depth can be calculated from the difference in frequency between the transmitted and received wave. FM systems have poor dynamic range, which is an important limitation. Since FM radars receive signals at the same time as it is transmitting, leakage signals between the antennas can obscure small reflections. Those two facts deviate the attention from FM systems to SF radars for ground applications.

An SF GPR uses a frequency synthesizer to go through a range of frequencies equally spaced by an interval Δ. At each step, a CW is send with a high stability and mixed with the received signal using a quadrature mixer. For each frequency, the amplitude and phase of the received signal are compared with the transmitted signal [5]. The reader can refer to Ref. [9] for a good overview of this technique. Stepped-frequency continuous-wave (SFCW) radars have some advantages over the others, including wider dynamic range and better signal-to-noise ratio (SNR) [3, 5, 6].

Besides the domain of operation, GPR antennas may be either monostatic (single emitting and receiving antenna), bistatic (different emitting and receiving antenna), or an array configuration of different antenna types and sizes. In the mine detection application, high depth resolution is needed and therefore ultra-wide band (UWB) antennas play an important role [8].

GPR design options				
Domain	**Time**		**Frequency**	
Modulation	Amplitude (mono cycle)	Carrier-free	Linear sweep	Stepped frequency

Table 2. Different types of GPR systems.

Vehicle-based systems generally use array antenna mode [10] in combination with other sensors such as metal detectors [11]. Laboratory prototypes of UWB GPR systems are built in bistatic mode [12, 13], and such a configuration is adopted for hand-held GPR-based systems [14, 15]. For all these configurations, different types of antennas such as horn, loop, spiral, Vivaldi, and combinations of them are used. An overview of their characteristics, advantages, and drawbacks for demining applications can be found in Refs. [5, 16].

2.3. UWB GPR laboratory prototypes

Landmine detection using GPR is a very particular problem. Commercial GPRs are mostly designed for geophysical applications and use central frequencies up to 1 GHz. As landmines are small objects, a large bandwidth is needed for a better depth resolution. Therefore, we have decided to build our own UWB system in the frame of the HUDEM (Humanitarian Demining) project (in collaboration with the Microwave Engineering and Applied Electromagnetism Department of the Catholic University of Louvain, UCL) and to use a GPR-based system (under the BEMAT (Belgian Mine Action Technologies) project) developed at the Environmental Sciences Institute of the UCL. These choices were made following five technical and practical requirements [5]:

1. The GPR system must be UWB (working in the frequency range 500 MHz to 4.5 GHz).
2. The GPR system must be used off ground (safety reasons and to increase mobility of the system, see point 3 below). Therefore, the antenna must be highly directive in order to couple sufficient energy into the ground for achieving a penetration depth of 15 cm in any soil [17].
3. The GPR system must guarantee a high degree of mobility, i.e., attention should be paid to dimensions and weight. Minefields have often rough surfaces, steep slopes and/or are covered with dense vegetation. Not all systems can guarantee a sufficient flexibility in such scenes.
4. The antenna properties must be independent of the ground properties.
5. The GPR system must be cheap in production to limit the overall cost of the sensor. This will always be asked for in the case of humanitarian demining.

2.3.1. Development of a transverse electromagnetic horn antenna

Based on the requirements listed before, many researchers have focused their attention into transverse electromagnetic (TEM) horn antennas, which have a high directivity, can work broadband and are nondispersive. A traveling wave TEM horn consists of a pair of triangular conductors forming a V structure (**Figure 8**), which can transmit and receive a fast transient pulse [18, 19]. It is presumed that the TEM horn conducts mainly the TEM mode within a selected frequency range by conserving constant characteristic impedance and that, by omitting the edge diffraction effect and fringe fields, a linearly polarized spherical wave is diffused.

This type of antenna is totally parameterized by three characteristics: L is the length of the antenna plates, φ_0 is the azimuth half-angle, and θ_0 is the elevation half-angle. The characteristic impedance of an infinite long TEM horn ($L = \infty$) only depends on the two angles φ_0 and

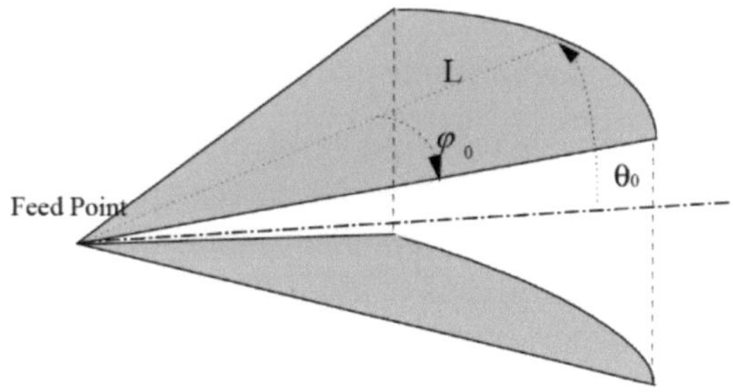

Figure 8. Conventional design of the TEM horn antenna characterized by three parameters: L the length of the antenna plates, φ_0 the azimuth half-angle, and θ_0 the elevation half-angle.

θ_0. In theory, a TEM mode does not have an upper cut-off frequency; however, this upper cut-off frequency will be practically restricted. The length L mainly determines the lower cut-off frequency.

For improving the directivity and reducing the physical antenna dimensions without diminishing the bandwidth, TEM horns can be filled with a silicone. A silicone characterized by a real relative permittivity ε_r of 2.89 and a loss tangent of 0.0084 at 1 GHz is used. As a result, the propagation speed of the TEM wave between the antenna plates will be divided by $\sqrt{\varepsilon_r}$ (or the electrical length of the antenna will be increased by a factor $\sqrt{\varepsilon_r}$) and the surge impedance will be reduced by a factor $\sqrt{\varepsilon_r}$ [20].

For preserving the same surge impedance as without the silicone, the angle θ_0 can be increased, which improves the directivity. In this case, the antenna impedance is chosen to match the 50 Ω driving cable. As a result, no mismatches will be found at the part of the transient traveling current that bounces back at the antenna aperture toward the excitation source, avoiding antenna ringing. The principle seems to work well for frequencies in the band of the antenna [20].

The antenna plates are replaced by printing sets of 41 wires on circuit boards (**Figure 9**). Since the distance between the wires is small, the antenna characteristics are preserved. In addition,

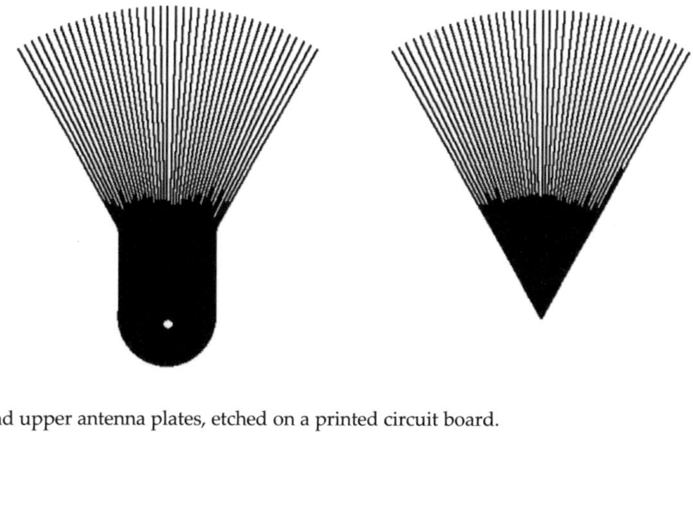

Figure 9. Lower and upper antenna plates, etched on a printed circuit board.

the currents are forced to be radial, limiting the surface of conducting metal. In this application, this is of great importance since GPRs are often used in combination with metal detectors.

In order to feed the TEM horn in its balanced configuration (**Figure 8**) with an unbalanced coaxial line, an UWB (frequency-independent) balun is required to prevent currents on the coax. Several realizations of a TEM horn were made, and, at first, measurements revealed an unbalanced current component on the coax exterior. This means that the coaxial feedline was reacting as an antenna. A common way of eliminating such currents is to add chokes (ferrite cylinders) around the feeding cable [21]. For this design, a new kind of balun is tested, which principle lays on an electrostatic reasoning described in Ref. [22]. The function of the taper in the bottom plate is to provide a gradual transition between the unbalanced upper antenna plates on a ground-plane, toward a balanced alignment with two symmetrical antenna plates. However, a slide change of the surge impedance along the antenna is introduced by such a transition. The surge impedance of one antenna plate on a ground-plane is half the value calculated using the wire model. Since θ_0 in the unbalanced configuration is measured from the ground plane, the resulting value will double the elevation half-angle of the balanced configuration. As a result, the surge impedance of the unbalanced part is found to be slightly inferior to the surge impedance of the balanced part of the antenna. Using the wire model and taking into account the reduction of the surge impedance due to the filling and the influence of the balun, an elevation half-angle θ_0 (defined in the balanced part) of 14.5° is calculated to match 50 Ω. Summarizing, the antenna parameters are $L = 12$ cm, $\varphi_0 = 30°$, and $\theta_0 = 14.5°$, which leads to a physical antenna aperture of 12 cm × 6 cm.

The dielectric-filled antennas were integrated in a laboratory UWB GPR. For this, a study was made in order to optimize the position and orientation of the Tx and Rx antennas. To reduce the coupling between the two TEM horns, they were put side by side with a common H-plane (E-field of both antennas is parallel to the interface). Generally speaking, antenna coupling is not critical since it could be neutralized. However, if the ringing between the two antennas lasts too long, it could obscure interesting parts of the returned signal. After some tests, the antennas were fixed at around 25 cm above the surface [12].

When configuring the Tx-Rx antenna system, it is important to consider the two antennas as one antenna. Therefore, the combined antenna pattern should be analyzed. The 3 dB beamwidth that results of this combination is normally function of the offset angle θ_1, as shown in **Figure 10**. When moving the GPR antennas over a fix target, in the backscattered signal displayed as function of time (B-scan) a hyperbolic structure will appear (**Figure 6**). A wider 3 dB beamwidth will produce wider hyperbolas in the B-scan and could therefore improve the detection of objects. The total energy found in these hyperbolic structures is considered as a measure for the optimization of the offset angle θ_1. The expected reflected energy of the object is represented by this total energy, after focusing the B-scan using an optimal migration methodology. In order to optimize the offset angle, different synthetic B-scans of a buried object at 6 cm depth are simulated for different values of θ_1 [12]. An offset angle of 20° is found to produce the hyperbola with the maximum of energy. In the configuration consider for this calculation (object at a depth of 6 cm), 20° corresponds to the angle that focuses the antennas on the object, considering the refraction. In a real scenario, the object is buried at an unknown depth, which is expected to be

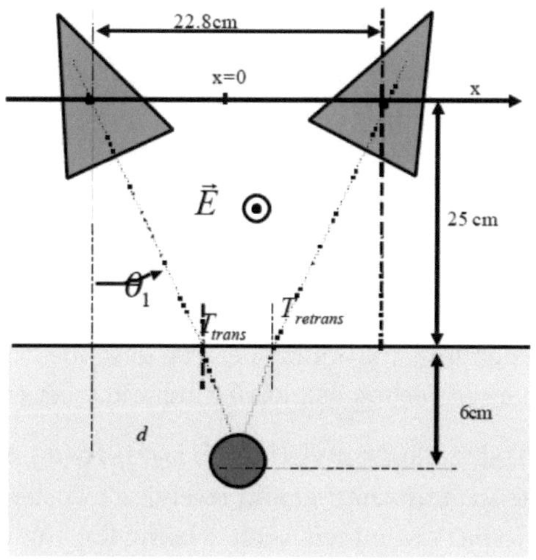

Figure 10. Dielectric-filled TEM horn antenna configuration in bistatic mode [12].

no deeper than 20 cm. Therefore, when building the laboratory UWB GPR, an angle θ_1 of 20° was selected.

Figure 11 represents a GPR image of a PMN AP landmine (12.5 cm diameter, plastic case), buried at 1 cm depth in dry sand. Data are acquired by an impulse UWB GPR emulated using a picosecond pulse labs step-generator type PSPL 4050B, followed by an impulse-forming-network and connected to two identical dielectric-filled TEM horn antennas. Data are taken by displacing the Tx and Rx antennas by steps of 1 cm (represented on the x-axis). In each antenna position, a short Gaussian impulse is radiated and the backscattered signal is recorded (y-axis).

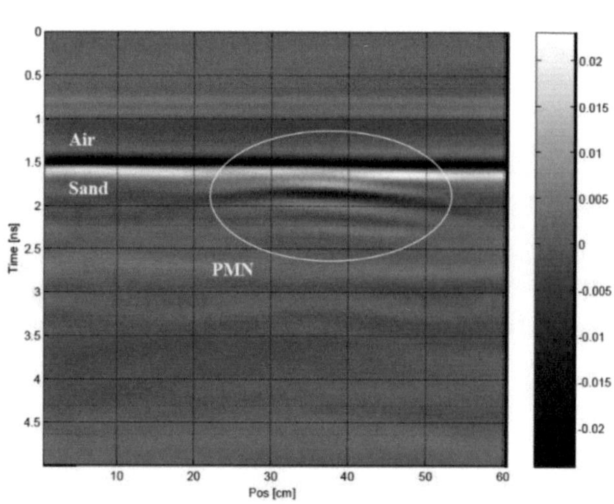

Figure 11. Time domain representation of the impulse UWB GPR measurements performed on sand for the PMN AP mine, buried at 1 cm.

2.3.2. UWB SFCW GPR in monostatic mode

An SFCW system is considered in the frame of the BEMAT project, which also covers several of the requirements mentioned before. It is based on the frequency domain radar-antenna-multi-layered medium model developed by Lambot et al. [23], which applies for SFCW radars operating off the ground in monostatic mode (in our case, a portable vector network analyzer (VNA) connected to a monostatic horn antenna, see **Figure 12**). In this approach, it is assumed that the distribution of the backscattered electric field measured by the antenna does not rely upon the elements of propagation (air and subsurface layers), i.e., only the amplitude and phase of the field change. Therefore, the antenna can be described by a model of linear transfer functions in series and in parallel, acting as global transmittances and reflectances (**Figure 13**).

In this model, $G(\omega)$ is the soil response and $H_i(\omega)$, $H_t(\omega)$, $H_r(\omega)$, $H_f(\omega)$ are, respectively, the antenna return loss, the antenna transmitting and receiving transfer functions, and the antenna feedback loss, which have to be determined, with ω being the angular frequency. The antenna equation can be written as

$$S(\omega) = H_i(\omega) + \frac{H_t(\omega)H_r(\omega)G(\omega)}{1 - H_f(\omega)G(\omega)} \tag{2}$$

The latter equation is represented in the frequency domain and the transfer functions are frequency dependent complex quantities. The multiple wave reflections occurring inside the antenna, which, as stated before, are a result of the impedance differences between the antenna feed point and the antenna aperture are represented by the return loss function $H_i(\omega)$. Such inner wave reflections are independent of the medium, $H_t(\omega)$ and $H_r(\omega)$ constitute the antenna gain and phase delay between the VNA calibration plane (the connection between the antenna

Figure 12. The UWB SFCW GPR system emulated using the hand-held VNA connected to the horn antenna via a 50-Ohm N-type coaxial cable.

http://dx.doi.org/10.5772/67007

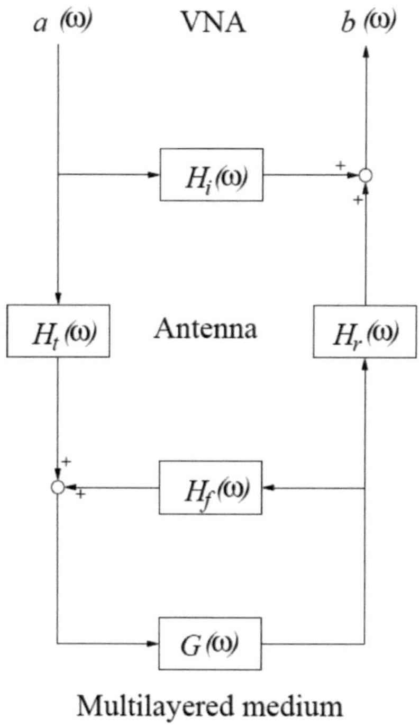

Figure 13. The VNA-antenna-multilayered medium system modeled as linear transfer functions in series and in parallel.

feed point and the VNA cable) and the source and receiver virtual point (antenna phase center). The feedback loss transfer function $H_f(\omega)$ represents multiple wave reflections between the antenna and the subsurface. $H_i(\omega)$ and $H_f(\omega)$ are global reflectances, and $H_t(\omega)$ and $H_r(\omega)$ are global transmittances. Finally, $G(\omega)$ accounts for the air-soil system [23] and is defined by a Green's function, solution of the 3D Maxwell's equations for electromagnetic waves propagating in horizontally multilayered media [24].

The antenna transfer function $H_i(\omega)$ can be determined from using a measurement in free space conditions, for which $G(\omega) = 0$. Therefore, from Eq. (2) $H_i(\omega) = S_{fe}(\omega)$. The antenna transfer functions $H_t(\omega)$, $H_r(\omega)$, and $H_f(\omega)$ are determined using measurements at two different heights above a planar metal sheet (considered as an infinite perfect electric conductor), for which $G(\omega)$ is known accurately, and correspond therefore to a specific distribution of the measured electromagnetic field. Consider $H(\omega) = H_t(\omega)\, H_r(\omega)$, then

$$H_f(\omega) = \frac{\frac{S^b(\omega)}{G^b(\omega)} - \frac{S^a(\omega)}{G^a(\omega)}}{S^b - S^a} \tag{3}$$

$$H(\omega) = \frac{S^a(\omega)}{G^a(\omega)} - S^a(\omega)H_f(\omega) \tag{4}$$

On the one hand, a necessary condition for this antenna model is to employ the antenna sufficiently far from the surface. On the other hand, it is necessary to minimize losses by spherical divergence in wave propagation, keep a high SNR, and ensure a high spatial resolution, conditions that are achieved by minimizing the distance between the antenna and the surface. In Ref. [25], laboratory experiments were carried out to characterize the transfer functions $H_i(\omega)$, $H(\omega)$, and $H_f(\omega)$ of a horn antenna (the antenna used in this study) as a function of height to determine to which extend they are still correct when approaching the soil surface. For this, electromagnetic measurements were carried out with the antenna aperture situated at 30 different heights h (from 1 to 28 cm) above a metal sheet of 2×2 m^2. Results shown small errors on the calculation of the transfer functions for $h > 12$ cm. Below $h \cong 12$ cm, these errors appear to increase, resulting in important errors in the calculation of the soil response $G(\omega)$ for field applications. These results indicate that, in the application of landmine detection, the independency of the received field distribution with respect to the medium is maintained when the antenna aperture is at minimum $h = 12$ cm.

In order to emulate the UWB SFCW radar, a low-cost hand-held VNA connected to a monostatic horn antenna via a 50 N-type coaxial cable is used. The VNA comprises a spectral analyzer (FSH6, Rohde and Schwarz), which uses a bridge and power divider (VSWR, Rohde and Schwarz) to give vector measurements. A linear polarized double-ridged broadband horn antenna (BBHA 9120A, Schwarzbeck Mess-Elektronik), with 22 cm length and 14×24 cm^2 aperture area, is used to collect data. For this application and from antenna characteristics, it can be considered as directive (3-dB beamwidth of 45° in the E-plane and 30°in the H-plane at 1 GHz and 27° in the E-plane and 22°in the H-plane at 2 GHz, when working in the 0.8–5.0 GHz frequency range).

This UWB SPCW radar is easy to use and affordable, and it covers several of the requirements listed before. It has a linear dynamic range of 60 dB, allowing detecting weak scatterers. Internal reflections inside the antenna are accurately calculated and included in the EM model described before. As a result, they do not influence negatively the signal-to-noise ratio of the system. Besides, amplitude drift (that can be due to mechanical or temperature changes on the connection point of the antenna) is limited by a precise and easy-to-do calibration method using a standard Open-Short-Match calibration kit. While gathering data, this calibration could be performed once more.

Figure 14 shows a time domain representation of radar measurements performed in a homogeneous sand for the PMN AP landmine (12.5 cm diameter, plastic case) buried at 10 cm. The horn antenna was displaced over the x-axis following constant steps of 2 cm. The height of the antenna aperture is 20 cm above the soil surface.

2.4. Hand-held dual-sensors

During the past 10 years, the development of GPR applied to landmine detection has evolved from the laboratory conditions and test fields to real minefields. Nowadays, systems using dual sensor technology combining MD and GPR (hand-held dual sensors) have enabled improved discrimination against small metal fragments to be demonstrated in live minefields. Some of them have reached the stage where they are being produced in large numbers. Such

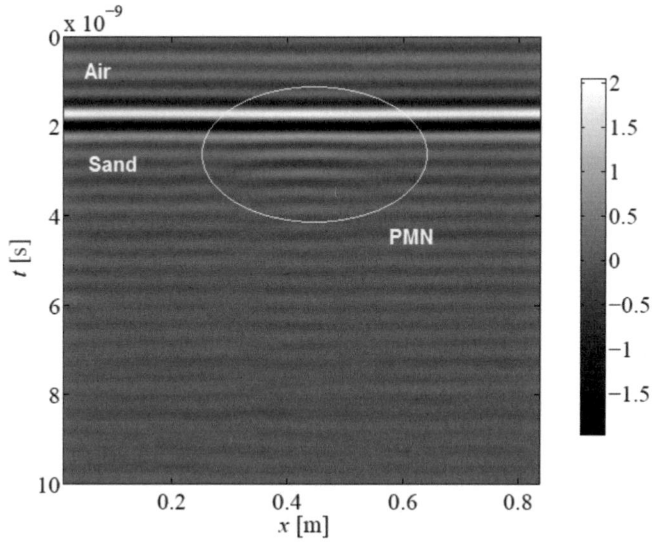

Figure 14. Time domain representation of the SFCW UWB GPR measurements performed on sand for the PMN AP mine, buried at 10 cm.

systems work with both MD and GPR, and they differ on the operating principle of GPR, the signal processing and the user interface. In this section, three of them are introduced.

MINEHOUND/VMR2 has been developed by ERA Tech., U.K., in collaboration with Vallon Gmbh, Germany. It combines a pulse induction MD and an impulse GPR. This dual-sensor transforms MD and GPR signals into two separated audio signals of different frequency of vibration and tone. MINEHOUND/VMR2 has been tested in different mine-affected countries including Afghanistan, Angola, Bosnia and Cambodia [26].

ALIS (Advanced Landmine Imaging System) is a Japanese detector developed at the Tohoku University. It incorporates an MD and a GPR, in combination with a sensor tracking system, which makes possible to analyze and visualize the data (after migration). Its hand-held version is equipped with a VNA-based GPR and a pulse induction MD. This dual-sensor provides two different user interfaces: audio for MD signal and images for both MD and GPR signals. Different trials have been performed with ALIS in mine-affected countries including Afghanistan, Cambodia, Croatia, and Egypt [27].

AN/PSS-14 (former HSTAMIDS, Hand-held Standoff Mine Detection System) was originally a project founded by the Defense Advanced Research Projects Agency, DARPA, U.S., and has been produced later by CyTerra Corp. (now L3 Communications). It is equipped with a pulse induction MD and a GPR based on a wide-band, SF radar. AN/PSS14 gives to the user an audio signal when a metallic object is detected. If there is a metal detection and the GPR system identifies other mine-like material as well, a second sound, of a different frequency of vibration and tonality, is played as aided target recognition. This system has been produced since 2006 for U.S. Army operations in Iraq [28].

3. GPR signal processing

GPR principal function is to detect differences on the electromagnetic (EM) properties in the soil-target medium. This permits to locate even low and nonmetallic landmines. Apart from the response from a potential target, the backscattered signal carries also undesirable effects from antenna coupling, system ringing and subsurface reflections, which hide the target signature [1–3]. These effects have to be filtered out from the signal to enhance landmine detection.

3.1. Clutter

As stated before, using GPR in landmine detection operations could be advantageous since the number of false alarms could be reduced and since low-metallic content landmines that are not detected by metal detection could be detected by GPR. However, extracting the landmine signature from GPR data is negatively affected by a list of influencing elements categorized as clutter, which can partly or totally obscure or deform the backscattered signal from a buried object. Mainly, these influencing elements are: (1) antenna reactions causing multiple reflections and signal deformation; (2) the subsoil EM characteristics and their spatiotemporal distribution controlling wave propagation velocity, attenuation, and surface and subsoil reflections; (3) the EM variation between the buried object and the subsoil influencing the strength of the landmine response; and (4) surface roughness causing diffuse scattering. Thus, it is necessary to investigate suitable techniques in order to reduce clutter while maintaining high landmine detection rates. This could be very challenging due to the complicated EM phenomena taking place in the theoretically unknown antenna-air-soil-mine system.

Different approaches are used to reduce this clutter and to recognize the landmine signal. Widely used approaches to reduce clutter are average and moving average background subtraction (BS). Other BS techniques are based on wavelet transform and system identification. Once these approaches suppress part of the clutter, the next expected procedure is to detect the buried object. In this step, different signal processing techniques for the identification of the target signal are applied, including advanced algorithms for hyperbola detection [29], convolutional models and migration techniques [5]. An overview of different signal processing techniques for landmine detection using GPR can be found in Refs. [1, 3].

3.2. Filtering antenna effects and soil surface reflection

In Ref. [30], we propose a method to filter out the antenna internal reflections and multiple reflections between the antenna and the ground, as well as the related distortion effects, by using the frequency-dependent linear transfer functions model developed in [23]. These functions also account for the antenna gain and wave propagation time, fixing time-zero at the antenna phase center. An example of data before and after filtering the antenna effects is presented in **Figure 15** (note that in the radar data represented in **Figure 14**, these effects were previously removed).

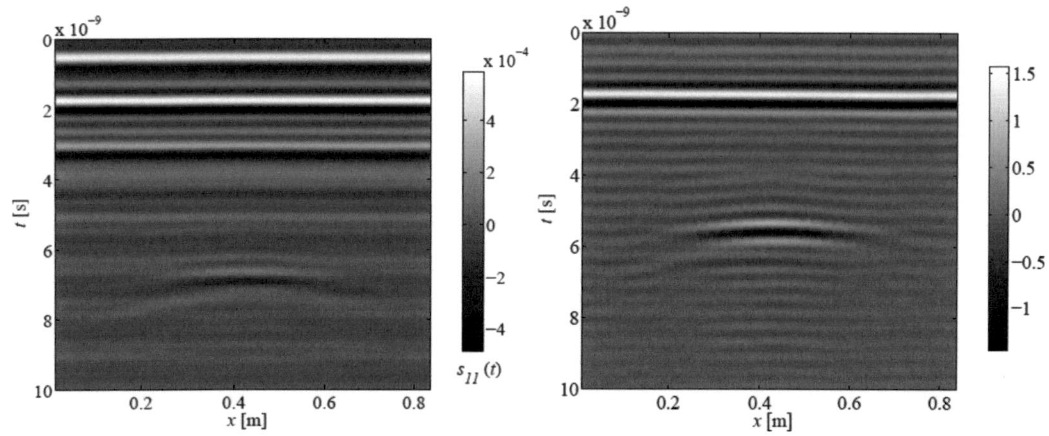

Figure 15. Time domain representation of the SFCW UWB GPR measurements performed on sand for an antitank (AT) mine, buried at 30 cm depth, before (left) and after (right) removing antenna effects.

On the left image we can observe the multiple reflections occurring in the antenna, between about 0 and 2 ns. These unwanted signals obscure the backscattered response when looking down in the time axis. At around 3.2 ns, the air-soil interface reflection appears. Then, a second-order reflection taking place between the antenna and the soil surface arrives at about 4.5 ns. Later in time, the backscattered signal coming from an antitank (AT) landmine appears around 6.5 ns (hyperbolic shape). Thanks to its relatively large size and metal content, as well as the strong electromagnetic contrast between this one and the sand, and to the low attenuation of the EM waves in the sand (the electric conductivity tends to zero), this landmine is easily noticeable to the naked eye. After filtering-out antenna effects (image on the right), the true time zero corresponds now to the antenna phase center, located at about 7 cm from the antenna aperture, inside the antenna, resulting in a time-shifting (note that the soil reflection and hyperbola originating from the AT landmine are shifted in time). The multiple antenna reflections have been taken away and the surface reflection appears clearer, around 1.9 ns, allowing the precise calculation of the antenna height. Moreover, the hyperbolic signature of the AT landmine is highlighted. There are some remaining oscillations which are still visible in the figure. These could be a consequence of the distinct suppositions postulated in the antenna model, mainly, (i) the condition of being in the far-field, (ii) the virtual Tx-Rx point of the antenna which is approximated at a fixed position (in reality, its position varies with frequency, the high frequencies being emitted nearer the feed point, and the low frequencies being emitted in the proportionally larger part of the antenna), and (iii) the fact that only the x-component of the electric field is supposed to be measured and that only an x-directed current source is available. These could also appear as a result of applying the inverse Fourier transform to data collected in a restricted frequency range. This chapter will show later that such oscillations are not a disadvantage for the detection and identification of AP landmines.

The surface reflection may be removed from a full GPR transect by subtracting from all measurements a mean measurement or an A-scan performed over a landmine-free area. In order to do this, the EM properties of the subsoil should be homogenous over the transect, all measurements should be performed with the antenna situated at a constant height above the ground, and the soil surface should be totally flat. In practice, real minefields cannot satisfy these requirements. For filtering the soil surface reflection we propose to subtract from the radar signal, preliminarily filtered for antenna effects, a computed Green's function $G'(\omega)$ describing wave propagation with the antenna in air above a single interface (mine-free soil surface). The Green's function is the solution of the three-dimensional (3D) Maxwell's equations for wave propagating in multilayered media. In our case, this function accounts for wave propagation in air, including spherical divergence, and the surface reflection. To compute that Green's function, two parameters should be known: (1) the soil surface relative dielectric permittivity ε_r and (2) the antenna height h. These two parameters are derived using full-wave inversion of the radar signal [31] for a measurement taken in a local landmine-free area. This approach has been validated and applied with success in laboratory and field conditions for identification of the soil relative dielectric permittivity. It is worth noting that the single measurement in a local mine-free area used to estimate the soil surface dielectric permittivity can be collected at any time in any surrounding, mine-free area. Therefore, the spatiotemporal variability of the soil properties can be partly taken into account to filter the surface reflection.

Figure 16 shows data after filtering the ground surface reflection. The effect of the AT landmine is isolated, which permits to differentiate the hyperbolic signature. The response of the target is then calculated by using Eqs. (2)–(4) and can be written as:

$$R_{obj}(\omega) = G(\omega) - G^{'}(\omega). \tag{5}$$

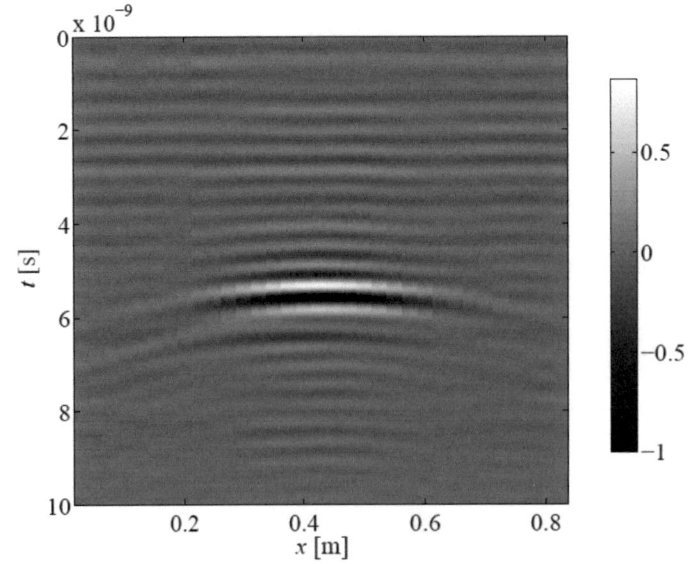

Figure 16. Results after filtering the soil surface reflection from radar data of **Figure 15** (AT landmine).

http://dx.doi.org/10.5772/67007

3.3. Focusing the radar image

The radar antenna transmits energy with a beamwidth pattern such that an object several centimeters away from the beam axis may be detected. As a result, objects of finite dimensions appear as hyperbolic reflectors on the B-scans. Migration techniques are used to reconstruct the reflecting structure present in the subsurface by focusing the reflections back into the true position of the object. We propose to filter out the effects of the antenna radiation pattern using the common Stolt's migration method [32], which applies a Fourier transform to back-propagate the scalar wave equation, extended in Ref. [30] for two media.

Consider the filtered signal R_{obj} as a 2D data set $R(x,\ z,\ \omega)$ with x being the distance along the scanning axis, z the depth and ω the angular frequency. Applying the Fourier transform with respect to the spatial distance x to the special frequency k_x yields an unfocused wavenumber data set

$$R(k_x, z, \omega) = \int R(x,\ z,\ \omega)e^{ik_x x}dx. \tag{6}$$

The Fourier transformation along the x coordinate makes sense only if the propagation velocity does not vary in this direction. This methodology permits variations of the propagation velocity in the z direction.

Considering the wavenumber k ($k\ =\ 2\pi/\lambda$, with λ being the wavelength in the ground) as the vector sum of k_x and k_z for one-way propagation, we have:

$$k = |\mathbf{k}| = \sqrt{k_x + k_z} = \frac{\omega}{v} \tag{7}$$

where v is the propagation velocity in the soil.

Assuming only upward coming waves and by introducing Eq. (7) into Eq. (6), the Fourier transform of the wavefront at depth z is done by

$$R(k_x, z, \omega) = R(k_x, 0, \omega)e^{-ik_z z}. \tag{8}$$

The migrated image will be the inverse Fourier transform of Eq. (8) at $t = 0$ as

$$\hat{r}(x, z) = r(x, z, 0) = \iint R(k_x,\ 0,\ \omega)e^{-(ik_x x - k_z z)}dk_x d\omega. \tag{9}$$

Figure 17 shows results after applying Slot's migration to the data set presented in **Figures 14** and **15**.

3.4. Migration by space-time deconvolution

As described above, the aim of migration is to focus target reflections in the recorded data back into their true position and physical shape. In this respect, migration can be seen as a form of spatial deconvolution that increases spatial resolution. It is a common practice not to include in

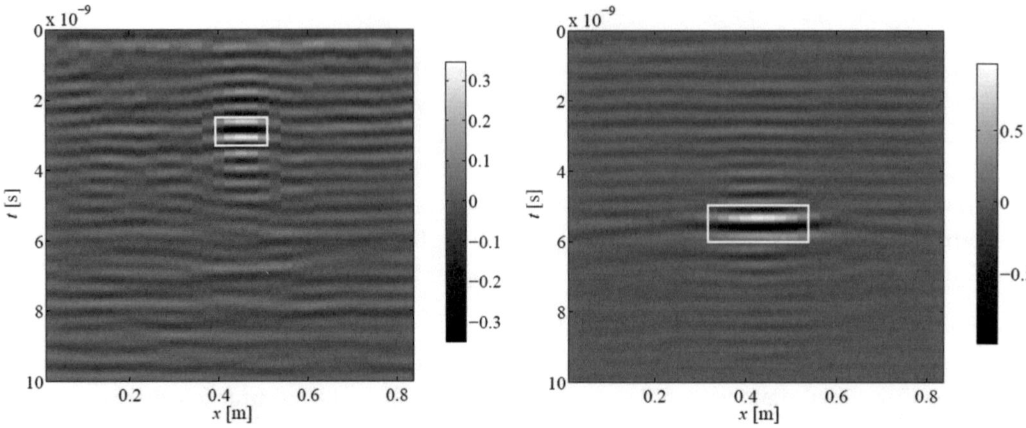

Figure 17. Results after applying the Slot's migration method for the datasets presented in **Figure 14** (PMN AP landmine, left) and **Figure 16** (AT landmine, right).

the migration approaches the characteristics of the radar system, e.g., antenna patterns, antenna impulse response, and source waveform. In this section, a migration approach that considers the system characteristics and to a certain extent, the ground characteristics, is presented [5]. Its strategy is based on the deconvolution of the collected data with the point-spread function of the radar system. As in most of migration algorithms, it is assumed that the interaction between the scatterers present in the medium is totally disregarded.

In order to perform the migration process by deconvolution, the data acquired by the UWB GPR have to be a convolution between the different layers and configurations present in the subsoil and the point-spread function of the system. This is valid under certain premises. In the interest of simplifying the analysis, a monostatic antenna configuration is taken as an example. The velocity of propagation through the propagation element can only vary in the groundward direction. The antennas are change location following a xy-plane at $z = 0$. The 3D data, $V_{rec}(x, y, z = 0, t)$ is collected following a rectangular grid with spacing Δx and Δy. Consider a point-target located in the subsoil at $\vec{r_o} = (x_o, y_o, z_o)$ in a first step, described by an IR $\Lambda_o(\vec{r_o}, t_o)$, independent of the incident direction and not necessarily a dirac impulse. For the antennas at any position $\vec{r_a} = (x_a, y_a, z = 0)$, the received voltage can be written as

$$V_{rec}(\vec{r_a}, t) = s(\vec{r_a}, \vec{r_o}, t) \otimes \Lambda_o(t) \tag{10}$$

Where $s(\vec{r_a}, \vec{r_o}, t)$ represents the voltage response of a point scatterer placed in $\vec{r_o}$ with IR $\delta\left(t - \frac{|\vec{r}_a - \vec{r}_o|}{v}\right)$, as a function of the antenna location $\vec{r_a}$ and time t. Additionally, for the antennas at $z = 0$ and the point scatterer at $z = z_o$, the response $s(\vec{r_a}, \vec{r_o}, t)$ is a function of $\vec{r_o}$ and $\vec{r_a}$ only by their difference. It is worth noting that the convolution present in Eq. (10) occurs in the time. Consider a target modeled by a group of independent point targets located in average at the same depth $z = z_o$, the resulting voltage $V_{rec}(x_a, y_a, t)$ can be calculated as a linear combination of the voltage responses of each individual point target as:

$$V_{rec}(x_a, y_a, t) = \iint_{x,y} \int_{\tau} s(x_a - x, y_a - y, z_o, t - \tau) \Lambda_{zo}(x, y, \tau) d\tau \, dxdy \tag{11}$$

Eq. (11) represents a space-time convolution along the co-ordinates x, y, and t, and can be written as

$$V_{rec}(x, y, t) = s_{z_o}(x, y, t) \otimes \Lambda_{zo}(x, y, t) \tag{12}$$

where $\Lambda_{zo}(x, y, t)$ is a 3D image including the feedbacks connected with the distributed targets at the positions (x, y, z_o). The 3D matrix $s_{z_o}(x, y, t)$ symbolizes the point-spread function of the UWB GPR system for a depth $z = z_o$ and can be approximated by forward modeling for different antenna positions $\vec{r_a}$ and a point scatterer with IR $\delta\left(t - \frac{|\vec{r}_a - \vec{r}_o|}{v}\right)$. An example of a 3D point-spread function for a UWB GPR system and a point scatterer at 6 cm of depth is given in **Figure 18**.

Even though the point-spread function $s_{z_o}(x, y, t)$ depends on the depth $z = z_o$, its contour will not be altered very much with changes in z. Basically, the point-spread function could be used for a broad depth range. Accordingly, the point-spread function $s_{z_o}(x, y, t)$ can be treated as space invariant. Therefore, the migrated image $M(x, y, t)$ can be calculated as a deconvolution of the recorded data $V_{rec}(\vec{r_a}, t)$ with the point-spread function $s_{z_o}(x, y, t)$, where z_o is chosen to be the most likely depth for an object. In the application of demining, z_o is taken 6 cm. The deconvolution can be performed in the frequency-wavenumber domain, using a Weiner filter, which makes the migration scheme very simple and not computational intensive. More details on the presented migration method can be found in Ref. [5].

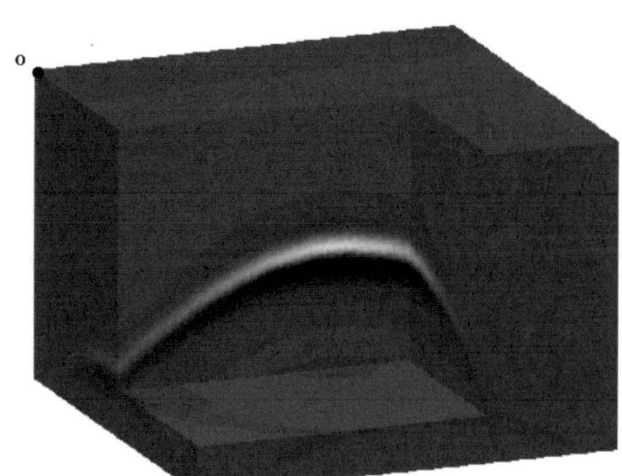

Figure 18. Synthetic C-scan of a fictive point scatterer at a depth of 6 cm below the air-ground interface, calculated by forward modeling.

Figures 19–21 show the results of the migration method on data taken by a laboratory UWB GPR described in Section 2.3, with the antennas mounted on an indoor *xy*-scanning table. The data are acquired over an area of 50 cm × 50 cm with a step of 1 cm in both *x*- and *y*-direction. Results are shown in **Figure 19** for a PMN mine buried at 5 cm of depth in sand, in **Figure 20** for a brick of dimensions 15 cm × 9 cm × 6 cm buried at the same depth, and in **Figure 21** for a piece of 20 cm barbed wire. There is a green, three-dimensional representation of the collected data in each of the figures, which is obtained after applying first a Hilbert transformation to the A-scan in order to calculate the cover for each A-scan. As a second step, the data are plotted by an isosurface 3D plot, accentuating all the pixel of a given intensity or higher. In each figure, the raw data are plotted the left. On the right, the migrated image is displayed. For clarity, the ground reflection is suppressed in **Figures 20** and **21**. When observing the targets from above, the rounded form of the PMN mine is clearly displayed, and the form of the brick is more box-like. **Figure 21b** shows how the form of the barbed wire can be easily distinguished from the

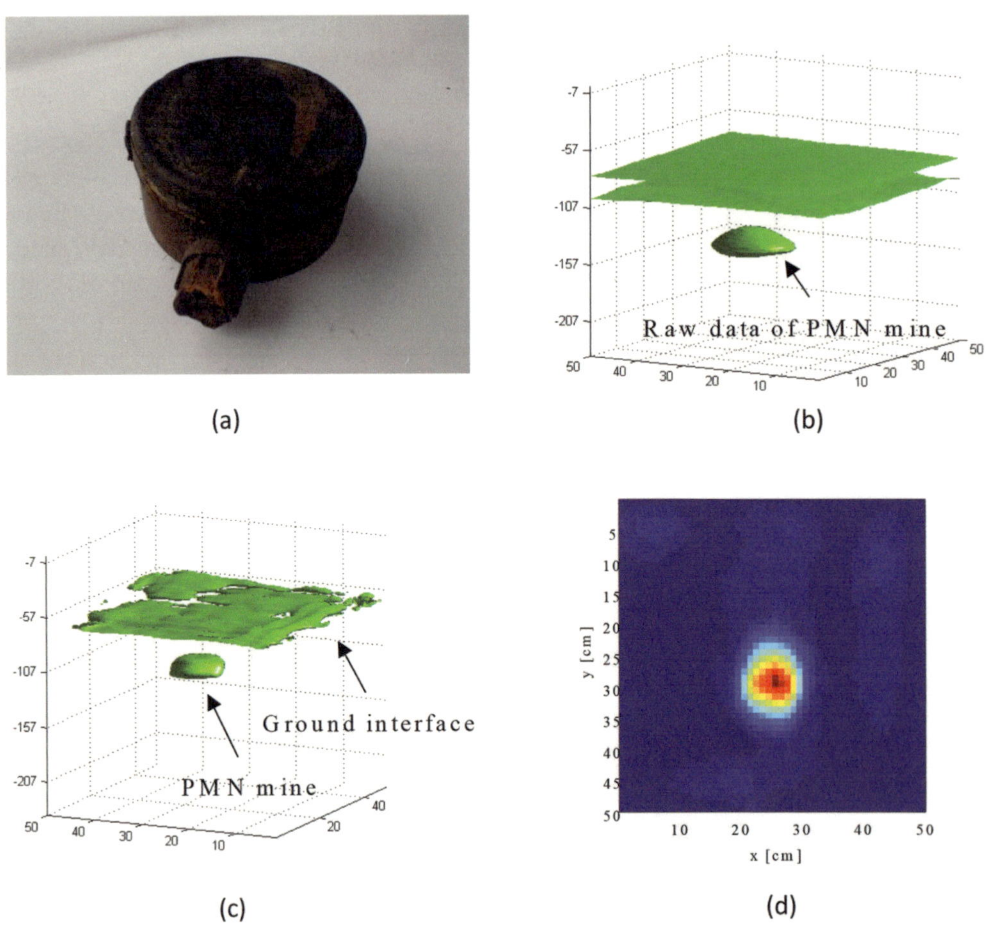

Figure 19. Results after applying deconvolution in order to focus the data collected on a PMN mine (diameter of 11 cm) located at 5 cm depth. (a) Photo of PMN mine ; (b) 3D C-scan view of raw data; (c) 3D C-scan view of migrated data; (d) 2D C-scan horizontal slide of migrated data.

http://dx.doi.org/10.5772/67007

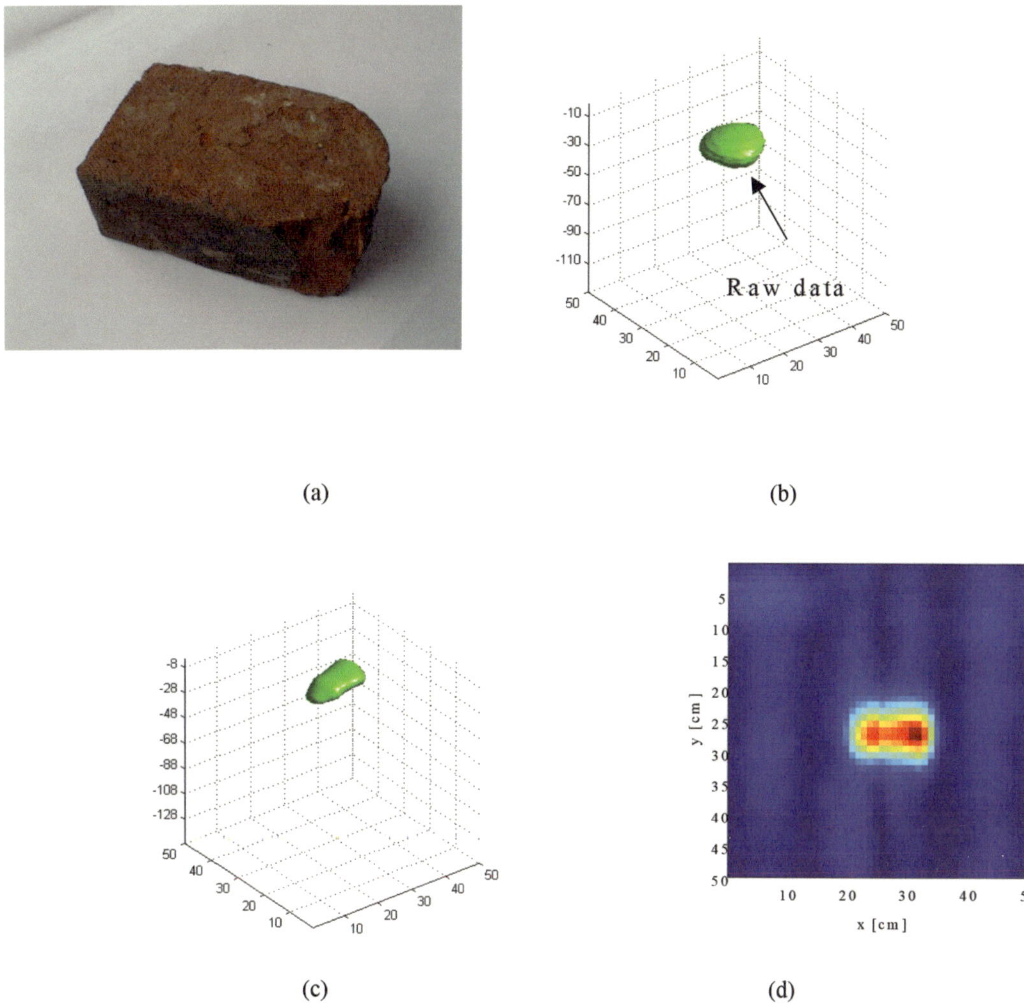

Figure 20. Results after applying deconvolution in order to focus the data collected on a brick (15cm × 9cm × 6cm) located at 5 cm depth. (a) Photo of the brick; (b) 3D C-scan view of raw data; (c) 3D C-scan view of migrated data; (d) 2D C-scan horizontal slide of migrated data.

other two forms and the three sets of pins present on the real wire are clearly noticeable. It is demonstrated with these three examples that the shape of a target in the subsoil can be extracted from the data gathered by the UWB GPR after applying the migration procedure explained above.

The aim of migration is not only to focus reflections on objects back into the true physical shape of the object but also into its true position. To illustrate the latter, an AP mine was buried under an angle of about 30° in dry sand, with the highest point of the mine at a depth of 5 cm. In the raw B-scan presented in **Figure 22a**, the strongest reflections on the mine are found in the lower right corner of the image, whereas in reality the mine (designated by the rectangular box) is situated in the middle of the image. This shift can be simply explained as follows. When

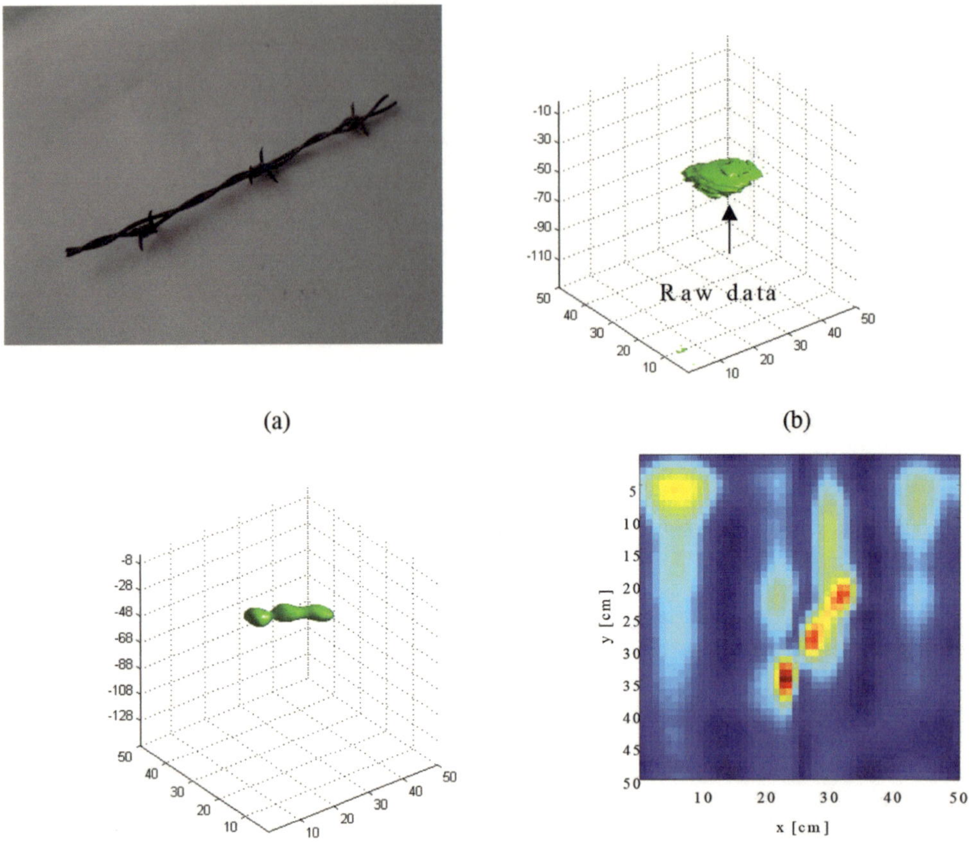

(a) (b) (c) (d)

Figure 21. Results after applying deconvolution in order to focus the data collected on barbed wire (approx. 20 cm length) located at 5 cm depth. (a) Photo of the barbed wire; (b) 3D C-scan view of raw data; (c) 3D C-scan view of migrated data; (d) 2D C-scan horizontal slide of migrated data.

the antennas are right above a tilted object, the latter will appear as a strong reflection in a direction away from the receiving antenna. For the antennas in the direction perpendicular to the flat top of such a target, the reflections going toward the receiving antenna will be stronger than in the case the antennas are right above the tilted object, causing a displacement of the target in the raw data. **Figure 22b** shows how, after applying the migration using the deconvolution approach presented before, the target is found in its actual position. Results of the migration not only show that the target was at the wrong position in the raw data, but also clearly show that the target is tilted. Because of the different backscatter structures in the target, its dimensions in the *z*-direction could be wrongly estimated. However, the position (in time or space) of the first reflection on the target (the specular reflection) will be accurate and reliable. As a consequence, the reconstruction of the envelope of buried objects will be correct. Comparison has been done with other migration methods, like Kirchhoff migration and Stolt migration, applied on the same data, which led to less satisfactory results than the

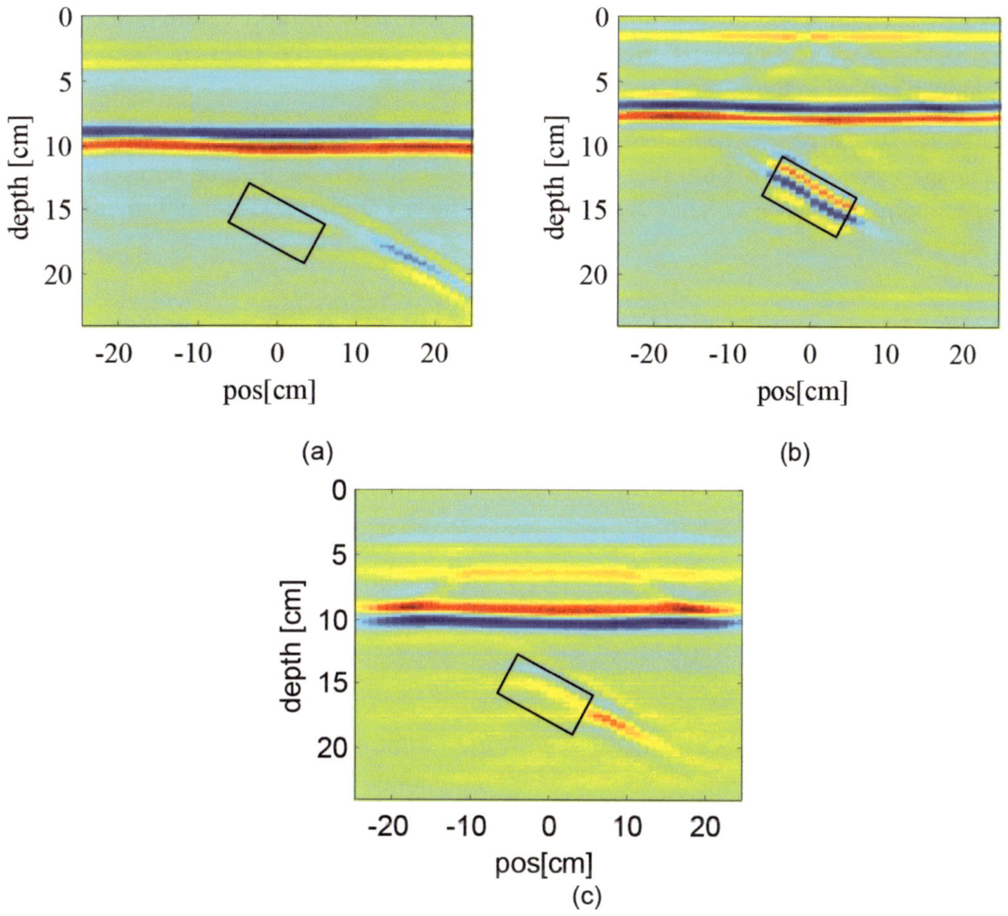

Figure 22. Oblique PMN mine under an angle of 30°. (a) Raw data, (b) image after migration by deconvolutions, and (c) image after Kirchhoff migration.

proposed deconvolution approach. In **Figure 22c**, the results after applying Kirchhoff migration can be seen. Although more information could be extracted from the migrated data, the Kirchhoff migration is not able to place the object completely back in its correct position.

3.5. Target classification

GPR may allow detecting buried objects such as metallic and nonmetallic AP landmines. However, this detection technique can be affected by false alarm rates as other reflectors (e.g., stones, metal fragments, roots) can produce similar echoes. In this regards, resonance features in backscattered signals are proposed here in order to identify unknown targets. These features can be studied in either the time-domain (TD) or the frequency-domain (FD). In Ref. [33], functions of both variables, time and frequency, are considered for this particular application. Time-frequency distributions, which are 2D functions, can reveal the time-varying frequency content of 1D signals. One of these 2D functions is the Wigner-Ville distribution (WVD), which is widely used for target recognition. This section shows how the application of the WVD on GPR data can yield to extract important information about the physical features of AP landmines located in the subsoil.

The WVD is one of the approaches of the time-frequency representations. It has a main advantage when compared to other representations such as the short-time Fourier transform or spectrogram, which is a higher time resolution. The WVD of a 1D signal $s(t)$, $s_{\mathrm{WVD}}(t,\ f)$, can be expressed as

$$s_{\mathrm{WVD}}(t,f) = \int s_a\left(t + \frac{\tau}{2}\right) + s_a^*\left(t - \frac{\tau}{2}\right) e^{-i2\pi f \tau} d\tau, \tag{13}$$

where

$$s_a = s(t) + i\hat{s}(t) \tag{14}$$

is the analytic signal consisting of the real signal $s(t)$ and its Hilbert transform

$$\hat{s}(t) = \frac{1}{\pi} \int \frac{s(\tau)}{1 - \tau} d\tau, \tag{15}$$

and $s_a^*(t)$ is its conjugate.

Figure 23 represents the WVD of one A-scan from **Figure 16** (AT landmine). The WVD is applied only to the A-scan containing the highest amount of energy backscattered from the target, calculated after filtering and migration. The dotted line in the left figure represents raw data acquired with the SFCW UWB GPR and filtered data are presented with the solid line.

At any time (frequency) point, the WVD can be considered as the summed spectrum (correlation) of the signal power at this point and the cross-power of two signals parts, spaced symmetrically with respect to the current time point.

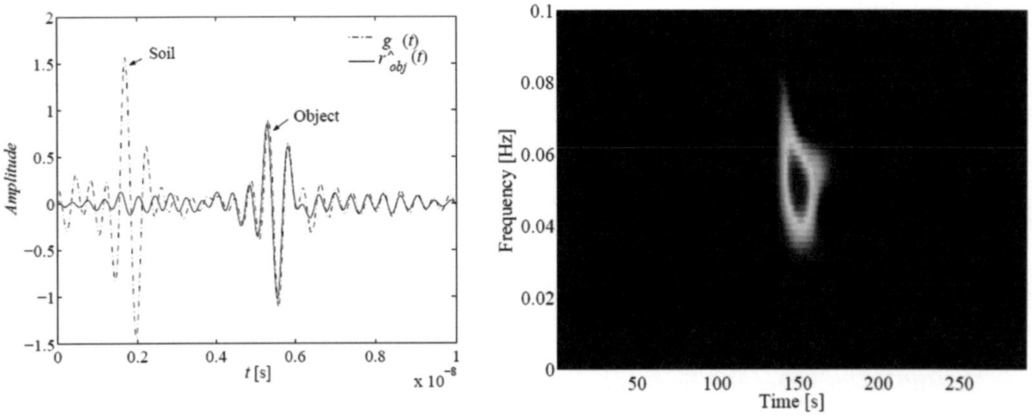

Figure 23. WVD (right) applied to a filtered and migrated A-scan data (left) for an antitank landmine buried in sand (from **Figure 16**).

http://dx.doi.org/10.5772/67007

As this distribution is a two-dimensional representation (matrix) of a one-dimensional signal, this transform inferred a given amount of redundancy. In order to confront this problem, Ref. [34] suggested the singular value decomposition (SVD), which is used here in conjunction with the concept of the center of mass (CM) to extract discriminant features [35].

SVD is intended for representing the WVD matrix **W**s with its singular values and singular vectors, which are unique for any matrix:

$$\mathbf{W}_s[m,\ n] \ = \ \mathbf{U}[m, m]\sum [m,\ n]\mathbf{V}[n,\ n] \tag{16}$$

where the matrixes **U** and **V** consist of the left-singular and right-singular vectors, respectively, and the matrix **Σ** consists of the singular values. Particularly, any matrix can be disintegrated into a number of singular triplets $\{uk,\ \sigma k,\ vk\}$, with $k \ = \ 1 \ldots \min(M, N)$, where each singular value σk can be considered as the square root of the corresponding triplet energy. Besides, the rows and columns of **W**s symbolized, respectively, the distribution of the energy in time (related to the singular vectors vk) and in frequency (related to the singular vectors uk).

Since the singular values and vectors are unique for any matrix, these triplets contain energy, time, and frequency features which help in discriminating different targets. Therefore, following Ref. [35], the CM (the center of mass is the strongest point in any distribution) of the singular vectors vk and uk is used as time and frequency features and the singular value σk as energy feature for each triplet $\{u_k,\ \sigma_k,\ v_k\}$,. After analyzing different triplets, we find that the following normalized features from the first triplet provide the best discriminant results:

$$\delta_t = \frac{1}{N}\sum_{n=1}^{N} n\mathbf{v}_1[n]^2 \tag{17}$$

$$\delta_f = \frac{1}{M}\sum_{m=1}^{M} m\mathbf{u}_1[m]^2 \tag{18}$$

$$\delta_e = \frac{\sigma_1}{\sum\limits_{k=1}^{\min(M,\ n)} \sigma_k^2} \tag{19}$$

This approach is applied to different targets, including landmines, improvised explosive devices (IED), and false alarms (FA), which are buried in different types of soils. A description of the targets is done in **Table 3**.

Figure 24 shows the calculated values from Eqs. (17) to (19) for targets of **Table 3**. The features of the AP landmines are well clustered and well separated from those of the metallic false alarms and stones. The false alarms are clearly separated as well. Classification between different AP landmines (with different shapes) and between different IEDs (with different materials) could be also obtained. Results also show that the extracted features could be independent of the target depth.

Target	Type	Shape	Diameter (cm)	Metal content
C3A1	Plastic AP	Irregular	5.1	Low
PMA	Plastic AP	Rectangular	15.2	Low
Stone1	FA	Irregular	12.0	No
Metallic can	FA	Irregular	12.0	High
IED5	PVC	Cylindrical	6.3	No
IED6	PVC	Cylindrical	6.3	Low
IED7	Glass	Cylindrical	5.5	No
IED8	Glass	Cylindrical	5.5	Low
Stone2	FA	Irregular	8.0	No
Metallic debris	FA	Irregular	6.0	High

Table 3. Some characteristics of the objects used.

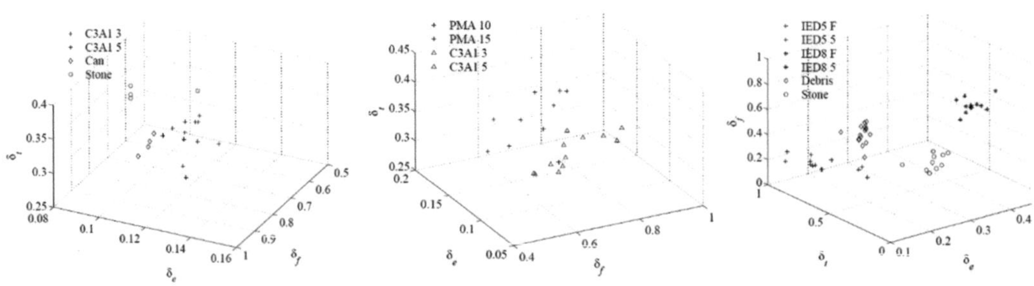

Figure 24. Extracted features from different targets buried at different depths.

4. Conclusions

In this chapter, two of the major challenges in the application of GPR in humanitarian demining operations are addressed: (i) development and testing of affordable and practical GPR-based systems, which can be used off-ground and (ii) development of robust GPR signal processing techniques for landmine detection and identification.

Different approaches developed at RMA in order to demonstrate the possibility of enhancing close-range landmine detection and identification using GPR under laboratory and outdoor conditions are summarized here. Raw GPR profiles give us a large quantity of information about the underground, and therefore performant signal processing techniques are needed to filter and improve the data quality in order to extract the right information. Data acquired using different affordable and practical GPR-based systems are used to validate a number of promising developments in signal processing techniques for target detection and identification. Removing undesirable reflections by filtering and focusing the data using migration algorithms are some of the techniques applied in image reconstruction which are introduced here. The proposed approaches have been validated with success for the imaging, detection and classification of buried objects in laboratory and outdoor conditions. Validation has been done for different scenarios, including AP, low-metal content landmines and IEDs, and real mine-affected soils.

Acknowledgements

The work presented in this chapter was carried out at the Signal and Image Centre, École Royale Militaire, Belgium, in collaboration with the Microwave Engineering and Applied Electromagnetism Department, the Telecommunications Laboratory and the Dept. of Environmental Sciences and Land Use Planning of the UCL (Belgium), and with the Dept. de Ingeniera Eléctrica y Electrónica of the Universidad de Los Andes (Colombia). They were funded by the Ministry of Defense of Belgium in the scope of HUDEM and BEMAT projects. The Instituto Colombiano para el Desarrollo de la Ciencia y la Tecnología—Colciencias (Colombia), the Batallón Baraya (Colombia) and the Escuela de Ingenieros Militares (Colombia) are gratefully acknowledged for their outstanding support during some trials.

Author details

Olga Lucia Lopera Tellez* and Bart Scheers

*Address all correspondence to: olopera@elec.rma.ac.be

CISS Department, Royal Military Academy, Brussels, Belgium

References

[1] MacDonald J, Loockwood J, Altshuler T, Broach T, Carin L, Harmon R, Rappaport C, Scott W, Weaver R. Alternatives for Landmine Detection. Santa Monica, CA: RAND; 2003

[2] Acheroy M. Mine action: Status of sensor technology for close-in and remote detection of anti-personnel mines. Near Surface Geophysics. 2007;**5**:43-55

[3] Daniels D. Ground Penetrating Radar. London: The Institution of Electrical Engineers; 2004

[4] Lopera O. An integrated detection and classification methodology applied to ground-penetrating radar for humanitarian demining applications [PhD thesis]. Catholic University of Louvain, Royal Military Academy; 2008

[5] Scheers B. Ultra-wideband ground-penetrating radar with application to the detection of antipersonnel landmines [PhD thesis]. Catholic University of Louvain, Royal Military Academy; 2001

[6] Nicolaescu I, van Genderen P, van Dongen K, van Heijenoort J, Hakkaart P. Stepped frequency continuous wave radar-data preprocessing. In: Proceedings of the 2nd International Workshop on Advanced GPR; 2003; The Netherlands: Delft. pp. 177-182

[7] Giannopoulos A. GprMax2D/3D V2.0 User Manual. Edinburgh: University of Edinburgh; 2005

[8] Yarovoy A, Kovalenko V, Fogar A. Impact of clutter on buried object detection by ground penetrating radar. In: International Geoscience and Remote Sensing Symposium; 2003; France. pp. 755-757

[9] Lagman A. The design of hardware and signal processing for a stepped frequency continuous wave ground penetrating radar [PhD Thesis]. Cape Town, South Africa: University of Cape Town; 2002

[10] Sato M, Fang G, Zeng Z. Landmine detection by a broadband GPR system. In: International Geoscience and Remote Sensing Symposium; 2003; France

[11] Chignell RJ, Hatef M. LOTUS—A real time integrated sensor suite for antipersonnel mine detection, incorporating the MINETEC GPR. In: Yarovoy, editor. Tenth International Conference on Ground Penetrating Radar; Delft, The Netherlands: TUDelft; 2004. pp. 689-692

[12] Scheers B, Plasman Y, Piette M, Acheroy M, Vorst AV. Laboratory UWB GPR system for landmine detection. In: Proceedings of the Eighth International Conference on Ground Penetrating Radar; SPIE; 2000. pp. 747-752

[13] Yarovoy A, Ligthart L, Schukin A, Kaploun I. Polarimetric video impulse radar for landmine detection. Subsurface Sensing Technologies and Applications. 2002;**3**:271-293

[14] Daniels D, Curtis P, Amin R, Dittmer J. An affordable humanitarian mine detector. In: SPIE, Detection and Remediation Technologies for Mines and Minlike Targets IX; 2004. pp. 1185-1193

[15] Sato M, Fujiwara J, Feng X, Zhou Z, Kobayashi T. Development of a hand-held GPR MD sensor system (ALIS). In: SPIE, Detection and Remediation Technologies for Mines and Minlike Targets X; 2005. pp. 1000–1007

[16] Pajewski L, Tosti F, Kusayanagi W. Antennas for GPR systems. In: Benedetto A, editors. Civil Engineer Applications of Ground Penetrating Radar. Springer International Publishing; Switzerland, 2015. pp. 41-67

[17] CEN Workshop Agreement. Humanitarian Mine Action—Test and Evaluation—Part 2: Soil Characterization for Metal Detector and Ground Penetrating Radar Performance. Available from: http://www.mineactionstandards.org/fileadmin/MAS/documents/cen-workshop-agreements/ 2008

[18] Kanda M. Time-Domain Measurements in Electromagnetics. New York: Van Nostrand Reinhold Company Inc.; 1986

[19] Klaasen JJA. An efficient method for the performance analysis of bounded-wave nuclear EMP Simulators. Sensor and Simulation Notes, 1992; note 345

[20] Scheers B, Piette M, Vander Vorst A. Development of dielectric-filled TEM horn antennas for UWB GPR. In: Millennium Conference on Antennas & Propagation AP-2000; Davos, Switzerland, vol. II, p. 187, April 2000

[21] Baum CE. Low-frequency-compensated TEM horn. Sensor and Simulation Notes. 1995; note 377

[22] Rumsey V. Frequency Independent Antennas. New York, USA: Academic Press Inc; 1966

[23] Lambot S, Slob E, van den Bosch I, Stockbroeckx B, Vanclooster M. Modeling of ground-penetrating radar for accurate characterization of the subsurface dielectric properties. IEEE Transactions on Geoscience and Remote Sensing. 2004;**42**:2555-2568

[24] Michalski K, Mosig J. Multilayered media Green's functions in integral equation formulations. IEEE Transactions on Antennas and Propagation. 1997;**45**:508-519

[25] Lambot S, Lopera O, Slob EC. Effect of the antenna height on the estimation of the soil electromagnetic properties using Full-Wave inverse modeling of GPR data. In: Daniels JJ, Chen CC, editors. Proceedings of the 11th International Conference on Ground Penetrating Radar; 2006; Columbus, Ohio, USA: The Ohio State University

[26] Daniels D, Curtis P. MINEHOUND trials in Cambodia, Bosnia and Angola. In: Broach JT, Harmon RS, Holloway, JH Jr., editors. Proceedings of the SPIE; 2006; Orlando, Florida USA. Vol. 6217. DOI: 10.1117/12.663537

[27] Sato M, Takahashi K, Fujiwara J. Development of the hand held dual sensor ALIS and its evaluation. In: International Workshop on Advanced GPR; 2007; Naples, Italy. IEEE; 2007

[28] Ritter K. AN/PSS-14 mine detection system offers improved countermine capability. Army AL&T Magazine. 2007;**1**:54-59

[29] Milisavljevic N. Analysis and fusion using belief functions theory of multisensor data for close-range humanitarian mine detection [PhD thesis]. Belgium: Ecole Nationale Superieure des Telecommunications, France, Royal Military Academy; 2001

[30] Lopera O, Slob E, Milisavljevic N, Lambot S. Filtering soil surface and antenna effects from GPR data to enhance landmine detection. IEEE Transactions on Geoscience and Remote Sensing. 2007;**45**(3):707-717

[31] Lambot S, Weihermüller L, Huisman JA, Vereecken H, Vanclooster M, Slob EC. Analysis of air-launched ground-penetrating radar techniques to measure the soil surface water content. Water Resources Research. 2006;**42**:W11, 403. DOI: 10.1029/2006WR005,097

[32] Ozdemir C, Demirci S, Yigit E. Practical algorithms to focus B-scan GPR images: Theory and application to real data. Progress in Electromagnetic Research B. 2008;**6**:109–122

[33] Lopera O, Milisavljevic N, Daniels D, Gauthier A, Macq B. A time frequency domain feature extraction algorithm for landmine identification from GPR data. Near Surface Geophysics. 2008;**6**:411-421

[34] Marinovic N, Eichmann G. An expansion of Wigner distribution and its applications. In: Proceedings of ICASSP'85; 26-29 April 1985; Tampa, Florida USA. IEEE. pp. 1021-1024

[35] Savelyev T, van Kempen L, Sahli H, Sachs J, Sato M. Investigation of time-domain frequency features for GPR landmine discrimination. IEEE Transactions on Geoscience and Remote Sensing. 2007;**45**:118-129

Chapter 4

Data Fusion for Close-Range Detection

Nada Milisavljevic

Additional information is available at the end of the chapter

http://dx.doi.org/10.5772/intechopen.68168

Abstract

Two approaches for combining humanitarian mine detection sensors are described in parallel, one based on belief functions and the other one based on possibility theory. In a first step, different measures are extracted from the sensor data. After that, based on prior information, mass functions and possibility distributions are derived. The combination of possibility degrees, as well as of masses, is performed in two steps. The first one applies to all measures derived from one sensor. The second one combines results obtained in the first step for all sensors used. Combination operators are chosen to account for different characteristics of the sensors. Comparison of the combination equations of the two approaches is performed as well. Furthermore, selection of the decision rules is discussed for both approaches. These approaches are illustrated on a set of real mines and non-dangerous objects and using three sensors: an infrared camera, an imaging metal detector and a ground-penetrating radar.

Keywords: close range antipersonnel mine detection, data fusion, belief functions, possibility theory

1. Introduction

Multi-sensor data fusion techniques prove to be useful for two main humanitarian mine action types: mined area reduction and close-range antipersonnel (AP) mine detection. In this chapter, data fusion for the latter mine action type is addressed. Close-range AP mine detection refers to detection of (sub-)surface anomalies that may be associated with mine presence (for instance, detection of differences in temperature thanks to an infrared camera (IR) or detection of metals by a metal detector (MD)) and/or to detection of explosive materials.

Efficient modelling and fusion of extracted features can improve the reliability and quality of single-sensor-based processing [1, 2]. Nevertheless, taking into account that there is a wide range of conditions and scenarios between minefields (such as mine types, structure of minefield and soil types) as well as within one minefield (e.g. burial depths and angles, moisture), there is no unique single-sensor solution, meaning that a high-enough performance of humanitarian mine action tools can be reached only using multi-sensor and sensor/data fusion approaches [3]. In addition, since the sensors used are, as a matter of fact, detectors of various anomalies, the classification and detection results can be improved by combining these complementary pieces of information. Last but not least, in order to take into account partial knowledge, intra- and inter-minefield variability, ambiguity and uncertainty, fuzzy set or possibility theory [4] and belief functions [5] within the framework of the Dempster-Shafer (DS) theory [6] prove to be beneficial.

The chapter is organized as follows. An analysis of modelling and of fusion of extracted features is performed. After that, two fusion approaches are presented, one of them being based on the belief function theory and the other one related to the possibility theory. These approaches are then illustrated using real data gathered within the Dutch project HOM-2000 [7], which are acquired using three intrinsically complementary sensors: infrared camera, metal detector and ground-penetrating radar (GPR). These results are obtained within two Belgian humanitarian demining projects: HUDEM and BEMAT. Importance of collateral information (knowledge about types of mines, mine records, etc.) is demonstrated.

2. About close-range detection

Due to a large variety of mine types as well as of conditions in which they can be found, no single sensor applied in close-range AP mine detection can obtain the necessarily high-detection rate in a wide range of possible situations/scenarios. Thus, a logical way towards deriving a solution consists in using several sensors that are complementary and taking the best out of their combination. To this end, an infrared camera (IR), a ground-penetrating radar (GPR) and an imaging metal detector (MD) present a very promising combination. In this chapter, we describe two approaches for combining these sensors, one based on the belief function theory and the other one on the possibility theory. These approaches can easily be adapted to other combinations of sensors.

An important part of the work performed in the field of fusion of dissimilar mine detection sensors is based on statistics [8, 9]. Examples of rare alternative approaches are [10] (neural networks) and [11] (fuzzy fusion of classifiers). The statistical approaches lead to good results for a particular scenario, but they ignore or just briefly mention that, once we look for more general solutions, several important problems have to be faced in this domain of application [12]. For instance, the data are variable, highly dependent on the conditions and on the context. Then, it is impossible to model every possible object (every mine or every other object that might be confused with mines). In addition, the data do not allow for a reliable statistical learning since they are not numerous enough. Finally, the data do not give precise information regarding the mine type, resulting in an ambiguity, typically between several mine types. Note that in the domain of humanitarian mine detection, a vast majority of the fusion attempts, for example, [13, 14], treat every alarm as a mine, and not as an object that could be a mine, but a false alarm as well.

In a previous work [15], a method based on the belief functions [6, 16, 17] has been proposed. In this chapter, we compare it with an alternative approach, based on the possibility theory, in order to take advantage of the flexibility in the choice of combination operators [18]. As shown in Ref. [2], this is exploited in order to account for the different characteristics of the sensors to be combined.

In this domain of application, to our knowledge, there is no work that applies the two fusion theories in parallel or that compares them. In other domains of application, some works on comparing the two theories are published, for example [19], where the qualitative possibility theory is opposed to the belief function theory and a fictitious example of assessing the value of a candidate is used as an illustration. On the contrary to that article, we use the quantitative possibility theory here.

3. Numerical information fusion using belief functions and possibility theory

3.1. Belief function fusion: overview

In the belief function theory or Dempster-Shafer (DS) evidence theory formalism [5, 6], both uncertainty and imprecision can be represented, using belief functions and plausibility obtained from a mass function. The mass allocated to a proposition A corresponds to a part of the initial unitary amount of belief, which supports that the solution is exactly in A. It is thus defined as a function m from 2^U into [0, 1], with U being the decision space, also called full set or frame of discernment. Usually, the following constraints are imposed:

$$m(\Theta) = 0, \tag{1}$$

$$\sum_{A \subseteq \Theta} m(A) = 1. \tag{2}$$

Not only the singletons of U but also any combination of possible propositions/decisions from the decision space can be quantified in this framework. This aspect represents one of the key advantages of the DS theory. As a matter of fact, this possibility allows for a rich and flexible modelling, which can fit to a wide range of situations, which are occurring typically in image fusion in particular. For example, the belief function theory can be successfully applied to situations that include partial or total ignorance, partial reliability, confusion between some classes (in only one or in several information sources), etc. [3, 15, 20–22].

In the DS framework, masses assigned by different sources (e.g. classifiers) are combined by the orthogonal rule of Dempster [6]:

$$m_{ij}(S) = \sum_{\substack{k,l \\ A_k \cap B_l = S}} m_i(A_k) \cdot m_j(B_l) \tag{3}$$

where S is any subset of the full set, while m_i and m_j are masses assigned by measures i and j, and their focal elements are $A_1, A_2,\ldots, A_p$ and $B_1, B_2,\ldots, B_q$, respectively [2].

As discussed in Ref. [2], Dempster's rule is commutative and associative, meaning that it can be applied repeatedly, until all measures are combined, and that the result does not depend on the order used in the combination. After the combination in this unnormalized form [23], the mass that is assigned to the empty set:

$$m_{ij}(\emptyset) = \sum_{\substack{k,l \\ A_k \cap B_l = \emptyset}} m_i(A_k) \cdot m_j(B_l) \tag{4}$$

can be interpreted as a measure of conflict between the sources. It can be directly taken into account in the combination as a normalization factor. It is very important to consider this value for evaluating the quality of the combination: when it is high (in the case of strong conflict), the normalized combination may not make sense and can lead to questionable decisions [24]. Several authors suggest not normalizing the combination result (e.g. [23]), which corresponds to Eq. (3).

This fusion operator has a conjunctive behaviour. This means that all imprecision on the data has to be introduced explicitly at the modelling level, in particular in the choice of the focal elements. For instance, ambiguity between two classes in one source of information has to be modelled using a disjunction of hypotheses, so that conflict with other sources can be limited and ambiguity can be possibly solved during the combination.

From a mass function, we can derive a belief function:

$$\forall A \in 2^{\Theta},\ Bel(A) = \sum_{B \subseteq A,\ B \neq \emptyset} m(B) \tag{5}$$

as well as a plausibility function:

$$\forall A \in 2^{\Theta},\ Pls(A) = \sum_{B \cap A \neq \emptyset} m(B). \tag{6}$$

After the combination, the final decision is usually taken in favour of a simple hypothesis using one of several rules [25]: for example, the maximum of plausibility (generally over simple hypotheses), the maximum of belief, the pignistic decision rule [26], etc.

For some applications, such as humanitarian demining, it may be necessary to give more importance to some classes (e.g. mines, since they must not be missed) at the decision level. Then maximum of plausibility can be used for the classes that should not be missed and maximum of belief for the others [27].

3.2. Fuzzy and possibilistic fusion: overview

In the framework of fuzzy sets and possibility theory [4, 28], the modelling step consists in defining a membership function to each class or hypothesis in each source, or a possibility distribution over the set of hypotheses in each source. Such models explicitly represent imprecision in the information, as well as possible ambiguity between classes or decisions.

For the combination step in the fusion process, the advantages of fuzzy sets and possibilities rely on the variety of combination operators, which may deal with heterogeneous information [18]. As stated in Ref. [2], among the main operators, we find t-norms, t-conorms, mean operators, symmetrical sums and operators taking into account conflict between sources or reliability of the sources. We do not detail all operators in this chapter, but they can be easily found in the literature, with a synthesis in Ref. [29].

We classify these operators with respect to their behaviour (in terms of conjunctive, disjunctive and compromise [18]), the possible control of this behaviour, their properties and their decisiveness, which proved to be useful for several applications [29]. It should be noted that, unlike other data fusion theories (e.g. Bayesian or Dempster-Shafer combination), fuzzy sets provide a great flexibility in the choice of the operator that can be adapted to any situation at hand. In particular, nothing prevents using different operators for different hypotheses or different sources of information.

An advantage of this approach is that it is able to combine heterogeneous information, which is usually the case in multi-source fusion (as in both examples developed in the next sections), and to avoid to define a more or less arbitrary and questionable metric between pieces of information issued from these images, since each piece of information is converted in membership functions or possibility distributions over the same decision space.

Decision is usually taken from the maximum of membership or possibility values after the combination step. Constraints can be added to this decision, typically for checking for the reliability of the decision (Is the obtained value high enough?) or for the discrimination power of the fusion (Is the difference between the two highest values high enough?). Local spatial context can be used to reinforce or modify decisions [2].

4. Close-range mine detection

4.1. Measures

From the data gathered by the sensors, a number of measures are extracted [15] and modelled using the two approaches [2]. These measures concern the following:

- the area and the shape (elongation and ellipse fitting) of the object observed using the IR sensor,
- the size of the metallic area in MD data and
- the propagation velocity (thus the type of material), the burial depth of the object observed using GPR and the ratio between object size and its scattering function.

Although the semantics are different, similar information can be modelled in both possibilistic and belief function models. The idea here is to design the possibility and mass functions as similar as possible and to concentrate on the comparison at the combination step.

The main difference relies in the modelling of ambiguity. The semantics of possibility leads to model ambiguity between two hypotheses with the same degrees of possibilities for these two hypotheses (e.g. Eqs. (7) and (12)). On the contrary, the reasoning on the power set of hypotheses in the belief function theory leads to assigning a mass to the union of these two hypotheses (e.g. Eqs. (9) and (14)).

Another distinction concerns the ignorance. It is explicitly modelled in the belief function theory, through a mass on the whole set (to guarantee the normalization of the mass function over the power set), while it is only expressed implicitly in the possibilistic model, through the absence of normalization constraint.

4.1.1. *IR measures*

The possibility degrees derived from elongation and ellipse-fitting measures are represented by π_{1I} and π_{2I}, respectively [2]. Being related to shape regularity, they are defined for a regular-shaped mine (*MR*), an irregular-shaped mine (*MI*), a regular-shaped non-dangerous (i.e. friendly) object (*FR*) and an irregular-shaped friendly object (*FI*).

In the belief function framework, the full set is: $\Theta = \{MR, MI, FR, FI\}$. As elongation and ellipse fitting aim at distinguishing regular and irregular shapes, masses assigned by these two measures, m_{1I} and m_{2I}, are split between $MR \cup FR$, $MI \cup FI$ and Θ.

Regarding elongation, we calculate r_1 as the ratio between minimum and maximum distances of bordering pixels from the centre of gravity (we work on threshold images) and r_2 as the ratio of minor and major axes obtained from second moment calculation. Using these two ratios, the following possibility degrees are derived:

$$\pi_{1I}(MR) = \pi_{1I}(FR) = \min(r_1, r_2), \tag{7}$$

$$\pi_{1I}(MI) = \pi_{1I}(FI) = 1 - \pi_{1I}(MR). \tag{8}$$

In the framework of belief functions, for this measure, masses are defined as follows:

$$m_{1I}(MR \cup FR) = \min(r_1, r_2), \tag{9}$$

$$m_{1I}(MI \cup FI) = |r_1 - r_2|, \tag{10}$$

and the full set takes the rest:

$$m_{1I}(\Theta) = 1 - \max(r_1, r_2). \tag{11}$$

In the case of ellipse fitting, let A_{oe} is the part of object area that belongs to the fitted ellipse as well, A_o is the object area and A_e is the ellipse area. Then we define:

$$\pi_{2I}(MR) = \pi_{2I}(FR) = \max\left(0, \min\left\{\frac{A_{oe} - 5}{A_o}, \frac{A_{oe} - 5}{A_e}\right\}\right), \tag{12}$$

$$\pi_{2I}(MI) = \pi_{2I}(FI) = 1 - \pi_{2I}(MR). \tag{13}$$

Masses for this measure are the following ones:

$$m_{2I}(MR \cup FR) = \max\left(0, \min\left\{\frac{A_{oe}-5}{A_o}, \frac{A_{oe}-5}{A_e}\right\}\right), \tag{14}$$

$$m_{2I}(MI \cup FI) = \max\left\{\frac{A_e - A_{oe}}{A_e}, \frac{A_o - A_{oe}}{A_o}\right\}, \tag{15}$$

$$m_{2I}(\Theta) = 1 - m_{2I}(MR \cup FR) - m_{2I}(MI \cup FI). \tag{16}$$

Note that in cases where it is sure that all mines have a regular shape, the possibility degrees of MR can be reassigned to mines of any shape ($M = MR \cup MI$) while the possibility degrees of MI can be reassigned to friendly objects of any shape ($F = FR \cup FI$). Similarly, masses given to $MR \cup FR$ can be reassigned to M, while masses given to $MI \cup FI$ can be reassigned to F [2].

The area directly provides a degree $\pi_{3I}(M)$ of being a mine. Namely, since the range of possible AP mine sizes is approximately known, the degree of possibility of being a mine is derived as a function of the measured size:

$$\pi_{3I}(M) = \frac{a_I}{a_I + 0.1 \cdot a_{I\min}} \cdot \exp\frac{-[a_I - 0.5 \cdot (a_{I\min} + a_{I\max})]^2}{0.5 \cdot (a_{I\max} - a_{I\min})^2}, \tag{17}$$

where a_I is the actual object area on the IR image, while the approximate range of expectable mine areas is between $a_{I\min}$ and $a_{I\max}$ (for AP mines, it is reasonable to set $a_{I\min} = 15$ cm^2 and $a_{I\max} = 225$ cm^2). On the contrary, friendly objects can be of any size, so the possibility degree is set to one whatever the value of the size:

$$\pi_{3I}(F) = 1. \tag{18}$$

The area/size mass assignment based on the above reasoning is given by

$$m_{3I}(\Theta) = \frac{a_I}{a_I + 0.1 \cdot a_{I\min}} \cdot \exp\frac{-[a_I - 0.5 \cdot (a_{I\min} + a_{I\max})]^2}{0.5 \cdot (a_{I\max} - a_{I\min})^2}, \tag{19}$$

$$m_{3I}(FR \cup FI) = 1 - m_{3I}(\Theta). \tag{20}$$

4.1.2. MD measures

In reality, as explained in Ref. [2], MD data are usually saturated and data gathering resolution in the cross-scanning direction is typically very poor, so the MD information used consists of only one measure, which is the width of the region in the scanning direction, w [cm]. As friendly objects can contain metal of any size, we define:

$$\pi_{\mathrm{MD}}(F) = 1. \tag{21}$$

If there is some knowledge on the expected sizes of metal in mines (for AP mines, this range is typically between 5 and 15 cm), we can assign possibilities to mines as, for example:

$$\pi_{\mathrm{MD}}(M) = \frac{w}{20} \cdot [1 - \exp(-0.2 \cdot w)] \cdot \exp\left(1 - \frac{w}{20}\right). \tag{22}$$

The corresponding mass functions are

$$m_{\mathrm{MD}}(\Theta) = \frac{w}{20} \cdot \left[1 - \exp(-0.2 \cdot w)\right] \cdot \exp\left(1 - \frac{w}{20}\right), \tag{23}$$

$$m_{\mathrm{MD}}(FR \cup FI) = 1 - m_{\mathrm{MD}}(\Theta). \tag{24}$$

4.1.3. GPR measures

All three GPR measures provide information about mines [2].

In the of burial depth information (D), friendly objects can be found at any depth, while it is known that there is some maximum depth up to which AP mines can be expected, mainly due to their activation principles. However, due to soil perturbations, erosions, etc., mines can, by time, go deeper or shallower than the depth at which they were initially buried. In any case, they can rarely be found buried below 25 cm ($D_{\max}$). Thus, for this GPR measure, possibility distributions π_{1G} for mines and friendly object can be modelled as follows:

$$\pi_{1G}(M) = \frac{1}{\cosh(D/D_{\max})^2}, \tag{25}$$

$$\pi_{1G}(F) = 1. \tag{26}$$

In terms of belief functions, the masses for this measure are

$$m_{1G}(\Theta) = \frac{1}{\cosh(D/D_{\max})^2}, \tag{27}$$

$$m_{1G}(FR \cup FI) = 1 - m_{1G}(\Theta). \tag{28}$$

Another GPR measure exploited here is the ratio between object size and its scattering function, *d*/*k*. Again, friendly objects can have any value of this measure, while for mines, there is a range of values that mines can have, and outside that range, the object is quite certainly not a mine:

$$\pi_{2G}(M) = \exp\left(-\frac{[(d/k) - m_d]^2}{2 \cdot p^2}\right), \tag{29}$$

$$\pi_{2G}(F) = 1, \tag{30}$$

where m_d is the *d*/*k* value at which the possibility distribution reaches its maximum value (here, $m_d = 700$, chosen based on prior information), and p is the width of the exponential function (here, $p = 400$).

Similarly, the mass assignments for this measure are

$$m_{2G}(\Theta) = \exp\left(-\frac{[(d/k) - m_d]^2}{2 \cdot p^2}\right), \tag{31}$$

$$m_{2G}(FR \cup FI) = 1 - m_{2G}(\Theta). \tag{32}$$

Finally, propagation velocity, v, can provide information about object identity. Here, we extract depth information on a different way than in the case of the burial depth measure [30] and we preserve the sign of the extracted depth. This information indicates whether a potential object is above the surface. If that is the case, the extracted v should be close to $c = 3 \times 10^8$ m/s, the propagation velocity in vacuum. Otherwise, if the sign indicates that the object is below the soil surface, the value of v should be around the values for the corresponding medium, for example, from 5.5×10^7 to 1.73×10^8 m/s in the case of sand:

$$\pi_{3G}(M) = \exp\left(-\frac{(v - v_{\max})^2}{2 \cdot h^2}\right), \tag{33}$$

where $v_{\max}$ is the value of velocity which is the most typical for the medium (here, for sand, it is $0.5 \times (5.5 \times 10^7 + 1.73 \times 10^8) = 1.14 \times 10^8$ m/s, and for air, it is equal to c), and h is the width of the exponential function (here, $h = 6 \times 10^7$ m/s). Once again, friendly objects can have any value of the velocity:

$$\pi_{3G}(F) = 1. \tag{34}$$

The corresponding mass functions are

$$m_{3G}(\Theta) = \exp\left(-\frac{(v - v_{\max})^2}{2 \cdot h^2}\right), \tag{35}$$

$$m_{3G}(FR \cup FI) - 1 - m_{3G}(\Theta). \tag{36}$$

4.2. Combination

The combination of possibility degrees, as well as of masses, is performed in two steps [2]. The first one applies to all measures derived from one sensor. The second one combines results obtained in the first step for all three sensors.

In the case of possibilities, only the combination rules related to mines are considered. The issue of combination rules for friendly objects is discussed in Section 4.4.

Let us first detail the first step for each sensor. For IR, since mines can be regular or irregular, the information about regularity on the level of each shape measure is combined using a disjunctive operator (here the max):

$$\pi_{1IM} = \max(\pi_{1I}(MR), \pi_{1I}(MI)), \tag{37}$$

$$\pi_{2IM} = \max(\pi_{2I}(MR), \pi_{2I}(MI)). \tag{38}$$

The choice of the maximum (the smallest disjunction and idempotent operator) as a t-conorm is related to the fact that the measures cannot be considered as completely independent from each other. Thus, there is no reason to reinforce the measures by using a larger t-conorm, and the idempotent one is preferable in such situations. These two shape constraints should be both satisfied to have a high degree of possibility of being a mine, so they are combined in a conjunctive way (using a product). Finally, the object is possibly a mine if it has a size in the expected range or if it satisfies the shape constraint, hence the final combination for IR is

$$\pi_I(M) = \pi_{3I}(M) + [1 - \pi_{3I}(M)] \cdot \pi_{1IM} \cdot \pi_{2IM}. \quad (39)$$

The conjunction in the second term guarantees that $\pi_I(M)$ is in [0,1].

In the case of GPR, it is possible to have a mine if the object is at shallow depths and its dimensions resemble a mine and the extracted propagation velocity is appropriate for the medium. Thus, the combination of the obtained possibilities for mines is performed using a t-norm, expressing the conjunction of all criteria. Here the product t-norm is used:

$$\pi_G(M) = \pi_{1G}(M) \cdot \pi_{2G}(M) \cdot \pi_{3G}(M). \quad (40)$$

For MD, as there is just one measure used, there is no first combination step and the possibility degrees obtained using Eqs. (21) and (22) are directly used.

In the case of possibilities, the second combination step is performed using the algebraic sum:

$$\pi(M) = \pi_I(M) + \pi_{MD}(M) + \pi_G(M) - \pi_I(M) \cdot \pi_{MD}(M) - \pi_I(M) \cdot \pi_G(M) - \pi_{MD}(M) \cdot \pi_G(M) + \pi_I(M) \cdot \pi_{MD}(M) \cdot \pi_G(M), \quad (41)$$

leading to a strong disjunction [18, 29], as the final possibility should be high if at least one sensor provides a high possibility, reflecting the fact that it is better to assign a friendly object to the mine class than to miss a mine [2].

In the belief function framework, for IR and GPR, masses assigned by the measures of each of the two sensors are combined by Dempster's rule in unnormalized form (Eq. (3)). A general idea for using the unnormalized form of this rule instead of more usual, the normalized form is to preserve conflict [27], i.e. mass assigned to the empty set, Eq. (4). Here, a high degree of conflict would indicate that either there are several objects and the sensors, as detectors of different physical phenomena, do not provide information on the same object, or some sources of information are not completely reliable. Our main interest is in the possibility that sensors do not refer to the same object, as the unreliability can be modelled and resolved through discounting factors [3]. After combining masses per sensor, the fusion of sensors is performed, using Eq. (3) again. If the mass of the empty set after combination of sensors is high, they should be clustered as they do not sense the same object.

4.3. Comparison of the combination equations

For IR, based on Eqs. (6)–(20) and (39), it can be shown that

$$Pl_I(M) \leq \pi_I(M). \quad (42)$$

This is in accordance with the least commitment principle used in the possibilistic model [2], as usually done in this framework.

As far as MD is concerned, there is no difference since it provides only one measure.

In the case of GPR, based on the comparison of Eqs. (25) and (27), Eqs. (29) and (31), as well as Eqs. (33) and (35), we can conclude that Eq. (40) can be rewritten as

$$\pi_G(M) = m_{1G}(\Theta) \cdot m_{2G}(\Theta)\, m_{3G}(\Theta). \tag{43}$$

Furthermore, the application of the Dempster's rule (Eq. (3)) to the mass assignments of the three GPR measures results in the fused mass of the full set for this sensor:

$$m_G(\Theta) = m_{1G}(\Theta) \cdot m_{2G}(\Theta)\, m_{3G}(\Theta) \tag{44}$$

which leads to

$$\pi_G(M) = m_G(\Theta). \tag{45}$$

This means that the ignorance is modelled as a mass on Θ in the belief function framework, while it privileges the class that should not be missed (M) in the possibilistic framework (i.e. the ignorance will lead to safely decide in favour of mines).

4.4. Decision

As the final decision about the identity of the object should be left to the deminer not only because his life is in danger but also because of his experience, the fusion output is a suggested decision together with confidence degrees [2].

In the case of possibilities, the final decision is obtained by thresholding the fusion result for M and providing the corresponding possibility degree as the confidence degree. As almost all possibility degrees obtained at the fusion output are either very low or very high, the selected regions having very low values of $\pi(M)$ (below 0.1) are classified as F, and the ones with very high values (above 0.7) are classified as M. Only a few regions exist at which the resulting possibility degree for M has an intermediary value and there, as mines must not be missed, the decision is M. In the following, this decision approach is referred to as *dec*1.

An alternative (*dec*2) for the final decision-making is to derive the combination rule for F as well, compare the final values for M and F and derive an adequate decision rule. Due to operation principles of GPR and MD, the measures of these two sensors can only give information where mines are possibly not. As they are non-informative with respect to friendly objects, it is not useful to combine their possibility degrees for F. Thus, for deriving the final combination rule for F, $\pi(F)$, we can rely only on IR, that is:

$$\pi(F) = \pi_I(F). \tag{46}$$

In the case of IR, since friendly objects can be regular or irregular, we apply a disjunctive operator (the max) for each of the shape constraints. In order to be cautious when deciding F, we combine the two shape constraints and the area measure using a conjunctive operator. Taking into account of Eq. (18), this reasoning results in

$$\pi(F) = \max(\pi_{1I}(F_R), \pi_{1I}(F_I)) \cdot \max(\pi_{2I}(F_R), \pi_{2I}(F_I)). \tag{47}$$

Thus, in this alternative way to derive decisions, in regions where IR gives an alarm, the decision rule chooses M or F depending on which one of the two has a higher possibility value, given by Eqs. (41) and (58), respectively. In other regions, at which IR does not give an alarm although at least one of the two other sensors gives an alarm, the decision is based on the fusion result for M, as in *dec*1.

In the case of belief functions, as shown in Ref. [15], usual decision rules based on beliefs, plausibilities [6] and pignistic probabilities [26] do not give useful results because there are no focal elements containing mines alone [27]. As a consequence, these usual decision rules would always favour friendly objects [2]. The underlying reason is that the humanitarian demining sensors are anomaly detectors and not mine detectors. In such a sensitive application, no mistakes are allowed so in the case of any ambiguity, much more importance should be given to mines. Hence, in Ref. [15], guesses $G(A)$ are defined, where $A \in \{M, F, \varnothing\}$:

$$G(M) = \sum_{M \cap B \neq \varnothing} m(B), \tag{48}$$

$$G(F) = \sum_{B \subseteq F,\, B \neq \varnothing} m(B), \tag{49}$$

$$G(\varnothing) = m(\varnothing). \tag{50}$$

In other words, the guess value of a mine is the sum of masses of all the focal elements containing mines, regardless their shape, and the guess of a friendly object is the sum of masses of all the focal elements containing nothing else but friendly objects of any shape, meaning that the guesses are a cautious way to estimate confidence degrees.

As the output of the belief function fusion module, the three possible outputs (M, F, conflict) are provided together with the guesses, for each of the sensors and for their combination.

For GPR, the focal elements are only F and Θ, so guesses for this sensor become simply:

$$G_G(M) = m_G(\Theta), \tag{51}$$

$$G_G(F) = m_G(F). \tag{52}$$

From Eqs. (45) and (51), we conclude that for GPR, the possibility degree of a mine is equal to the guess of a mine:

$$\pi_G(M) = G_G(M). \tag{53}$$

Furthermore, Eqs. (6) and (48) show that the guess of a mine is equal to its plausibility, while Eqs. (5) and (49) show that the guess of a friendly object is equal to its belief. This means that the relation given by Eq. (42) shows, actually, that for IR:

$$G_I(M) \leq \pi_I(M). \tag{54}$$

4.5. Results

The proposed approach has been applied to a set of known objects, buried in sand, leading to 36 alarmed regions in total [2]: 21 mines (M), 7 placed false alarms (PF, friendly objects) and 8 false alarms caused by clutter (FN, with no object).

The results of the possibilistic fusion are very promising, since all mines are classified correctly with the proposed approach, as can be seen in **Table 1**. The numbers given in the parenthesis indicate the number of regions selected in the pre-processing step for further analysis, that is, measure extraction and classification. Regarding the combination operators, the results given in this table are based on the combination proposed in Section 4.2. (Eqs. (39)–(41)). The second fusion step is important, since a decision taken after the first step provides only 18 mines for IR, nine for MD and 13 for GPR. This illustrates the interest of combining heterogeneous sensors.

The two decision rules, *dec*1 and *dec*2, give the same results for mines and friendly objects caused by clutter [2]. In the case of placed false alarms, two are correctly classified in the case of *dec*2, which is a slight improvement with respect to *dec*1 and the same result as for the belief function fusion, shown in **Table 2**. It is not surprising that the placed false alarms are not so well detected by any of the methods, since our model is designed in order to favour the detection of mines. This is also the type of results expected from deminers. Regarding correct classification of mines, the results of the possibilistic fusion are slightly better than those obtained using the belief function method (19 mines detected, **Table 2**). This is due to the increased flexibility at the combination level. False alarms with no objects are correctly identified by the belief function method (six out of eight), and it is the same result as for the two possibilistic decision rules. This result shows that a power of our methods is in decreasing the number of clutter-caused false alarms without decreasing the result of mine detection, thanks to knowledge inclusion.

All results have been obtained with the models proposed in Section 4.1., with the same parameters. Note that although the general shapes of the possibility distributions are important and have been designed based on prior knowledge, they do not need to be estimated very precisely, and the results are robust to small changes in these functions. What is important is that the functions are not crisp (no thresholding approach is used) and that the rank is preserved (e.g. an object with a measure value outside of the usual range should have a lower possibility degree than an object with a typical measure value). Two main reasons explain the experienced robustness: (i) these possibility distributions are used to model imprecise information, so they do not have to be precise themselves and (ii) each of them is combined in the fusion process (Section 4.2.) with other pieces of information, which diminishes the importance and the influence of each of them.

Analysis regarding the robustness of the choice of the operator is also performed (within a class corresponding to the type of reasoning we want to achieve) [2]. Different operators within the same family have been tested, leading to the maximization and minimization of the possibility degrees of mines, thus being the safest and the least safe situations from the point of view of mine detection. The results obtained show that the model is robust indeed: all mines are detected in the second step, for all fusion schemes.

Classified correctly, possibility theory	Sensors			Fusion	
	IR	MD	GPR	dec1	dec2
M (total: 21)	18 (18)	9 (9)	13 (13)	21 (21)	21 (21)
PF (total: 7)	0 (4)	0 (4)	2 (6)	1 (7)	2 (7)
FN (total: 8)	0 (1)	0 (0)	6 (7)	6 (8)	6 (8)

Table 1. Correct classification results, possibilistic fusion.

Classified correctly, belief functions	Sensors			Fusion
	IR	MD	GPR	
M (total: 21)	10 (18)	9 (9)	13 (13)	19 (21)
PF (total: 7)	3 (4)	0 (4)	1 (6)	2 (7)
FN (total: 8)	0 (1)	0 (0)	6 (7)	6 (8)

Table 2. Correct classification results, belief functions.

Differences between the results of **Tables 1** and **2** can be formally explained as discussed in Section 4.3. For GPR, Eq. (53) explains why the results are the same for the two fusion approaches. In the case of IR, Eq. (54) indicates that the possibilistic approach would favour mines more than the belief function approach, which is indeed the case here.

5. Conclusion

Fusion approaches for close-range humanitarian mine detection are presented and compared. These approaches are based on the belief functions as well as on the fuzzy/possibility theory. The differences at the combination step are mainly highlighted in this comparison. The modelling step is performed according to the semantics of each framework, but the designed functions are as similar as possible, so as to enhance the combination step. Different fusion operators are tested, depending on the information and its characteristics. An appropriate modelling of the data along with their combination in a possibilistic framework leads to a better differentiation between mines and friendly objects. The decision rule is designed to detect all mines, at the price of a few confusions with friendly objects. This is a requirement of this sensitive application domain (mines must not be missed). Still the number of false alarms remains limited in our results. The robustness of the choice of the operator is also tested, and all mines are detected for all fusion schemes. The proposed modelling is flexible enough to be easily adapted to the introduction of new pieces of information about the types of objects and their characteristics, as well as of new sensors.

The work shown in this chapter is useful in many other applications, even in quite different domains, and constitutes thus a large set of methods and tools for both research and applicative work. The developed schemes have a noticeable variety and richness and constitute a real improvement over existing tools.

Acknowledgements

The author of this chapter would like to thank Prof. Isabelle Bloch (Télécom ParisTech, Paris, France) for her long-term help and support. The author also thanks the Belgian Ministry of Defence for its financial support and the TNO Physics and Electronics Laboratory (The Hague, the Netherlands) for the permission to work on the data gathered on their test facilities within the Dutch HOM-2000 project. Parts of this chapter were previously published in Milisavljevic et al. [2].

Author details

Nada Milisavljevic

Address all correspondence to: nmilisav@yahoo.co.uk

Royal Military Academy, Signal and Image Centre, Belgium

References

[1] Acheroy M. Mine action technologies: Problems and recommendations. Journal for Mine Action. 2003;**7**(3):48, 49 & 64

[2] Milisavljevic N, Bloch I, Acheroy M. Multi-sensor data fusion based on belief functions and possibility theory: Close range antipersonnel mine detection and remote sensing mined area reduction. In: Habib MK, editor. Humanitarian Demining. Vienna, Austria: InTech; 2008. pp. 95–120. Ch. 4. DOI: 10.5772/5410

[3] Milisavljević N, Bloch I. Improving mine recognition through processing and Dempster-Shafer fusion of multisensor data. In: Sarfraz M, editor. Computer-Aided Intelligent Recognition, Techniques and Applications. New York: John Wiley; 2005. pp. 319–343. Ch. 17. ISBN: 0-470-09414-1

[4] Dubois D, Prade H. Fuzzy Sets and Systems: Theory and Applications, Collection: Mathematics in Science and Engineering. Vol. 144. New York, London: Academic Press; 1980

[5] Smets P. The combination of evidence in the transferable belief model. IEEE Transactions on Pattern Analysis and Machine Intelligence. 1990;**12**(5):447–458

[6] Shafer G. A Mathematical Theory of Evidence. Princeton, NJ: Princeton University Press; 1976

[7] de Jong W, Lensen HA, Janssen YHL. Sophisticated test facility to detect landmines. In: Proceeding of SPIE Conference, Detection and Remediation Technologies for Mines and Minelike Targets. Vol. 3710. Orlando, USA; 1999. pp. 1409–1418

[8] Cremer F, Schutte K, Schavemaker JGM, den Breejen E. A comparison of decision-level sensor-fusion methods for anti-personnel landmine detection. Information Fusion. 2001;**3** (2):187–208

[9] Yee ML. Multisensor probabilistic fusion for mine detection. In: Proceeding of SPIE Conference, Detection and Remediation Technologies for Mines and Minelike Targets. Vol. 4394. Orlando, USA; 2001. pp. 959–969,

[10] Stanley RJ, Gader PD, Ho KC. Feature and decision level sensor fusion of electromagnetic induction and ground penetrating radar sensors for landmine detection with hand-held units. Information Fusion. 2002;**3**(3):215–223

[11] Auephanwiriyakul S, Keller JM, Gader PD. Generalized Choquet fuzzy integral fusion. Information Fusion. 2002;**3**(1):69–85

[12] Milisavljević N, Acheroy M. An approach to the use of the Bayesian rule in decision level fusion for multisensor mine detection. In: Proceedings of Physics in Signal and Image Processing (PSIP). Paris, France; 1999. pp. 261–266

[13] den Breejen E, Schutte K, Cremer F. Sensor fusion for anti personnel landmine detection. In: Proceedings of SPIE Conference, Detection and Remediation Technologies for Mines and Minelike Targets. Vol. 3710. Orlando, USA; April 1999. pp. 1235–1245

[14] Perrin S. Contribution à l'algorithmique multicapteur pour la détection de mines antipersonnel [PhD dissertation]. France: Ecole Centrale de Lille, USTL; 2001

[15] Milisavljević N, Bloch I. Sensor fusion in anti-personnel mine detection using a two-level belief function model. IEEE Transactions on Systems, Man and Cybernetics, Part C. 2003;**33**(2):269–283

[16] Smets P, Kennes R. The transferable belief model. Artificial Intelligence. 1994;**66**:191–234

[17] Smets P. What is Dempster-Shafer's model? In: Yager RR, Fedrizzi M, Kacprzyk J, editors. Advances in the Dempster-Shafer Theory of Evidence. New York: Wiley; 1994. pp. 5–34

[18] Dubois D, Prade H. A review of fuzzy set aggregation connectives. Information Sciences. 1985;**36**:85–121

[19] Dubois D, Grabisch M, Prade H, Smets P. Using the transferable belief model and a qualitative possibility theory approach on an illustrative example: The assessment of the value of a candidate. International Journal of Intelligent Systems. 2001;**16**(11):1245–1272

[20] van Cleynenbreugel I, Osinga SA, Fierens F, Suetens P, Oosterlinck A. Road extraction from multi-temporal satellite images by an evidential reasoning approach. Pattern Recognition Letters. 1991;**12**:371–380

[21] Mascle S, Bloch I, Vidal-Madjar D. Application of Dempster-Shafer evidence theory to unsupervised classification in multisource remote sensing. IEEE Transactions on Geoscience and Remote Sensing. 1997;**35**(4):1018–1031

[22] Le Hégarat-Mascle S, Bloch I, Vidal-Madjar D. Introduction of neighborhood information in evidence theory and application to data fusion of radar and optical images with partial cloud cover. Pattern Recognition. 1998;**31**(11):1811–1823

[23] Smets Ph. Belief functions: The disjunctive rule of combination and the generalized Bayesian theorem. International Journal of Approximate Reasoning. 1993;**9**:1–35

[24] Dubois D, Prade H. Representation and combination of uncertainty with belief functions and possibility measures. Computational Intelligence. 1988;**4**:244–264

[25] Denœux T. A k-nearest neighbor classification rule based on Dempster-Shafer theory. IEEE Transactions on Systems, Man and Cybernetics. 1995;**25**(5):804–813

[26] Smets P. Constructing the pignistic probability function in a context of uncertainty. In: Kanal LN, Henrion M, Shachter RD, Lemmer JF, editors. Uncertainty in Artificial Intelligence 5. New York: Elsevier; 1990. pp. 29–39

[27] Milisavljević N, Bloch I. A two-level approach for modeling and fusion of humanitarian mine detection sensors within the belief function framework. In: Proceedings of Applied Stochastic Models and Data Analysis. Vol. 2. Compiègne, France; 2001. pp. 743–748

[28] Zadeh L. Fuzzy sets. Information and Control. 1965;**8**(3):338–353

[29] Bloch I. Information combination operators for data fusion: A comparative review with classification. IEEE Transactions on Systems, Man and Cybernetics. 1996;**26**:52–67

[30] Milisavljević N, Bloch I, van den Broek SP, Acheroy M. Improving mine recognition through processing and Dempster-Shafer fusion of ground-penetrating radar data. Pattern Recognition. 2003;**36**(5):1233–1250

Chapter 5

Remote Sensing for Non-Technical Survey

Yann Yvinec, Nada Milisavljevic, Charles Beumier,
Idrissa Mahamadou, Dirk Borghys,
Michal Shimoni and Vinciane Lacroix

Additional information is available at the end of the chapter

http://dx.doi.org/10.5772/66691

Any sufficiently advanced technology is indistinguishable from magic.

—Arthur C. Clarke

Abstract

This chapter describes the research activities of the Royal Military Academy on remote sensing applied to mine action. Remote sensing can be used to detect specific features that could lead to the suspicion of the presence, or absence, of mines. Work on the automatic detection of trenches and craters is presented here. Land cover can be extracted and is quite useful to help mine action. We present here a classification method based on Gabor filters. The relief of a region helps analysts to understand where mines could have been laid. Methods to be a digital terrain model from a digital surface model are explained. The special case of multi-spectral classification is also addressed in this chapter. Discussion about data fusion is also given. Hyper-spectral data are also addressed with a change detection method. Synthetic aperture radar data and its fusion with optical data have been studied. Radar interferometry and polarimetry are also addressed.

Keywords: multi-spectral, hyper-spectral, radar, interferometry, polarimetry

1. Problem statement

Experience shows that a lot of effort is sometimes spent to clear areas that are not actually mines [1]. It is therefore paramount to have a good assessment of the areas that are contaminated so that clearance is done on actually mined areas. Remote sensing can be used for that. It provides information on large areas that surveyors cannot enter because of the suspicion of mine contamination.

In the early days of research, the main objective was to find ways to detect mines by airborne survey by developing the right sensors. Nowadays, the goal is to use remote sensing to collect the right data about an area, to merge it with information from other sources, for instance expert knowledge and then improve awareness on the suspicious areas to update our assessment of their contamination.

This chapter describes some techniques developed to use remote sensing to help this activity.

A key objective is the detection of elements in the remote sensing data that may indicate the possible presence of mines. Trenches are such indicators. Section 2 describes the detection of long trenches on visible images.

Another example of the detection of indicators is the detection of craters as in Section 3.

The texture of an area as seen in airborne data may be an indication of the land use and therefore may help understand the contamination. A good example is the use of Gabor filters as in Section 4.

Remote sensing can also be used to provide a three-dimensional (3D) representation of an area. This information proves to be invaluable to photo-interpreters when they want to understand a certain area. A method to provide this surface, or terrain, model is in Section 5.

Classification of the land use may benefit from the multi-spectral sensors as described in Section 6.

Section 7 explains how to fuse the data from several sensors to obtain a common and improved understanding of an area.

Hyper-spectral sensors can bring a lot of added information for instance for change detection as explained in Section 8.

Radars are less often used in mine action but prove to be quite useful. A good example can be found in Section 9.

In order to help the mine action community to use remote-sensing data, a geoinformation platform has been created and is available on the TIRAMISU website: http://www.fp7-tiramisu.eu/.

2. Long trench detection (Vinciane Lacroix)

Using aerial photographs for individual mine recognition has been proposed in Humanitarian Demining since 1998 (e.g., see Ref. [2]) although with little success. Maathuis then introduced the concept of indirect minefield indicators in Ref. [3]. This concept has been further developed in the SMART project (e.g., see Ref. [4]) and used in AIDSS, an operational system designed to find indicators of mine presence (IMP) and indicators of mine absence (IMA) (e.g., see Ref. [5]). In the scope of the EU FP7 TIRAMISU project, new methods to detect and map IMAs and IMPs on aerial and satellite images were developed and tested [6–9]. The choice of indicators results from an analysis involving photo-interpreters and possibly former members of the military who took part in the conflict that caused the contamination.

In the case of military conflicts, trenches are examples of IMPs. The first step for detecting an IMP/IMA consists in translating the indicator in terms of basic image features. As trenches can be detected thanks to their long straight linear shadows, an automatic dark line detection tool is proposed for their extraction.

The low occurrence of trenches in the huge amount of aerial photos collected during a flight over a suspected hazardous area (SHA) makes the detection task overwhelming for a photo-interpreter. **Figure 1** gives an idea of the amount of data collected during a typical campaign. The line in **Figure 1** shows the path of a plane, which performed an aerial campaign over the SHAs of Bihać, Bosnia and Herzegovina, in 2010.

Five stripes of 17 colour photographs of size 4288 × 2848 were taken using Nikon D90 camera. On the right of the figure, the corresponding photographs of one of the airplane tracks are overlaid.

Various scenarios using the dark line detector were proposed to extract some suspicious photographs from the whole set. The result of such a scenario is shown in **Figure 2**. Note that the area is covered by forest producing tree shadows generating many other dark lines.

The trench detector based on dark line detection is described in Ref. [8] and summarized hereafter.

Figure 1. Example of a partial aerial campaign over Bihać region in Bosnia and Herzegovina. Left: the partial fight path. Right: a stripe made of 17 colour photographs taken during the flight.

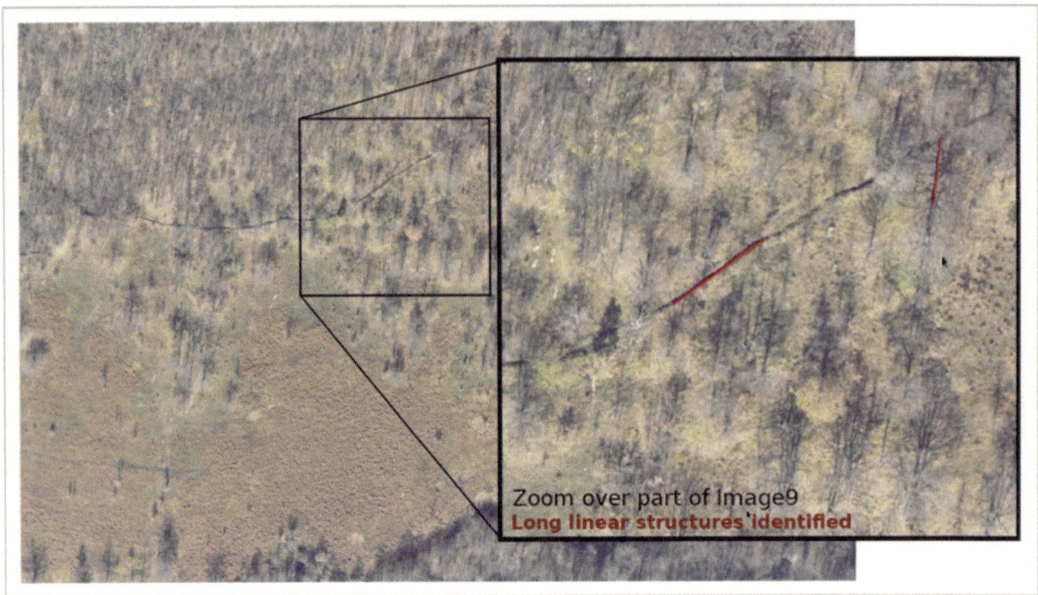

Figure 2. Extraction of the most suspicious image thanks to the long trench detection tool: Image #9 has been identified as having the longest dark linear structures.

The success of line detection relies on (i) a line filter producing the contrast of the line and its direction at each pixel, (ii) an efficient non-maximum suppression to extract the line axis and (iii) an appropriate linking of line elements. In order to use the line detection process for trench detection, a set of relevant local properties enabling to discriminate a trench from other elements should then be computed.

The line filter is based on the gradient line detector, which exploits the change in gradient direction at each side of a line [10]. Each pixel p is considered as a potential line element: in an eight-neighbourhood, dp, the dot product of the gradient of the intensity at pixels arranged symmetrically around p is computed (see **Scheme 1**); in the presence of a line, dp is negative for some of the four pairs, and |dp|, its absolute value, is the highest for the pair lying in the direction perpendicular to the line (quantized direction). A more precise direction is provided by the difference of the gradient vectors G located at the pixels belonging to this pair. The direction of each of these gradient vectors enables to discriminate between dark and bright lines (see **Scheme 1** (centre) and (right)), while the square root of |dp| provides quantitative information related to the contrast in the neighbourhood of p.

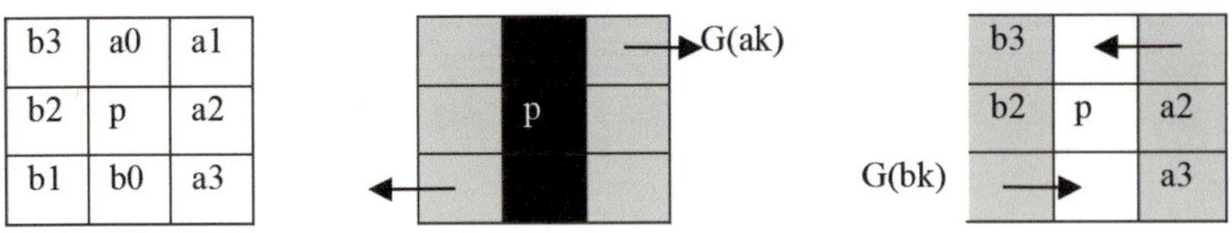

Scheme 1. Local neighbourhood at p (left); dark line at p (centre); bright line at p (right).

A line filter providing a vector field is thus available over the image, based on the above computation. Non-maxima suppression is then used to obtain the line median axis. Local line width is then computed by associating the line axis with its borders; the gradient norm at these borders enables to compute the local line contrast, while linking axis elements enables to obtain various average attributes over line segments: contrast, direction and width.

Red, green, blue and possibly panchromatic channels may be available for trench detection. As linear shadows should be detected, the channel offering the highest dark line contrast is kept for line extraction. This will most probably be the red channel. The IMP (i.e., trenches) should then be translated into image features according to image resolution and scene characteristics. In this case, the minimum length (in pixels) of the line segments and the minimum average width (in pixels) should be provided. Several scenarios were considered, all of them analysing the series of images taken during the aerial campaign displayed in **Figure 1**. In the first one, information about potential trench orientation is provided thanks to old annotated scanned maps, focusing the detection on a specific orientation. In another one, the detection in a specific orientation will be avoided; this could be an option in areas where tree shadows may generate a lot of dark lines. In the most general scenario, no constraint on the orientation is provided. In all cases, line features satisfying the constraints are provided as a list vectors ranked based on the average line contrast. The vectors may be superimposed on the image or imported in a geographic information system. The most suspicious image is the one providing the larger total length of valid segments.

When looking at the results, the photo-interpreter may consider the image as suspicious and note it for further visual analysis, object-based image analysis [9] or further processing such as ortho-photo production.

The result of launching the dark line detector with a minimum length of 180 pixels and a minimum width of five pixels (the image resolution is about 10 cm) with no orientation constraint provides only three suspected photographs, all of which contain trenches. Part of the most suspected one is displayed together with the superimposed detected linear objects.

3. Crater detection (Vinciane Lacroix)

3.1. Introduction

In the scope of the TIRAMISU project, image-processing tools to help non-technical survey were developed and tested (see Section 2). In this framework, the Cambodian Mine Action Centre (CMAC) expressed the need for having the mapping of craters in the eastern part of Cambodia as they might provide indication on the presence of unexploded ordnance (UXO). The presence of UXOs resulting from the US bombing during the late 1960s and 1970s is still preventing the use of the land in Cambodia. When dropped, the bombs produced craters that may still exist today. Many of them are filled with water so that they appear as circular objects on satellite images.

Similar work was made by Hatfield-Consultants [11] for Laos. The authors used historic Corona satellite images; they computed differences between the original image and its smoothed version and used these differences in an unsupervised K-means fuzzy classifier.

We rather used data acquired by the WorldView-2 (WV2) instrument. We used circularity of the craters with the assumption of water or bare soil inside.

This initial circle detection method applied to the panchromatic image is described in Ref. [12] and has been validated as a *circle detection technique* on various sets of images used for circle detection [13] and on artificial images made of controlled shapes (circles, ellipse, squares and triangles). The method called CGC (constrained gradient for circle) has been enhanced and further applied to the pan-sharpened Band 8 (covering near-infrared (NIR) range from 860 to 900 nm) and to various normalized difference indices. In order to validate the method as a *crater detection tool*, since ground truth was not available, a visual crater detection has been performed by an independent photo-interpreter at IGEAT (Institut de Gestion de l'Environnement et d'Aménagement du Territoire). A summary of the whole process is provided in this section.

3.2. Circle detection overview

The state of the art in circle detection is provided in Ref. [12]. The most existing approaches use the fact that some basic elements (pixels, edge elements or 'edgels', or connected segments) are part of a circumference so that they can be combined to generate a centre-radius pair hypothesis. To our knowledge, none of these methods *directly* uses a centre-radius hypothesis. We designed a 'Roundness Index' filter providing at each pixel a radius and a contrast corresponding to the most contrasted circle having the current pixel as centre. The current design of the filter is such that a unique circle is assigned to a pixel (i.e., if the pixel is the centre of several circles, only the most contrasted one will be kept). The filtering involves a computation on pre-defined digital circles of radius spanning the radius range to detect (see **Figure 3**, left). The Roundness Index computation described in the next section is based on the average flux and circulation of the gradient of the intensity along the tested circle and a *gradient angle compatibility* test. Indeed, in

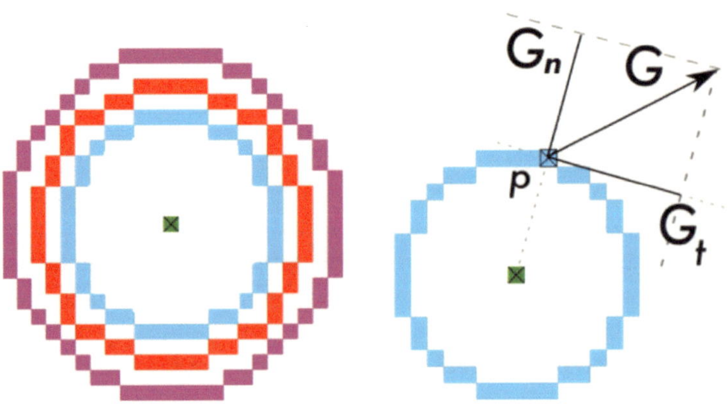

Figure 3. Left: Various digital circles of odd radius (from 7 to 11) generated by Bresenham's circle algorithm. Right: local computation at *p* located on the digital circle of radius 7, involving projection of *G*, the gradient of the intensity, along two perpendicular directions (normal and tangent to the circle).

the case of a perfect circle, all gradient directions of pixels located on (and nearby) the circumference should point along the normal—that is, the line joining the current point to the centre of the circle; in the case of an imperfect circle, a small difference α should be tolerated. If the test is positive for a minimum proportion of pixels on the considered circumference, the pixel is considered as a potential circle centre candidate, and the radius providing the highest contrast is kept. The local maxima of the contrast are then selected. A final test on the variation of the gradient on the identified digital circle is performed in order to keep circular shapes only. A reasonable range for the tolerated gradient difference α is [0.06, 0.14] radian. And a reasonable range for the fraction of pixels satisfying the gradient angle compatibility is [0.7, 1]. The method should not be used for too small radius (i.e., above or equal to 6 pixels) because the digitalization generates errors on the gradient direction.

3.3. Round index computation

The computation of the Gaussian gradient G and angle A is a prerequisite to the filtering process. Let r_0 and r_n be, respectively, the minimum and maximum radius of the circles to detect. Let $C = \{C_0, \ldots, C_k, \ldots, C_n\}$ be the list of digital circles located at the origin, corresponding to the increasing radii $r_0, \ldots, r_k, \ldots, r_n$. In order to initialize the process, the following information about each C_k is stored: Z the number of pixels in the circle and, for each of these pixels, the integer Cartesian coordinates i and j, and the angle in polar coordinates θ. Bresenham's circle algorithm may be used to compute i and j. Finally, let α be the tolerance angle, that is, the maximum angle difference between the gradient and the normal to the assumed circle, and let b represent the minimum portion of pixels satisfying the gradient angle compatibility test.

Two rasters, R, the local radius, and F, the filtered image providing the 'Roundness Index', are initialized. The image is scanned; each pixel c is considered as a potential circle centre candidate; each digitized circle in C is translated at c; the average flux $\overline{|G_n|}$ (absolute value of the projection of G along the normal) and average circulation $\overline{|G_t|}$ (absolute value of the projection of G along the tangent) over all pixels of the translated circle is computed (see an example in **Figure 3**, right). If $\overline{|G_n|} > \overline{|G_t|}$ and if the gradient angle compatibility test are satisfied for at least bz pixels, c is accepted as a circle centre candidate. The compatibility test at pixel P located on circle C_j is $|\theta - A_p| < \alpha$, where A_p represents the gradient angle at p translated by c and θ is the polar angle of P in C_j. The maximum average flux $\overline{|G_n|}$ computed over all digital circles and the radius r of the circle providing that maximum values are stored in R and F, respectively. Bright circles are discriminated from dark circles based on the sign of the gradient projection (bright circles are such that the gradient is pointing towards the centre), so that the average flux is computed for dark and bright circles separately. For practical purpose, dark circles are stored as negative values. The result of the filtering process holds in these two rasters R and F.

3.4. Crater detection

The main assumption underlying the crater detection is that craters are circular. Moreover, we assume that they are either filled with water or made of bare soil. This is probably too restrictive but limits the false-positive rate especially in the presence of trees whose crowns are circular. Therefore, we have identified Band 8 covering near-infrared (NIR) range from 860 to 900 nm as ideal band to detect dark circles. Water has indeed a low reflectivity in this part of the spectrum. A threshold at 0.6 on the normalized difference vegetation index (NDVI) computed from WV2 data has been added: only circles having their average NDVI equal to or lower than threshold are considered as potential craters. NDVI is computed on WV2 as the normalized difference between Band 6 B_6 and Band 5 , that is, $\frac{B_6-B_5}{B_6+B_5}$. All potential normalized differences have also been considered as a basis for dark circle detection.

We purchased WV2 data taken on 26 November 2011, covering 100 km^2 of the eastern part of Cambodia near the border with Vietnam, in a rural zone (Choam Kravien), with a spatial resolution of about 0.5 and 2 m for panchromatic band and multi-spectral band, respectively, and a pixel depth at acquisition of 11 bits. The dark circle detection has been performed on the data after ortho-rectification and atmospheric correction.

3.5. Validation

In order to validate the crater detection algorithm, as ground truth was not available, a photo-interpreter analysed five zones of 4000 × 4000 pixels, each one covering 4 km^2, based on a coloured composition (red = Band 7, green = Band 5, blue = Band 3). The photointerpreter provided for each detected crater the centre, the radius, the land cover, the circularity expressed in percentage (100% representing perfect circles) and a confidence measurement also expressed in percentage. She found 334 craters over the five zones (20 km^2). In the following, 'visual crater' and 'CGC circle' denote a crater identified by the photo-interpreter and a circle detected using the automatic method, respectively. **Table 1** summarizes some statistics derived from the visual crater detection over the five zones.

The following remarks can be drawn from **Table 1**:

- As far as the land cover is concerned, the assumption of bare soil or water inside crater makes sense as only four visual craters over 334 (less than 1%) have another land cover according to the photo-interpreter.
- The circularity assumption (i.e., assuming that the crater is almost a circle) is far from true: 152 visual craters only (45%) have a circularity coefficient larger than 60.
- The radius of the visual crater can be very small: 71 (21%) have a radius lower than 6 pixels, which are the minimum radius recommended for our circle detection technique.
- Only 211 visual craters (63%) are detected with a confidence larger than 50.

Confidence	#	Circularity	#	Radius	#	Land cover	#
[75, 100]	102	[80, 100]	32	Larger than 12	12	water	187
[50, 75]	109	[60, 80]	122	[9, 12]	80	Bare soil	143
[25, 50]	86	[40, 60]	109	[6, 9]	171	Vegetation	2
[0, 25]	37	[20, 40]	58	[3, 6]	69	Unknown	2
		[0, 20]	13	Lower than 3	2		
Total	**334**		**334**		**334**		**334**

Table 1. Statistics on visual crater detection: confidence, circularity, radius and land cover.

http://dx.doi.org/10.5772/66691

Given these remarks, it seems *a priori* difficult to detect the craters based on the CGC method: less than 1/3 (33%) satisfies the required assumption of a radius larger or equal to 6 and circularity larger or equal to 70%. The percentage drops to 28% if one considers only the craters with a confidence larger or equal to 75%. Nevertheless, the comparison between the visual craters and the CGC circles has been performed. In order to obtain quantitative results, precision p and recall r have been computed over the five zones; p and r are defined as $p = \frac{TP}{TP+FP}$ and $r = \frac{TP}{TP+FN}$ where TP, FP and FN are the number of true positive, false positive and false negative, respectively. A false positive is a CGC circle, which is not identified as a visual crater. A true positive is a CGC circle, which corresponds to a visual crater. A false negative is a visual crater that has been missed by the automatic method. Precision is also referred as the positive-predictive value while recall is also called true-positive rate, sensitivity or hit rate. These definitions are better understood in view of **Table 2**, where it is assumed that visual detection corresponds to 'Ground Truth'.

		Ground Truth			
		Crater	**No Crater**	**Sum**	**Ratio**
Detection	Circle	TP	FP	**Detected**	P = TP/**(TP + FP)**
	No Circle	FN	TN		
	Sum	**Existing**			
	Ratio	r = TP/**(TP + FN)**			

Table 2. Precision p and recall r definitions.

The circle detection has been performed on Band 8 (with NDVI of >0.6), and on various normalized difference indices: ND_{83}, ND_{84}, ND_{85}, ND_{73}, ND_{74}, ND_{75} and ND_{65}, where ND_{ij} refers to the normalized difference between B_i and B_j, that is,$ND_{ij} = \frac{B_i - B_j}{B_i + B_j}$. A true positive is counted if the centre of the CGC circle is falling inside a visual crater. Fusion using logical operator 'OR' has been considered. Several precisions and recalls (summed up on **Figure 4**) have been computed in different ways:

1. All visual craters have been considered with an equal weight.

2. Visual craters (TP and FN) are counted according to the confidence assigned by the photo-interpreter: the weight assigned is 25th of the confidence so that a crater of confidence 100 has a weight of 4 compared to a crater with the lowest confidence having a weight of 1. False positives have an equal weight of 1.

3. Only visual craters having a radius larger than or equal to 6 and circularity $c \geq 70$ are equally considered. Nevertheless, CGC circles corresponding to visual craters having radius or circularity outside these bounds are not considered as FP.

4. Only visual craters having a radius larger than or equal to 6 and circularity $c \geq 70$ are considered and weighted according to their confidence. Similarly, CGC circles corresponding to visual craters having a radius or circularity outside these bounds are not considered as FP.

Selected results among these four sets of experiments are presented in **Figure 5**. The selection is based on the 'best precision', the 'best recall' and the 'best average of precision and recall', taking each considered image independently and fusion (OR) of all of possible results.

As expected, results are less good if all the visual craters are considered in the computation, independently of their circularity and radius. Indeed, the CGC method assumes a minimum circularity of the shape and should not be used for detecting too small circles. As long as the craters satisfy the working hypothesis, the detection of craters showing much evidence is very good (more than 80% for the best fusion result).

Weighting the craters according to the confidence given by the photo-interpreter always improves the results, which means that the CGC circles mainly correspond to visual craters with high confidence.

In general, recall is better than precision, which means that they are more FP than FN, which could be interpreted as CGC method generating too many FP. This statement should be qualified; indeed, there are 116 FP for the experiment providing the best precision, which corresponds to a density of 5.8 FP per km^2 or 1.45 FP for 10^8 pixels, which is quite reasonable compared to the 334 visual craters corresponding to a density of 16.7 craters per km^2or 4.17 visual craters for 10^8 pixels. For the experiment providing the best recall, they were 738 FP (36.9 per km^2 or 9.22 for 10^8 pixels) which is still acceptable.

The distribution of FP is not uniform: in particular, one zone presents a lot of false positives, which are generated either by the shadow of small-squared houses (on Band 8) or by the house themselves (on ND_{74}) as can be seen in **Figure 5**.

The best precision is always obtained for the ND of Band 7 and Band 4—which means it is the best image to consider if low false-positive rate is searched for. CGC method applied to that image provided 161 FP and 116 TP.

The best recall is a fusion of results obtained on different images. If all visual craters are considered, the fusion involves Band 8, ND_{83}, ND_{84}, ND_{75} and ND_{65}. On the other hand, if craters

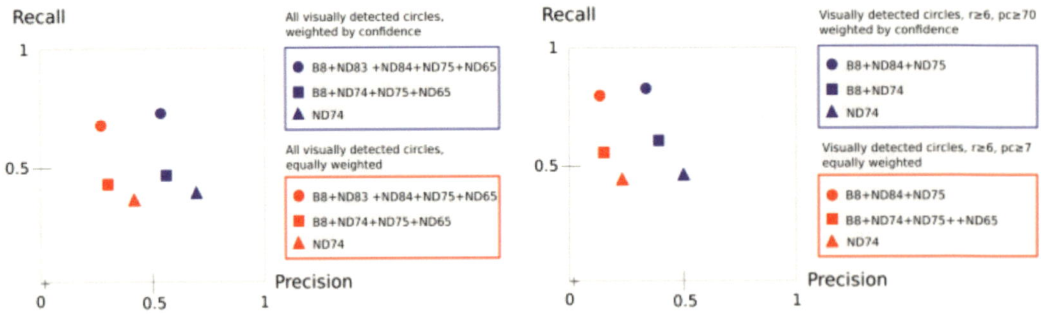

Figure 4. Precision and recall of the CGC automatic crater detection method.

http://dx.doi.org/10.5772/66691

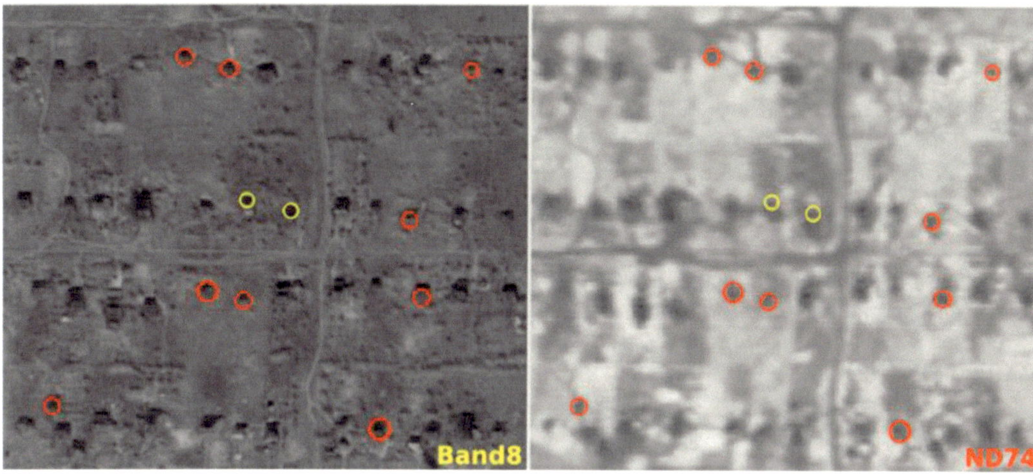

Figure 5. Examples of FP superimposed on part of Band 8 (left) and part of ND_{74} (right). Bright and dark circles represent CGC circles detected on band 8 and ND_{74} respectively.

outside the domain of CGC are filtered out, Band 8 fused with ND_{84} and ND_{75} provides the best results. That experiment provided 738 FP and 223 TP.

The CGC circles have been presented to the photo-interpreter. While some circles (FP) have been easily identified as house shadow or house themselves (see **Figure 5**), others were recognized by the photo-interpreter as potential craters that could have been missed. Indeed, this visual inspection is quite tedious and is prone to errors. Other examples of TP, FP and FN are shown in **Figure 6**.

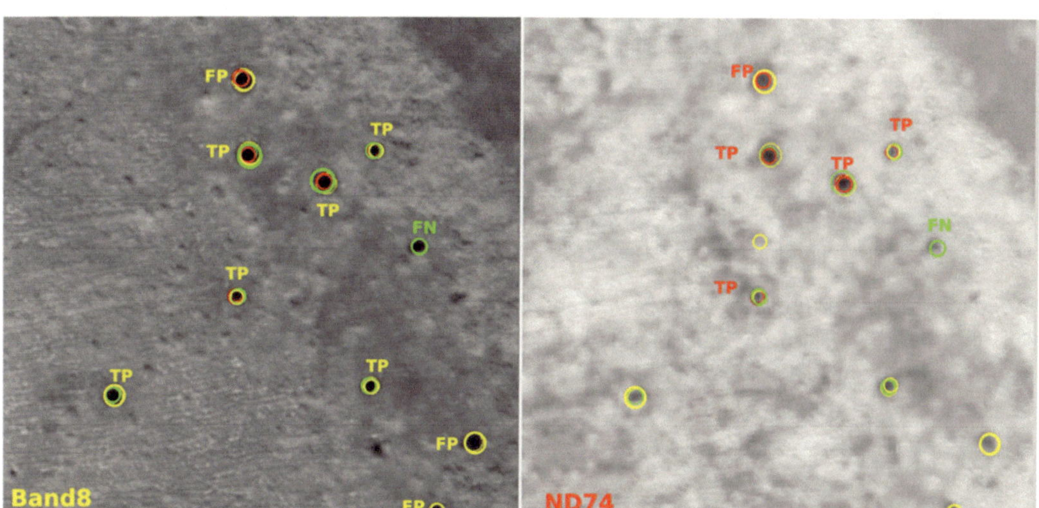

Figure 6. Examples of TP, FP and FN over part of Band 8 (left) and ND_{74} (right).

3.6. Conclusions

This study showed that, according to a photo-interpreter, the naive idea of a crater being seen as perfect circle on satellite image is not true; indeed, only 45% of the craters visually detected can be considered as such. Nevertheless, the CGC method used over Band 8 and fused with various normalized difference indices enables to detect at least half of them, mainly the ones having a radius larger than 6 pixels, and the ones for which the photo-interpreter gives a larger confidence, all this with a reasonable amount of false positives. The fusion method providing the best recall should probably be used in the current context as it should be more important not missing a crater than having more false positives.

The performances would be enhanced if the method could be adapted to small circle detection, for example, by making the detection parameter dependant on the radius.

4. Segmentation with Gabor filters (Mahamadou Idrissa)

4.1. Introduction

In satellite image interpretation, classification is the operation by which an operator would like to detect different kinds of region like forest, urban zone, waterways, and so on. As the scene is a set of points (pixels) with intensity in grey-scale values in several bands, most methods use these grey-scale values to determine the kind of terrain. However, a single ground cover usually occupies a number of pixels with some variability in their grey-scale values. A more satisfactory interpretation of the scene should thus include textural aspects of regions.

In general, texture is characterized by invariance of certain local attributes that are periodically or quasi-periodically distributed over a region. There are many approaches for analysing image texture. Haralick [14] proposed a set of features (energy, entropy, maximum probability, correlation, etc.) based on grey level co-occurrence matrices (GLCMs). Some statistical techniques use Markov Random Field models to characterize textures [15]. The spectral approach [16, 17] to texture analysis is referred to as the multi-channel filtering approach. Textures are characterized by their responses to filters, each one being selective in spatial frequency and orientation.

We present a spectral approach to extract texture features. The textured input image is decomposed into feature images using a bank of Gabor filters. These feature images are used to form feature vectors and each of them corresponds to one dimension of the feature space. Then, we present the Fuzzy c-means-clustering algorithm used for unsupervised classification of the input pixels based on their associated feature vector. This method considers clusters as fuzzy sets, while membership function measures the possibility that each feature vector belongs to a cluster. At last, we present methods for evaluating how well different textures are separated in feature space, as well as measuring classification performance. In most applications, the number of classes is unknown. Here, we propose a method for choosing the best number of classes and we apply it to a real texture representation problem.

http://dx.doi.org/10.5772/66691

4.2. Gabor filters

Gabor filters perform a local Fourier analysis thanks to sine and cosine functions modulated by a Gaussian window. In the complex space, these filters are defined as follows:

$$G(x, y|X, Y, k_x, k_y) = e^{\frac{-((x-X)^2+(y-Y)^2)}{2\sigma^2}}.e^{j(k_x x+k_y y)} \quad (1)$$

where *x,y* represent the spatial coordinates while represent the frequency coordinates. *X* and *Y* are the spatial localization of the Gaussian window.

As signal is discrete, two simplifications are proposed [18]. The first one makes use of short-time Fourier transform (STFT) enabling a fixed window size independently of the filter frequency. The second introduces binomial window as approximation of the Gaussian. The basis functions of this decomposition are as follows:

$$\begin{aligned} C_{k,l}(n, m) &= W(n, m) \times \cos 2\pi\left(\frac{kn}{N+1} + \frac{lm}{M+1}\right) \\ S_{k,l}(n, m) &= W(n, m) \times \sin 2\pi\left(\frac{kn}{N+1} + \frac{lm}{M+1}\right) \end{aligned} \quad (2)$$

where $W^2(n, m) = \frac{1}{2^{(N+M)}} C_N^{(\frac{N}{2}+n)} C_M^{(\frac{M}{2}+m)}$ is the $(N + 1) \times (M + 1)$ binomial window. The coefficients are given by

$$C_j^i = \left\{ \begin{array}{ll} \frac{j!}{i!(j-i)!} & 0 \leq i \leq j \\ 0 & \text{elsewhere} \end{array} \right\} \quad (3)$$

The indexes $n = -\frac{N}{2}, \ldots, \frac{N}{2}$ and $m = -\frac{M}{2}, \ldots, \frac{M}{2}$ are the window spatial coordinates with *N* and *M* even integers. The filters selectivity in frequency (expressed in number of cycles per pixel) and orientation (in radian) is derived from the following equation:

$$\begin{aligned} f &= 2\pi\sqrt{\left(\frac{k}{N+1}\right)^2 + \left(\frac{l}{M+1}\right)^2} \\ \theta &= \arctan\frac{k}{N+1}\frac{M+1}{l} \end{aligned} \quad (4)$$

with $k = -\frac{K}{2}, \ldots, \frac{K}{2}$, $l = -\frac{L}{2}, \ldots, \frac{L}{2}$ and $K \leq N$, $L \leq M$. **Figure 7** shows a subset of filter functions (their corresponding sine and cosine terms) for a 31 × 31 binomial window with $K = 4$ and $L = 4$, while **Table 3** shows their corresponding sine and cosine terms.

A set of feature images is obtained by convolving the image *I* with each filter providing a local 'Energy' given by

$$E_{k,l}(n, m) = [(C_{k,l} \quad I)(n, m)]^2 + [(S_{k,l} \quad I)(n, m)]^2 \quad (5)$$

Here, ⊗ denotes the convolution and *k,l* the frequency-orientation coordinates.

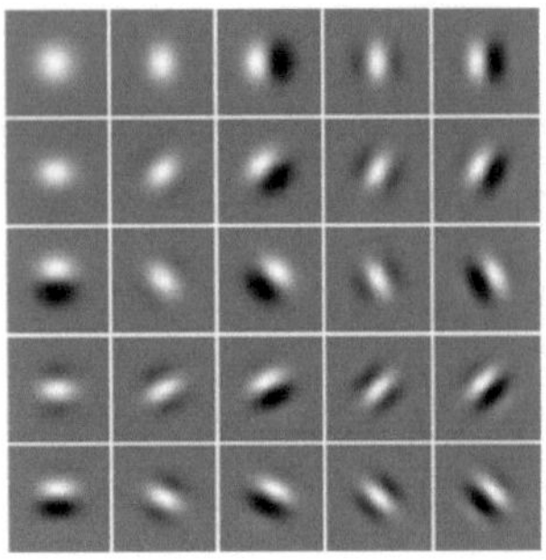

Figure 7. Filter bank for N = M = 30 (grey = 0, white = positive values, black = negative values).

C0,0	C0,1	S0,1	C0,2	S0,2
C1,0	C1,1	S1,1	C1,2	S1,2
S1,0	C−1,1	S−1,1	C−1,2	S−1,2
C2,0	C2,1	S2,1	C2,2	S2,2
S2,0	C−2,1	S−2,1	C−2,2	S−2,2

Table 3. Filters legend: respectively, for the sine and cosine terms.

4.3. Classification

Many algorithms have been developed for supervised and unsupervised classification. In supervised classification, training sets are needed whereas unsupervised classification classifies images automatically by finding clusters in the feature space. One of the unsupervised data-clustering methods is the hard k-means-clustering algorithm. It assigns each sample (feature vector) to one and only one of the clusters. This method assumes that boundaries between clusters are well defined. The model does not reflect real data where samples are in general mixed. In fuzzy c-means algorithm [19], unlike the k-means algorithm, samples belong to all clusters with various membership degrees. Membership degree is defined as a function of the distance between sample and the *i*th-cluster centre. The fuzzy c-means may be seen as a generalization of the k-means method with membership values in the interval [0,1] rather than in the set {0,1}.

4.4. Classification validity

Classification of data should be of high quality, that is, all samples should have a large membership degree for at least one cluster. This problem is related to how many classes there are in the data. In fuzzy c-means algorithm, the number of clusters is required though in many applications this information is unknown. A method for measuring performance is needed to compare the goodness of different classification results. Gath has defined an 'optimal partition' of the data into subgroups as follows [20]:

1. Clear separation between the resulting clusters.
2. Minimum volume of the clusters.
3. Maximum number of data points concentrated around the cluster centre.

In literature, this is known as cluster validity problem; there exists a wide variety of cluster validity parameters. Gath has proposed the partition density *PD* more related to the geometry of the dataset. The method accounts for variability in cluster shapes and the number of data points in each of the subsets:

$$PD = \frac{S}{FHV} \tag{6}$$

where $s = \sum_{i=1}^{C} \sum_{j=1}^{n} u_{ij}$ is the fuzzy cardinality of dataset and $F_{HV} = \sum_{i=1}^{C} [\det\left(\sum_i\right)^{1/2}$ the fuzzy hyper-volume of dataset.

This parameter is only sensitive to well-compact substructures in the dataset and does not take the overlapping between them into account.

For this work, we have defined a cluster validity parameter, which combines Gath partition density *PD* (Eq. (6)) and a resemblance function *r* defined by Bloch in Ref. [21]. This function *r* is used as separation measurement between clusters

$$r = \frac{f(A \cap B)}{f(A \cup B)} \tag{7}$$

where *A* and *B* are two fuzzy sets; *f*(*X*) is the fuzzy cardinality of a set *X*. The intersection $A \cap B$ is the largest fuzzy set, which is contained in both *A* and *B*; the union is the smallest fuzzy set containing both *A* and *B*. The new definition of Compactness and Separation coefficient is

$$nCS = \frac{PD}{\max(r)} \tag{8}$$

The factor *max*(*r*) corresponds to the maximum resemblance value measured between two clusters. This normalized factor always takes a value close to zero as soon as the clusters are disjoint; the higher resemblance value is close to one. It can also be interpreted as a punishing factor applied to the partition density *PD* to avoid the over-segmentation problem.

With this definition, the procedure to select the best number of classes is to cluster the dataset for many different numbers of classes; the best one should give the maximum *nCS* index.

4.5. Experiments

We now apply this texture classification method to a grey-scale image (**Figure 8**) taken in November 1998 during an airborne minefields survey in Mozambique [22]. In the surveyed area, minefields are homogeneous regions with no human activity inside and surrounded by agricultural fields. Since the vegetation in these minefields is different from the vegetation outside, texture-based classification is a good approach for the interpretation of this image. Unfortunately, exact class maps or texture models are not available. Thus, the use of the proposed cluster validity parameter will help us to find the optimal class number, which maximizes the final classification accuracy.

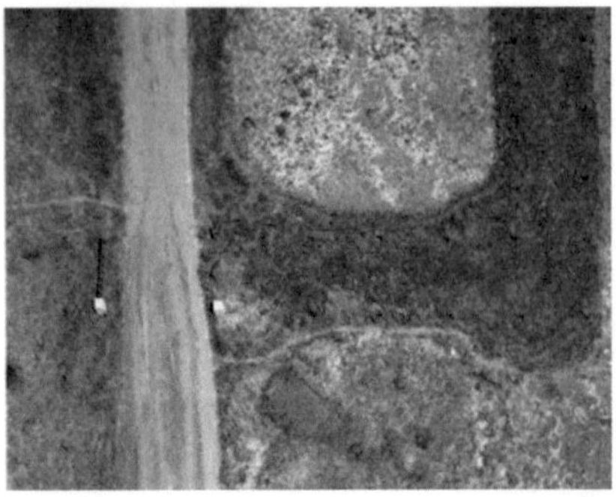

Figure 8. Aerial image of a minefield area in Mozambique.

Gabor filters are computed with an 11 × 11 binomial window and five discrete values of frequency coordinates k, l {−2,−1,0,1,2} (empirically determined). We have used only 9 features selected from the 13 feature images. The feature selection scheme is based on a simple method, which consists in sorting the feature images based on their amount of energy and pick up as many features as needed to achieve 95% of the total energy.

The optimal number of clusters that we find for this image is 5 (**Figure 9**), and the classification result is presented in **Figure 10**. As we can see in **Figure 11**, the dark area corresponding to the suspected minefield is determined and the other types of vegetation as well.

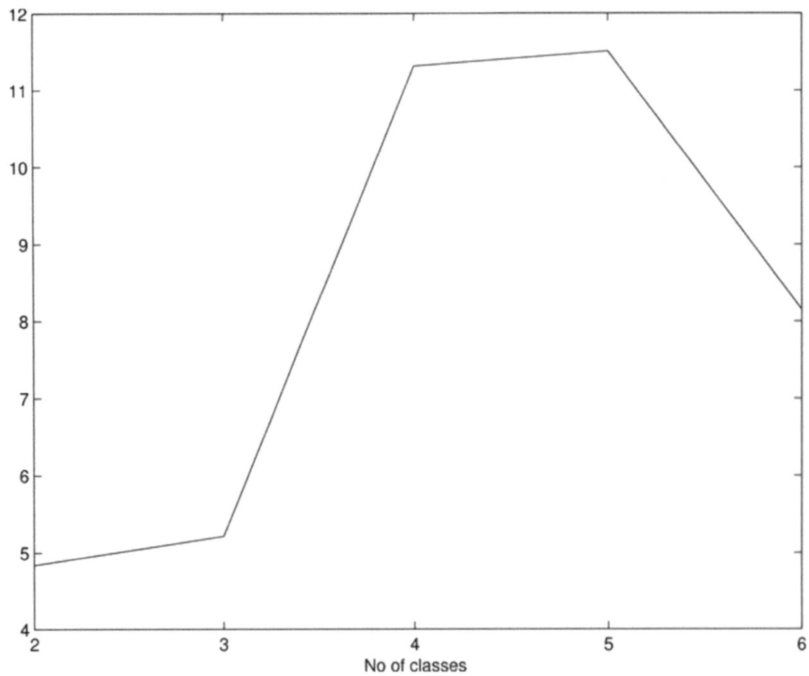

Figure 9. Compactness-separation function (nCS) *for the aerial image.*

http://dx.doi.org/10.5772/66691

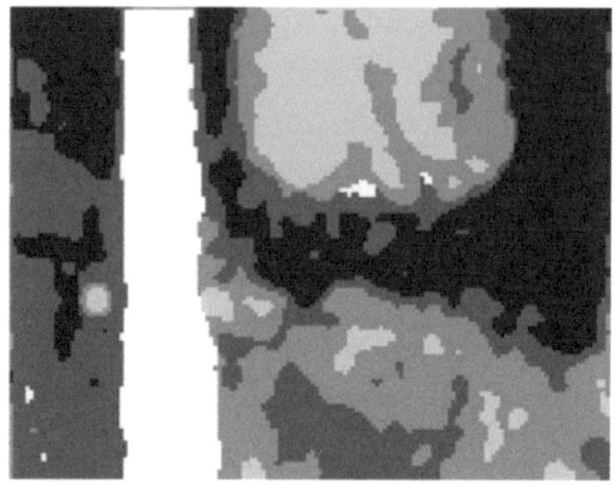

Figure 10. Classification in five classes.

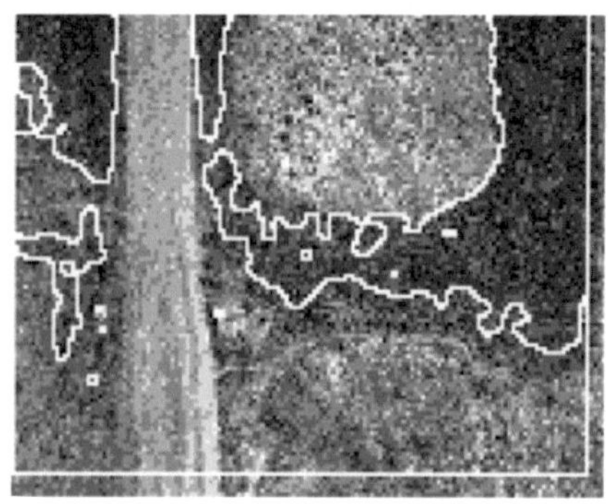

Figure 11. Minefield region classification.

4.6. Conclusions

A fuzzy-clustering approach to textured image classification has been presented. The texture features are extracted using a set of Gabor filters with different frequencies and orientations.

The fuzzy c-means algorithm has been successfully used for discriminating different types of textured image, but the drawback is that one has to specify the number of clusters. We thus discussed the use of cluster validity parameters. A modification of Compactness and Separation validity function is proposed as index to estimate the optimal number of texture categories, which have proven adequate for texture discrimination.

5. From DSM to DTM (Charles Beumier)

The bare ground is of interest for applications such as flood-risk evaluation, environment planning or terrestrial navigation. In the specific case of minefield delineation, a terrain model is a precious piece of information to spot where camps could have been settled and hence where mines could have possibly been laid or buried.

In a mine context, the earth surface model is ideally captured by a remote-sensing method, resulting in a fast and cost-effective solution if the area of interest may be flown over. LIDAR and photogrammetry are the two classic options to deliver a dense digital surface model (DSM) with high resolution. Since it is taken from the air, this model contains elevated objects such as trees and buildings. These usually complicate the automatic extraction of features of the bare soil useful for detection.

By contrast, a digital terrain model (DTM) is a representation of the bare ground surface. It is advantageously derived from the DSM by automatic or semi-automatic procedures. Most of them filter elevated points and fill filtered areas by interpolation between remaining points.

In the specific case of the TRAMISU project, gigantic DSMs (one or several giga-pixels) were produced with a spatial resolution of 0.3 m. Fortunately, a rougher level of detail for the desired DTM is sufficient and we opted for 3-m resolution, leading to tractable images in terms of speed and memory.

We implemented a DTM from DSM approach based on the sate-of-the-art solution 'lasground' distributed by rapidlasso Gmbh in the LASTOOLS suite, and inspired from the work of Ref. [23] about LIDAR point filtering. The method filters out recursively locally elevated points from a point cloud and returns the reduced cloud normally made of ground points.

We first converted the DSM raster to a point cloud after low-pass filtering and regularly sub-sampling the raster to get a reasonable file size. The lasground programme was then run to deliver a reduced point cloud, normally made of ground points. The main parameter ('step') was set to 10 to accommodate for rather mountainous/forest scenes. Finally, a raster was created from the resulting point cloud and filled in by interpolation thanks to the las2dem of LASTOOLS.

Figure 12. Digital surface model and digital terrain model.

The project TIRAMISU successfully employed the produced DTM. By subtracting it to the DSM, obtaining the so-called normalized DSM (nDSM), the vegetation and buildings were easily localized. **Figure 12** shows perspective views for one example of extracted DTM and the corresponding nDSM.

6. Multi-spectral classification (Nada Milisavljevic)

Daedalus, the multi-spectral scanner used in the project SMART [24], records the electromagnetic spectrum in 12 channels ranging from visible blue to thermal infrared, with the spatial resolution of 1 m. The appearance of objects in the different Daedalus channels may vary widely, depending on the molecular reflection and absorption characteristics of the matter of the objects.

Generally speaking, multi-spectral remote sensing takes benefit from the fact that the way that different types of material or of land cover absorb and reflect solar radiation depends on the wavelength. This means that any type of land cover or material has its own multi-spectral signature. This signature can be represented as a plot of the fraction of reflected radiation in function of the incident wavelength. For instance, the reflectance of bare soil increases with the increasing wavelength. On the contrary, clear water generally has a low reflectance, but it reaches the maximum at the blue region of the spectrum and decreases with the increasing wavelength. Finally, the reflectance of vegetation is minimal in blue and red parts of the spectrum; the reflectance increases as the wavelength increases towards the near-infrared part of the spectrum, and becomes significantly higher than in the visible part. Apart from these general tendencies of soil, water and vegetation, differences exist in the spectral signature (i.e., shape of the reflectance spectrum) between different types of the same class. In other words, the exact shape of the signature of bare soil is a function of the type of soil, its moisture, composition, and so on. The main idea behind our work with the Deadalus data is based on the fact illustrated by these examples. Namely, in order to perform land-cover classification, we analyse the shape of the multi-spectral signatures and then assign pixels to classes based on their similar signatures.

There are two main types of multi-spectral classification. In supervised classification (applied when the knowledge about the data and types of land-cover classes is good), classes are related to known features on the ground. In unsupervised classification, the pixels are analysed in an automatic way and divided in spectrally distinct classes (i.e., accumulations of points in a multidimensional feature space). The unsupervised classification is used when we have a low knowledge about the data before classification, and the classification result is given to the image analyst, who should then give meaning to the obtained classes.

In our case, there is no reliable information on the minefield indicators (as a matter of fact, one goal of our work is to determine useful minefield indicators). In addition, we have to be careful, taking into account the context, and avoid any misclassification that could have dangerous consequences. Consequently, our plan is to proceed as follows:

- as a first step, to perform an unsupervised classification which should provide us with classes of different land cover;

- then, based on the available ground-truth information, we attach meaning to typical and wide-spread classes of land cover (such as forest, water, grass and cornfield);

- finally, the remaining regions that are not labelled could be related to land-cover anomalies, which means that they are possibly dangerous and should be analysed further.

Note that in order not to discard, misclassify or miss any mined area, the first two steps have to be performed slowly and carefully. Another reason for being cautious lies in the fact that the mine density is often low in mine-affected regions of Croatia. Furthermore, combining these results with data coming from synthetic aperture radar (SAR) (Section 7), we can improve the reliability of the obtained results.

During the application of this unsupervised classification, we take advantage of another piece of information related to the multi-spectral signature analysis—the information provided by comparing the reflectance in the red part (r) and in the near-infrared of the spectrum (n), the so-called vegetation index. NDVI, that is, the normalized difference vegetation index is calculated as $\frac{n-r}{n+r}$ and

- is below 0 for snow and water;

- is close to 0 for bare soil and rocks;

- is between 0 and 1/3 in case of lower vegetation;

- increases above 1/3 with the vitality and density of the plant canopy (e.g., for fresh vegetation and forests).

Water and forests are not potential minefield indicators. Thus, before we apply a land-cover classification, we can mask areas with NDVI below 0 and above 1/3, taking into account that there is no fresh biomass in several year-old pioneer vegetations of the mine-suspected fields.

Thus, the way our unsupervised classification proceeds can be summarized in the following way. As a first step, the 12 Daedalus channels are analysed pixel per pixel. A first analysed pixel is assigned to a first class, and its multi-spectral signature (i.e., its values $p_{1,1}, p_{1,2}, \ldots, p_{1,12}$in the 12 channels) is stored as the representative of that class, noted as $c_{1,1}, c_{1,2}, \ldots, c_{1,12}$. The following pixel, which has values $p_{2,1}, p_{2,2}, \ldots, p_{2,12}$in the 12 Daedalus channels, is compared with the representative of the first class based on their multi-spectral distance t, as $t = \sqrt{\sum_{i=1}^{12} \left(p_{2,i} - c_{2,i}\right)^2}$.

The analysed pixel is grouped in the first class if t is lower than or equal to a tolerance set in advance (k), and the representative multi-spectral signature of that class is then updated (replaced by averaging the signature of the prior existing pixel in the class and the newly found pixel). If t is greater than k, the analysed pixel is the founder of a new class, which means that its values initialize the multi-spectral representative of that class. For all pixels having NDVI > 1 or NDVI < 0, a separate rejection class is formed. This procedure is repeated so that each pixel signature is compared with all existing classes. Each pixel is always grouped to the class having the minimum distance t or it creates a new class if t is greater than k. Once all pixels are analysed, we obtain a first output of our unsupervised classification, as well as the final multi-spectral signature representative for each of the classes. At this step, there are two possibilities: our method can either end or it can go on for a new iteration, where the results of the last iteration are used as a priori classes. As a matter of fact, in the first iteration, the process of finding the multi-spectral signatures of the classes consists in updating the class representatives each time a new class member is found, which means that misclassifications for pixels belonging to similar vegetation types are possible. After the first iteration, the class representatives are fixed, so they can be used to reclassify the pixels and correct most misclassifications (if not all). Furthermore, in such a way, a strong influence of the choice of the tolerance, k, that influences the number of classes is significantly diminished.

The described method is illustrated for the mine-suspected area shown in **Figure 13**. **Figure 14** contains results of the method for $k = 19$ (20 classes) and $k = 22$ (10 resulting classes), while results obtained for a larger area are given in **Figure 15**. If the number of classes increases, both

Figure 13. Daedalus geocoded images of a mine-suspected area in Croatia. Left: visible, right: infrared.

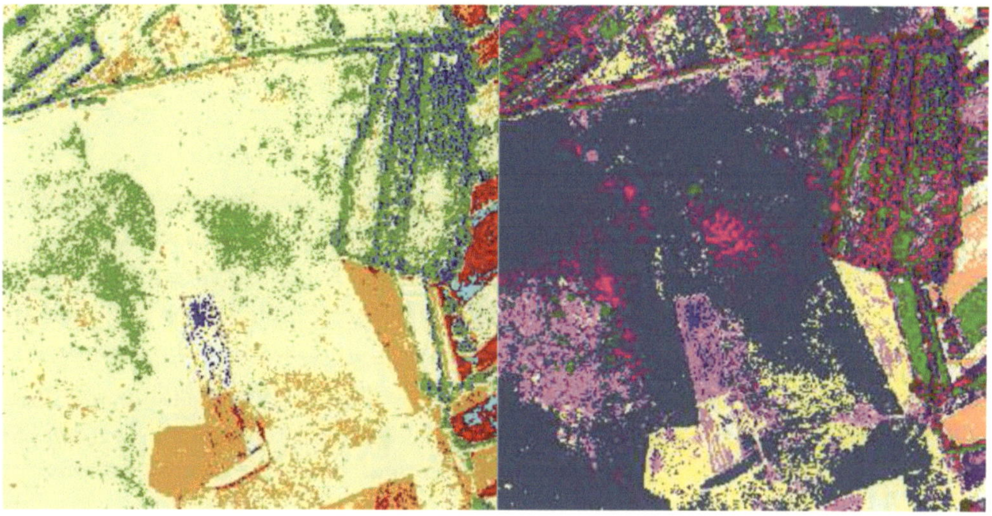

Figure 14. Results of an unsupervised classification of Daedalus data for the test area in Croatia: 10 classes (left), 20 classes (right).

the level of details and noise (i.e., isolated pixels that are grouped with other classes than their surroundings) increase as well, where the former could result in identifying possible anomalies in land cover. Although the resulting classified data are relatively noisy, we do not smooth the image in order to preserve possible minefield indicators.

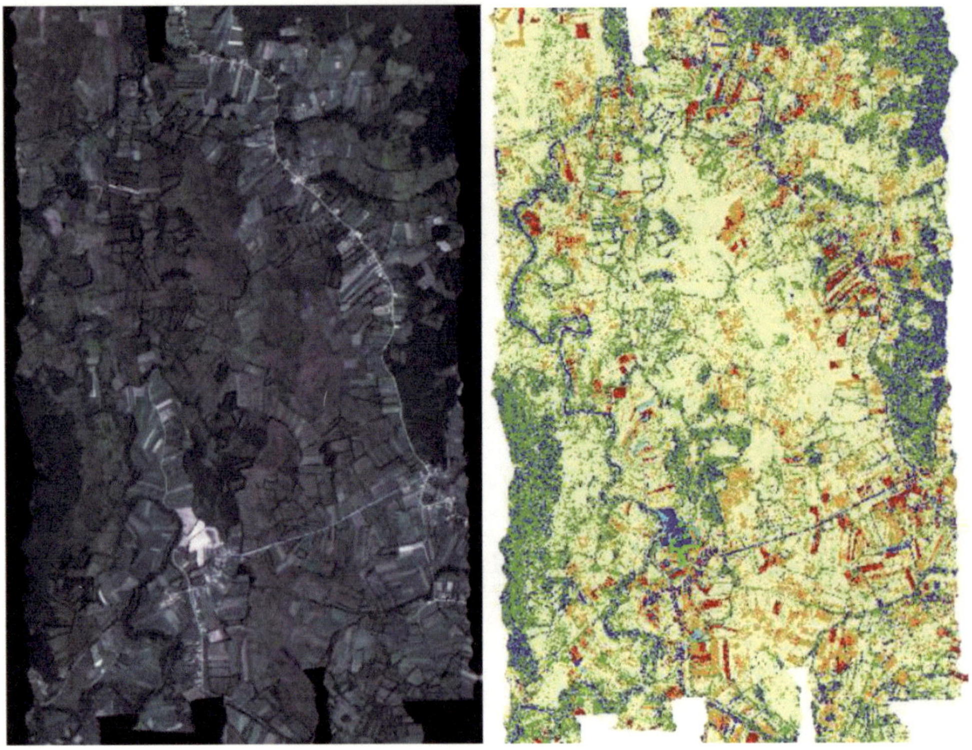

Figure 15. Left: RGB composite of geocoded Daedalus channels of a larger area in Croatia; r,g,b = ch5, ch3, ch2. Right: result of an unsupervised classification of Daedalus data, 10 classes.

7. Multisensor data fusion (Nada Milisavljevic)

As shown in detail in Ref. [25], in order to find indicators of mine absence and mine presence as well as to provide image analysts with adequate tools to analyse and interpret potentially mined zones during the process of area reduction, various pieces of information obtained using Daedalus and airborne full-polarimetric SAR, together with the existing context information (all integrated in a geographical information system), are classified and then combined.

As far as the data fusion module is concerned, we have worked on several numerical fusion methods. The proposed methods are to a large extent original thus representing a result of the project SMART. Results are obtained on three test sites using the three most promising methods: fusion using belief functions [26] and having a global discounting factor, fusion using belief functions and using confusion matrices for a more specific discounting, and fusion using fuzzy logic. Then, in the next step, we have shown how the results obtained on this basic level can be improved by introducing additional knowledge in the fusion process (e.g., knowledge on the width of the roads, on the existence of rivers, etc.). Finally, as the third step, a spatial regularization at a regional level further improves the results, based on the idea that isolated pixels of one class quite unlikely appear in another class. Note that the results are at least as good as those obtained for each class using the best classifier for that particular class. In other words, they are globally better than any separate input detector or classifier, which proves the improvement that is brought by fusion. The user can influence the choice of the classifiers as well as the choice of some parameters (some supervision is still required in the choice of parameters for the fuzzy fusion approach in particular).

As an illustration, **Figure 16** contains results with fuzzy fusion. For more detailed information, please refer to Ref. [25].

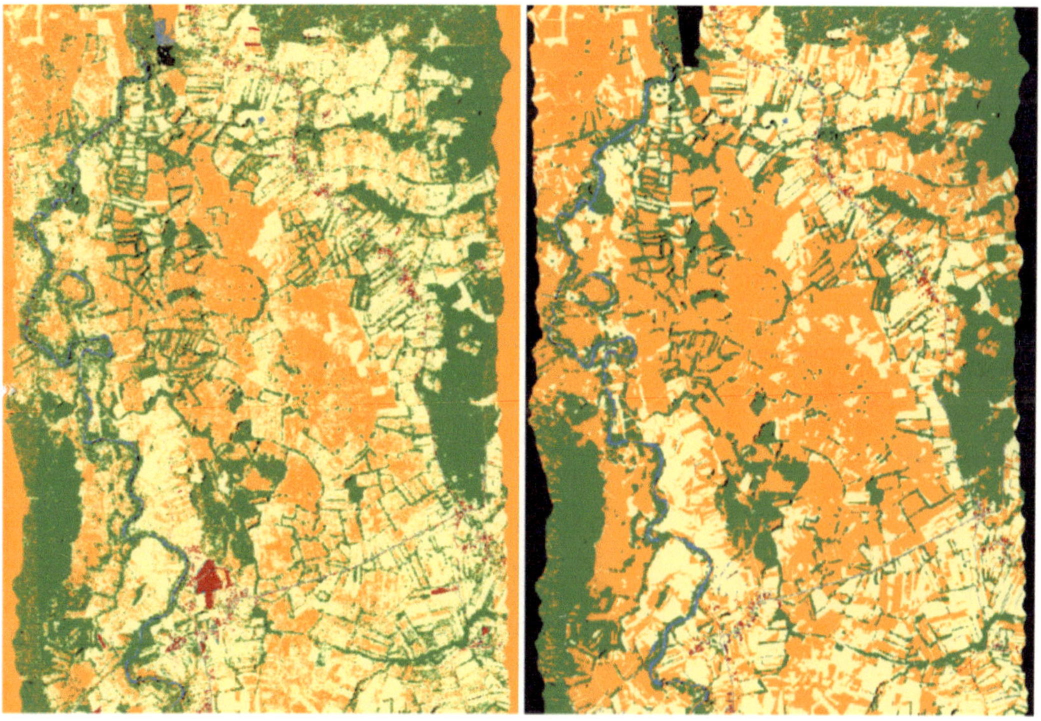

Figure 16. Results with fuzzy logic fusion: basic version (left), after knowledge inclusion and spatial regularization (right).

8. Hyper-spectral change detection (Michal Shimoni)

8.1. Introduction

Hyper-spectral imaging collects information from the visible to the microwave regions of the electromagnetic spectrum by taking a series of images over a range of narrow and continuous wavelengths. This results in the creation of a three-dimensional hyper-spectral imaging cube, where the x- and y-axes present the spatial dimension of the image and the z-axis denotes the spectrum as dependent on the number of bands. For the visible to short-wave infrared (SWIR) wavelengths, a continuous reflectance spectrum for each pixel is formed, whereas for the mid-wave infrared (MWIR) to long-wave infrared (LWIR) wavelengths, an emittance spectrum is created. These wave ranges are, respectively, named in the literature as the reflective and the emittance domains.

Hyper-spectral imagery has been applied with success to many security and defence applications including the detection of buried mines and occluded improvised explosive devices (IEDs) [27]. In the reflective domain, spectral-based methods were applied for the detection of surface landmines using anomaly detection [28], matched filtering [29], the constrained energy minimization (CEM) [30] and the discrete wavelet transform (DWT) [31]. The presence of buried landmines was investigated using the changes in the spectrum of the covered vegetation or soil upon burying the devices [32–37].

In the radiative domain, the detection of disturbed soil exploits the 'reststrahlen effect', in which the emissivity absorption of a particular soil is changed within the 8–10-μm spectral range [38]. However, the reststrahlen effect has been shown to be quite variable depending on the geographic location and the soil composition, and it requires the use of higher-order decision fusion (i.e., fusion of thermal, textural and spectral information) to achieve target/clutter discrimination from a single thermal hyper-spectral image [34]. Our study tries to overcome this limitation by applying spectral-based change detection of buried IED using temporal thermal hyper-spectral scenes. Our study assesses the detection of small buried objects using multi-temporal thermal hyper-spectral images and by applying two change detection methods based on multivariate statistical algorithms: the Chronochrome (CC) and the class-conditional change detector (QCC).

8.2. Dataset

A measurement campaign took place on the 21 October 2009 in the location of a quarry in the periphery of Quebec City (Canada).

The TELOPS imager (Hyper-Cam) [39] was located at a distance of 2.5 m away from the area of interest. The sensor scanned the area site looking downwards at an angle of 26° and with an instantaneous field of view of 1.4 μrad (**Figure 17**). The spectral images were collected using 660 bands in the wavelengths 800–1350 cm^{-1}. A magnifying telescope was installed for improving the spatial resolution. The detector array was set to 220 × 200 pixels, resulting in an FOV of 17.6° by 16°.

Four aluminium plates consisting of 6.5 × 8 × 0.25 inches (**Figures 17** and **18**) were buried in a sandy soil and were framed with several rocks (diameter of 10–25 cm). The plates were buried at different depths: 2, 5, 10 and 15 cm (**Figure 18**) the day prior to the experiment on 20 October 2009, between 12h00 and 13h00. Part of the site was undisturbed as indicated at the upper part of **Figure 18**.

Images were collected from 6h40 am on the 21 October 2009 in nine sequences consisted with the following:

http://dx.doi.org/10.5772/66691

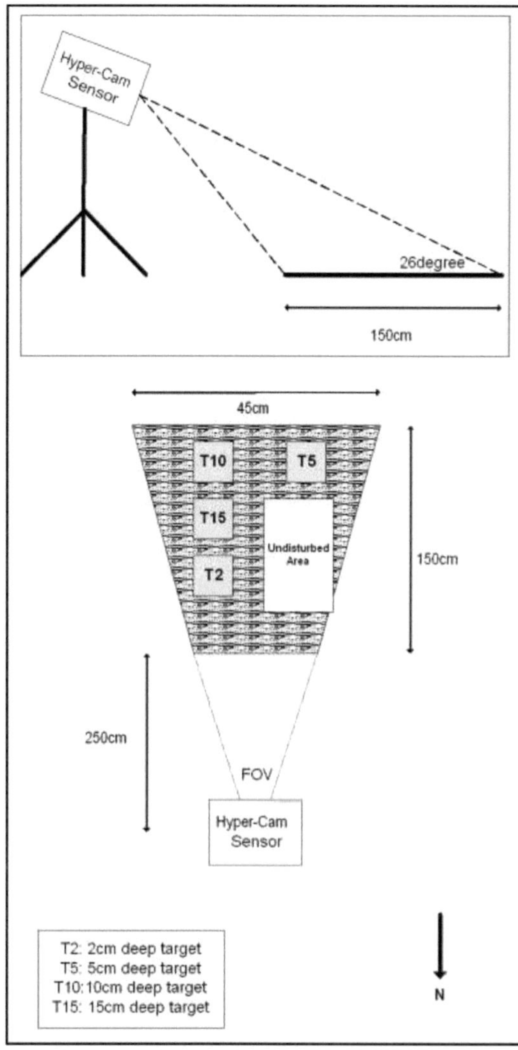

Figure 17. View of the experimental layout showing the Hyper-Cam field of view, buried targets, and undisturbed area.

- 50 acquisitions of a blackbody at a given temperature;
- 100 acquisitions of the target site;
- 50 acquisitions of a blackbody at a second given temperature.

Table 4 summarizes the environmental conditions in the nine measured sequences.

8.3. Change detection methods

For automatic target detection of the buried objects, two change detection methods were applied: the Chronochrome (CC) and the class-conditional CC (QCC).

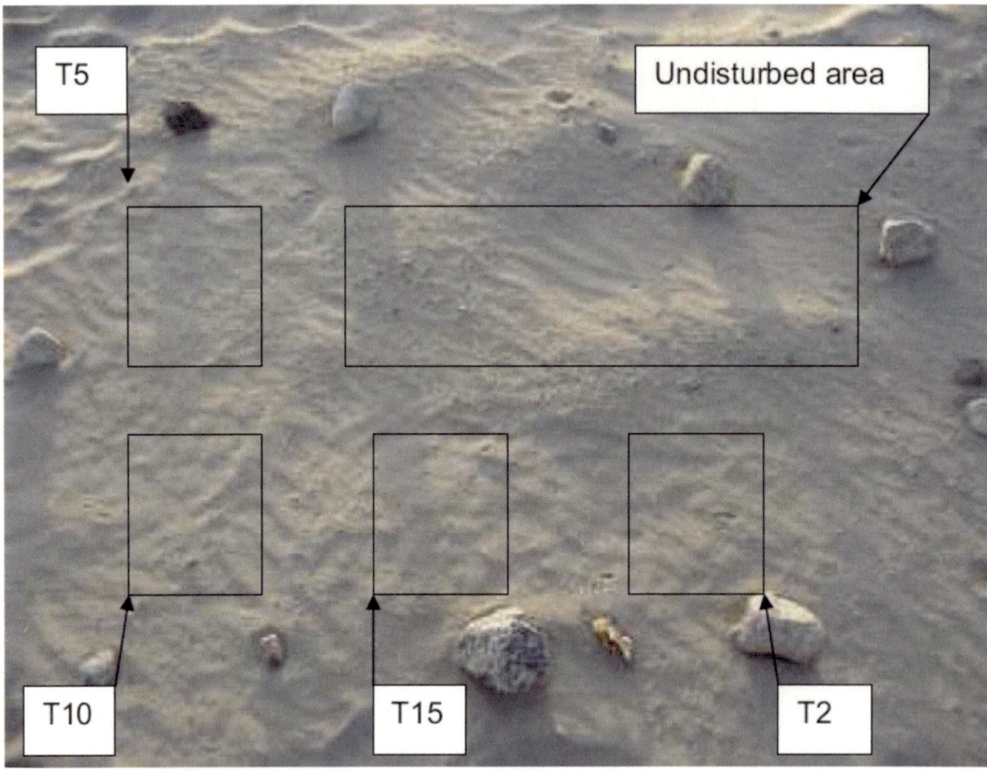

Figure 18. Target site showing the location of the buried targets.

Sequence	Start time	Ambient temp °C	Humidity %	Ground temp ($E = 0.98$)
IED#1	06:40	1.3	80	−7.3
IED#2	07:55	1.8	79	−2.5
IED#3	09:10	3.1	78	7
IED#4	10:50	6.5	57	8.5
IED#5	00:00	9.9	45	16
IED#6	13:16	9.1	45	12–14
IED#7	14:24	9.3	44	9.5–11
IED#8	15:34	7.9	44	3.5–4.5
IED#9	17:24	4.4	54	−1.3

Table 4. Experiment log.

8.3.1. Chronochrome (CC)

For the two hyper-spectral matrices x and y, the diagonal matrices T_x and T_y are written as follows [40]:

$$x_i = T_x\rho_i + d_x$$
$$y_i = T_y\rho_i + d_y \quad (9)$$

where ρ_i is the spectral reflectance and d_x and d_y are the changes between the observations. The spatial position index on the vector is dropped for notation convenience:

$$\begin{aligned} x &= T_{xy}y + d_{xy} \\ T_{xy} &= T_x T_{y^{-1}} \\ d_{xy} &= d_x - T_{xy}d_y \\ \hat{x} &= \hat{T}_{xy}y + \hat{d}_{xy} \end{aligned} \quad (10)$$

And the change residual image (δ) is defined as follows:

$$\delta = x - \hat{x} - \left(\hat{T}_{xy}y + \hat{d}_{xy}\right) \quad (11)$$

Using second-order statistics, the transformation parameters $\hat{T}_{xy}$ and $\hat{d}_{xy}$ can be estimated using the mean vectors m_x and m_y and the covariance matrices C_x and C_y. Following, the covariance matrices are diagonalized:

$$\begin{aligned} C_x &= V_x D_x V_{x^T} \\ C_y &= V_y D_y V_{y^T} \end{aligned} \quad (12)$$

where V_x and V_y are the eigenvector matrices, and D_x and D_y are the diagonalized covariance matrices. The Chronochrome (CC) change detection method is written as

$$\begin{aligned} \hat{T}_{xy}^{(CC)} &= C_{xy}C_{yy}^{-1} \\ \hat{d}_{xy}^{(CC)} &= m_x - \hat{T}_{xy}^{(CC)} m_y \end{aligned} \quad (13)$$

8.3.2. Class-conditional CC (QCC)

The QCC [41, 42] is an image with a normal mixture model that allows the transformation parameters $\hat{T}_{xy}$ and $\hat{d}_{xy}$ to vary between spectral classes. In QCC, each spectrum (x) is defined by the class index q (where $q = 1, 2, \ldots, Q$) with a prior probability $P(q)$ to belong to each respective class. In this method, a class-conditional probability function $p(x|q)$ is assigned to the transformation parameters in Eq. (2):

$$\hat{x}|q = \hat{T}_{xy}|q.y + \hat{d}_{xy}|q \quad (14)$$

And the change residual image (δ_1) is defined as

$$\delta_1 = x - \hat{x}|q = x - \left(\hat{T}_{xy}|q.y + \hat{d}_{xy}|q\right) \quad (15)$$

As a classification method, we applied the stochastic expectation maximization (SEM) [43] after principal component analysis on the reference image (x). The SEM is a quadratic-clustering method that addresses the bias against overlapping class-conditional probability density functions and uses a Monte-Carlo class assignment. Because in thermal imagery, the temperature variations will cause a large non-linear radiance offset [43], we selected a stochastic model rather than a linear-mixing model. We assume that local thermal variation occurs due to changes in the upwelling radiance and the down-welling illumination from nearby object emissions.

8.4. Results

The nine sequenced acquisitions were resulted in 72 pairs of images to which the CC was applied. From **Figure 19**, one can learn that the obtained changes are spectral-based and not thermal-based. Apart from the illumination and the shadow effects, the obtained changes did not vary along the day. Nevertheless, because the thermal spectra are influenced by the down-welling illumination, clear changes were recorded only from sequence #3 (i.e., 09h10 am). When the up-spectra increase in the afternoon, the spectral changes between the disturbed soil and its surrounding are reduced.

In **Figure 19**, the disturbed soil is clearly detected using the sequenced images #3, #4 and #5 (i.e., image x = #3). It is also clearly detected using the short temporal sequences (i.e., $x = 3$; $y = 4$) as the long temporal sequences (i.e., $x = 1$; $y = 5$).

Further, the detection of the buried objects was applied using the QCC method. Following the results acquired using the CC method, the transformation parameter-based spectral classes were obtained only using the hyper-spectral images x = #3, #4, #5. We implemented the QCC with Q = QSEM (i.e., the maximum number of classes Q may be obtained using the expectation maximization). In our case study, the $Q = 5$ classes: three different types of soils (included the disturbed soil), stones and shadows (**Figure 20**). The spectral doublet features of quartz that are located at 1176 and 1123 cm^{-1} (8.5 and 8.9 μm) are clearly identified in the spectra of the undisturbed soils shown in **Figure 21**. On the other hand, the spectral features of the quartz were suppressed in the recently disturbed soil.

Figure 22 presents the results of the QCC. In comparison to the CC method, the disturbed soil is clearly identified in the different sequence images when using the QCC. Using the two applied methods, the changes were clearly detected using the short temporal sequences as the long temporal sequences.

By merging the QCC results from different sequences (**Figure 22**) and after applying a threshold filter, the detection of the disturbed soil became significant.

8.5. Conclusions

This research found that spectral-based change detection is a valuable method for the detection of buried IEDs under disturbed soil condition. The change detection methods (Chronochrome

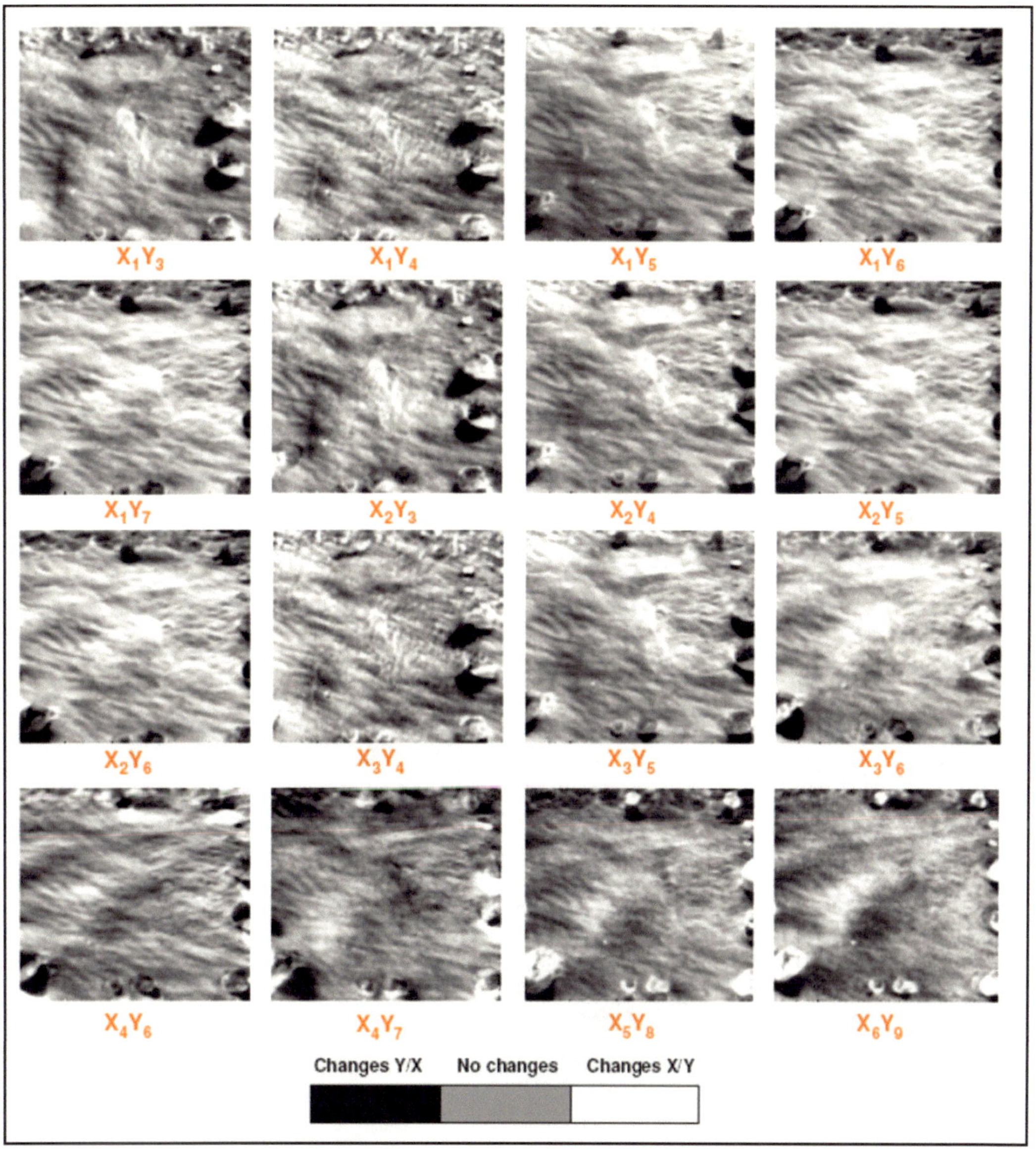

Figure 19. Several CC change detection results.

(CC)) and the class-conditional (QCC) were able to detect changes even using short temporal sequences. The efficiency of these methods increases using pair of images from large daily temporal differences (e.g., morning and evening). A thermal hyper-spectral sensor with very high spectral resolution (as the TELOPS Hyper-Cam) was found to be a sensitive tool for the detection of small changes in the upper layer of the soil that were caused by the buried small objects.

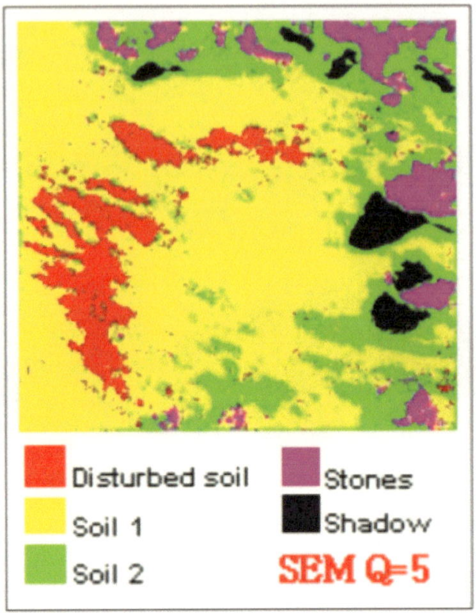

Figure 20. SEM results using $Q = 5$.

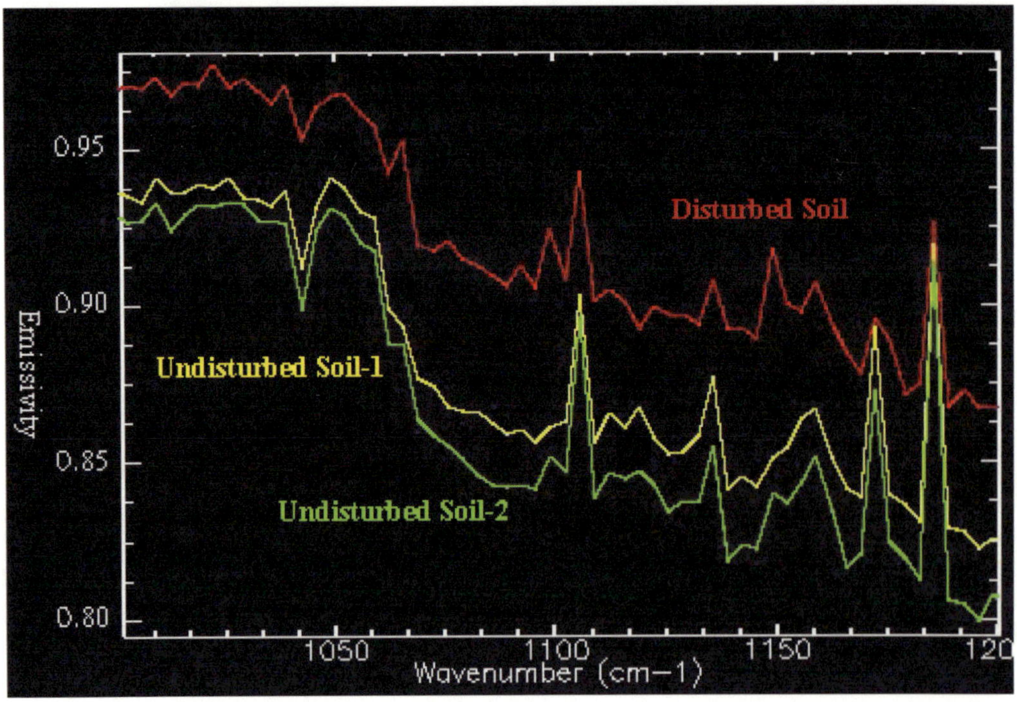

Figure 21. Spectra of the three detected soils using SEM.

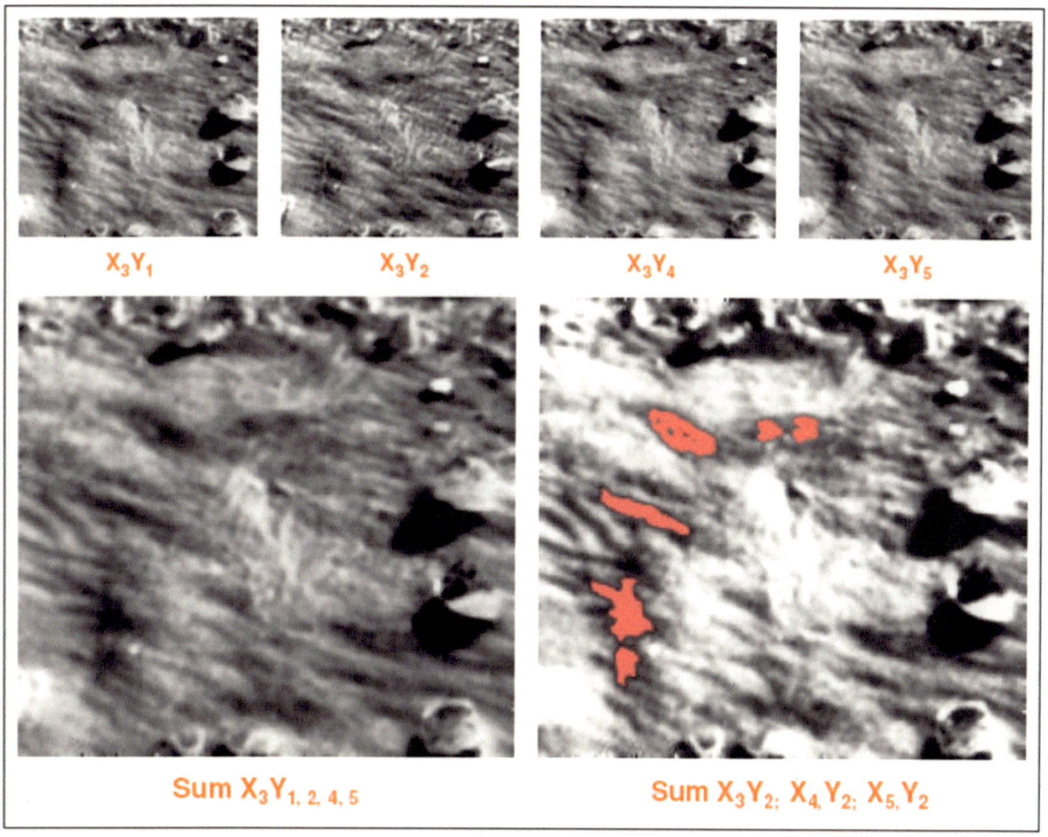

Figure 22. The QCC results.

9. Synthetic aperture radar (Dirk Borghys and Michal Shimoni)

9.1. Introduction

Remote sensing by synthetic aperture radar (SAR) systems provides information that is complementary to information available from optical images. Moreover, in areas under risk where rapid land-cover mapping and suspected mined area reduction are required, the advantages of SAR offering cloud penetration and day/night acquisition are evident in comparison to optical data.

The current chapter summarizes results obtained by using different SAR modalities for helping in mined-area reduction.

All the results presented in this chapter were obtained by automatic image processing. They could be further improved by introducing human intervention in a semi-automatic approach.

9.2. Properties of SAR images

The aim of this paragraph is to provide a very brief overview of the principal properties of SAR images. For a more detailed and rigorous explanation, the interested reader is referred to the plethora of books, websites and online tutorials on SAR, for example, [44–48].

In SAR imaging, an electromagnetic wave is emitted towards the earth surface where it is scattered by the elements on that surface. The energy scattered into the direction of the sensor is then measured and converted into an image. Usually, the emitter and the receiver are located on the same platform (airplane, satellite or UAV). This is called mono-static SAR, and in this case, the radar backscattering is measured.

The backscattered energy depends on properties of the scattering surface (geometrical properties, dielectric constant, etc.) as well as on the characteristics of the SAR system (wavelength, incidence angle, used polarization, imaging mode, etc.). SAR is operated in the microwave at different frequencies (**Table 5**). In general, each frequency has its advantage and the longer the wavelength, the higher the penetration depth of the SAR energy.

The geometrical properties determine the type of scattering that occurs on the surface. For a given SAR system, the type of scattering and the dielectric constant of the scattering surface (determined by the material type and humidity) determine the amount of backscattered energy and thus the intensity of the image. For a given type of land cover, the type of backscattering depends on the frequency of the SAR system.

In a SAR system, the emitted and received radar waves are polarized. Usually, the SAR system emits either horizontally (H) or vertically polarized waves (V), and the receiver is able to capture also either H or V waves, or both. This leads to four possible polarization configurations denoted HH, HV, VH or VV. A fully polarimetric system (PolSAR system) can emit/receive these four modes simultaneously. In PolSAR, the tight relation between the physical properties of natural media and their polarimetric behaviour leads to highly descriptive results that can be interpreted by analysing underlying scattering mechanisms [49, 50].

Different SAR images acquired from a slightly different location can, under certain conditions, be combined into a set of interferometric couples. SAR interferometry provides information about the 3D structure of the terrain and can be used for generating digital elevation models (DEMs) [51]. Interferometry can also be used for detecting subtle changes by a method called coherent change detection (CCD). A complete chapter of this book is devoted to the subject.

Band	Wavelength (cm)	Frequency (GHz)	Main applications
X	2.4–3.8	12.5–8	Urban
C	3.8–7.5	8–4	Land cover, soil moisture and ice
S	7.5–15	4–2	Security, manmade targets
L	15–30	2–1	Forest and soil moisture
P	30–100	1–0.3	Tomography

Table 5. Wavelength/frequency bands used in SAR remote sensing.

The combination of interferometric and polarimetric information (PolInSAR) provides information on scattering type and 3D structure of the image objects [52, 53]. PolInSAR data provide information about the structure and the complexity of the observed objects. When utilized concurrently, these different capabilities allow substantial improvements in land-cover determination [54, 55].

Table 6 gives an overview of past, current and planned civilian SAR remote-sensing satellites used for applications over land. The table shows that most satellites operate in C- or X-band, that since 2007 metric resolution satellite SAR images are available and that several SAR systems offer polarimetric capabilities. The column labelled 'Min. SR' gives the best spatial resolution available from that satellite. Usually, this best spatial resolution is obtained only for single-band spotlight mode images. The best obtainable spatial resolution in full-polarimetric mode is usually three times less good. The images from some of the satellites mentioned in the table are freely available (PALSAR, Sentinel-1). Furthermore, in an attempt to have a larger community benefit from satellite SAR images, space agencies freely offer SAR toolboxes that include most of the relevant processing tools and are able to process most of the available SAR satellite images [56–59].

9.3. Problem statement, dataset and region of interest

The SAR dataset used in this chapter was acquired in the frame of the SMART project [4]. SMART was dedicated to help mined area reduction by providing methods and integrated tools to help the use of airborne and space data.

In the project, a large dataset of airborne images was acquired over four test areas in Croatia. The flight campaign took place in August 2001 and was performed by the German Aerospace Agency (DLR). Metric resolution multi-spectral images (Daedalus line scanner with 12 bands in VIS-LWIR), very high resolution (5-cm) panchromatic visual images and metric resolution

Satellite	Lifetime	Agency	Band	Polarization	Min. SR
ERS-1	1991–2000	ESA (EU)	C	VV	25 m
JERS	1992–1998	NASDA (JP)	L	HH	25 m
ERS-2	1995–2012	ESA (EU)	C	VV	25 m
RADARSAT-1	1995–	CSA (CA)	C	HH	10 m
ENVISAT-ASAR	2002–2012	ESA (EU)	C	HH/HV/VV	15 m
ALOS-PALSAR	2006–	JAXA (JP)	L	PolSAR	10 m
TerraSAR-X	2007–	DLR (GE)	X	PolSAR	1 m
Tandem-X	2008–	DLR (GE)	C	PolSAR	1 m
Cosmo-Skymed constellation	2007–	eGEOS (IT)	X	PolSAR	1 m
RADARSAT-2	2008–	MDA (CA)	C	PolSAR	1 m
Sentinel-1	2014–	ESA (EU)	C	Dual-Pol	5 m
NovaSAR-S	2015–2022	SSTL-EADS (UK)	S	PolSAR	6 m
Radarsat constellation	2018 (planned)	CSA (CA)	C	PolSAR	3 m

Table 6. Overview of civilian SAR remote-sensing satellites.

multi-frequency SAR data from the E-SAR system of the DLR were acquired. The SAR data consist of PolSAR data in L- and P-band as well as HH- and VV-polarized data in X- and C-band (**Table 7**).

The land-cover mapping serves for 'suspected area' reduction and not for mine detection. The end users defined nine land-cover classes of interest: residential areas, roads, forests, wheat and corn fields, pastures, abandoned areas, bare soil, and rivers. These remote-sensing data were mainly used for identifying and detecting indicators of mine presence or mine absence. A list of such indicators was compiled in the project and the different available data were investigated for their adequacy for detecting the defined indicators. **Table 8** shows the indicators for which SAR data were found to improve the detection capability. The table shows

E-SAR parameter	X-band	L-band	P-band
Central frequency	9.6 GHz	1.3 GHz	450 MHz
Polarization	VV	HH-VV-HV-VH	HH-VV-HV-VH
Nr of looks	1	1	1
Spatial resolution	1.5 m	2 m	4 m
Altitude above sea level	3160 m	3160 m	3160 m
Radiometric resolution	<2 dB	<2 dB	<2 dB
Incidence angle	50°	50°	55°
Azimuth beam width	17°	18°	30°
Elevation beam width	30°	35°	60°

Table 7. Characteristics of the E-SAR data (©DLR) acquired in the SMART project.

Class	Indicators of mine presence	Methods/tools
C1	Cultivated land (A)	Classification
C2	Asphalted roads (A)	Classification and detection of linear features
C3	Agricultural areas that are no longer in use (P)	Classification and change detection
C4	Edges of forests (P)	Classification
C5	Trenches and man-made embankments (P)	Detection of linear features
		Interactive enhancement and extraction
C6	Soft edges of hardtop roads (P)	Classification and detection of linear features
C7	Crossroads, especially crossings of main roads with tracks no longer in use (P)	Detection of linear features, classification
C8	River banks (P)	Detection of linear features, classification
C9	Power supply poles (P)	Dedicated detector based on polarimetric features
C10	Hilltops and elevated plateaus (P)	DEM generation SAR interferometry

Table 8. Indicators of mine presence (P) or absence (A) for which SAR data can contribute to the detection.

that a combination of image classification and detection of linear features contributes to the detection of most indicators. For detecting power supply poles, a dedicated detector was developed by DLR, which exploits the specific effect of dihedral scattering on PolSAR images. Note that this is an example where objects much smaller than the resolution cell can be detected using SAR image processing. The relative 3D structure of the terrain can be easily extracted from interferometric image couples [51].

The remainder of the current chapter will discuss how the different SAR modalities contribute to increased classification accuracy for the other indicators. The research area is located in Glinska Poljana (Croatia), which is a post-war land mine-affected zone (**Figure 23**). The Croatian Mine Action Centre found the SAR to be the most valuable remote-sensing source due to the high frequency of its acquisition and its independency on cloud covers. Training and validation datasets have been constructed using the collected ground-truth data. Due to the risk incurred in mine-suspected areas, the ground-truth survey could be applied in only a small part of the research zone and its peripheries. **Figure 23** presents the locations of the training and the validation sets superimposed on the L-band SAR image.

Based on the confusion matrix determined using the validation set, four statistics are calculated for the validation: global Kappa coefficient (κ) [60], overall accuracy (OA), producer's accuracy (PA) and user's accuracy (UA) [61].

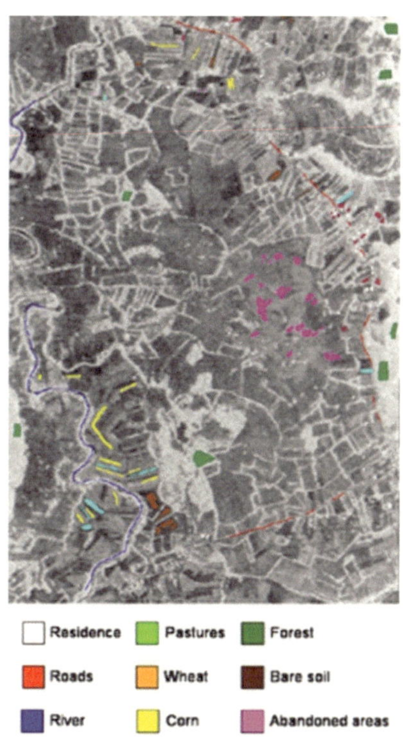

Figure 23. Training and validation dataset locations superimposed on the L-band E-SAR.

9.4. Fusion of single-channel SAR and optical data

For investigating the potential of 'future' satellite remote-sensing data for mine-area reduction-related land-cover classification, the available airborne data were used for simulating these satellite data. The Daedalus images were sub-sampled to the spectral bands of a typical optical satellite (Pleiades). A four-channel optical image was thus simulated (B1: 0.45–0.52 μm, B2: 0.52–0.60 μm, B3: 0.63–0.69 μm, B4: 0.76–0.90 μm). For the SAR data, VV-polarized X-band data were used.

For the land-cover classification from the single-channel SAR data, a feature-based supervised classification method was developed and the information available from the single-channel X-band SAR image was augmented by combining intensity information with spatial information. **Figure 24** presents a schematic overview of the developed approach.

The introduced spatial information consisted of a combination of textural features derived from grey level co-occurrence matrices (GLCMs) [62] and the results from a detector of bright and dark lines [63]. The line detector was applied after speckle reduction [64]. Feature-level fusion based on stepwise logistic regression (LR) [65] was applied to obtain the classification map. The stepwise LR implicitly performs a feature selection for each class to be detected [66].

Figure 25 shows the speckle-reduced X-band VV-polarized SAR amplitude image of part of the region of interest. The results of the two line detectors and of the extraction of textural parameters are shown in **Figures 26** and **27**, respectively. **Figure 28** shows the classification results obtained from the SAR image. From the classification results, it appears that the method based on only a single channel allows detecting the location of forests and edges as well as large parts of the roads and some of the buildings. It also provides an idea of the location of abandoned land and fields in use with vegetation; however, the distinction between

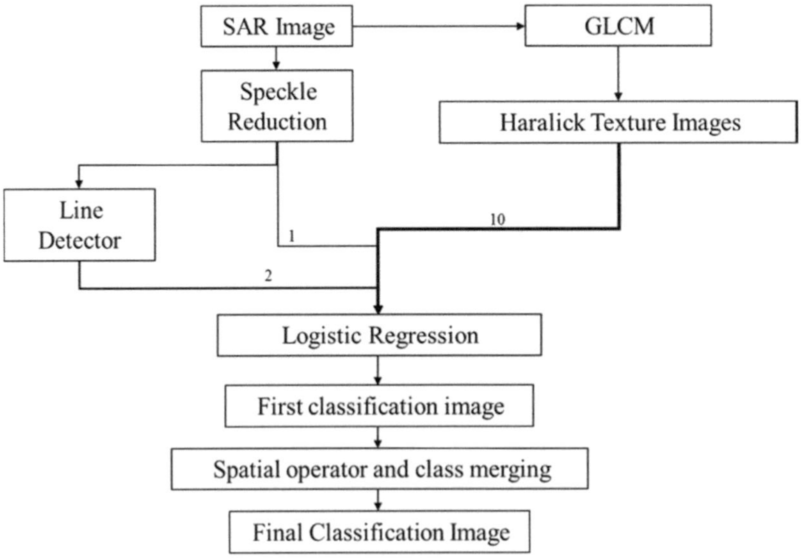

Figure 24. Processing scheme for land-cover classification from single-channel SAR data.

them is difficult because the two classes are mixed. Nevertheless, using only the single-band SAR image the distinction between water and radar shadows is very difficult, because both surface types result in a very low backscattering. In **Table 9**, the global classification accuracy as well as the user and producer accuracy per class is presented.

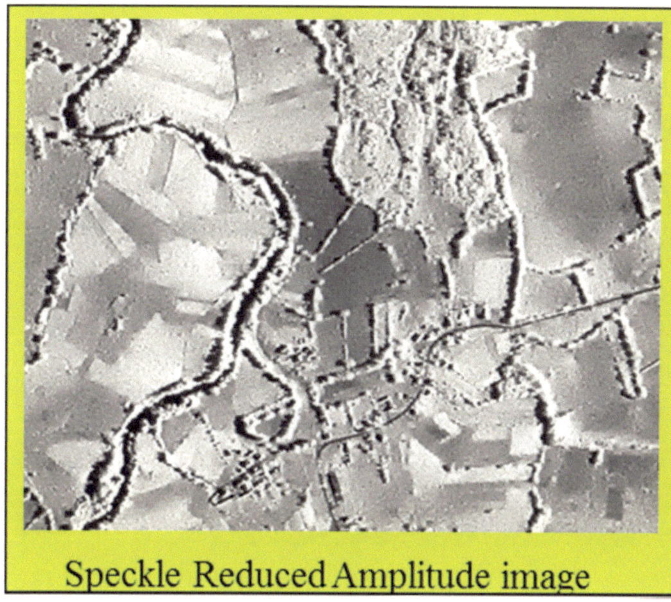

Figure 25. Speckle-reduced VV-polarized, X-band E-SAR image (©DLR).

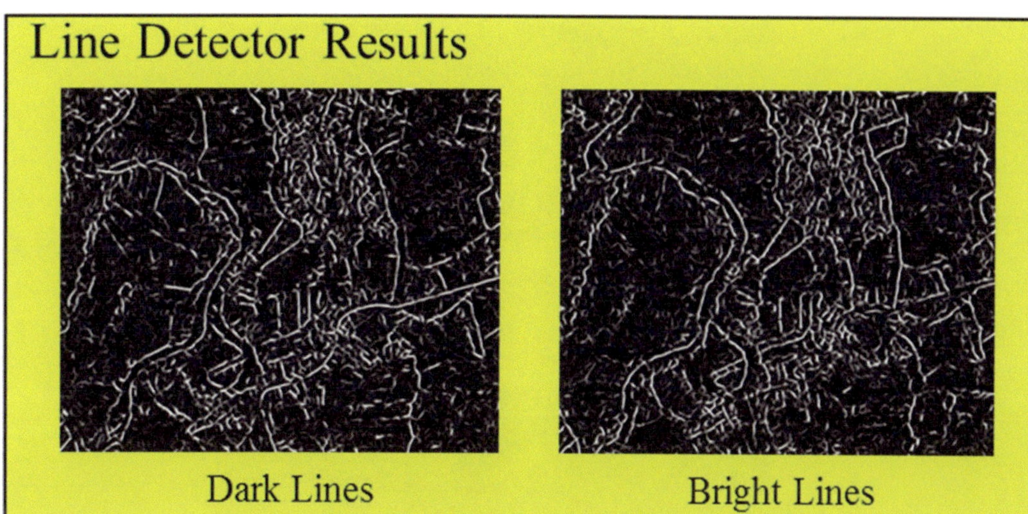

Figure 26. Results of the dark and bright line detectors.

For building a classification map from the optical data, a minimum distance classifier was applied. Decision-level fusion was used for combining the classification from SAR and optical data. Classification results obtained from the optical data and by fusion of optical and SAR are also presented in **Table 9**. The table shows that while the optical image leads to better classification results than the single-channel SAR image, the fusion of both results improves further the classification accuracy. This is particularly the case for the class 'abandoned land', which is a very important indicator for possible mine presence.

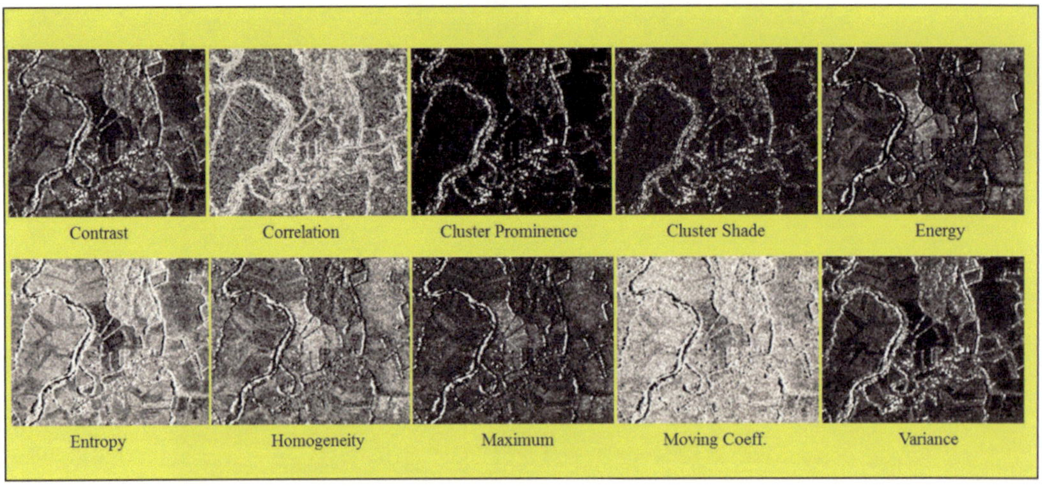

Figure 27. Images of the various textural features.

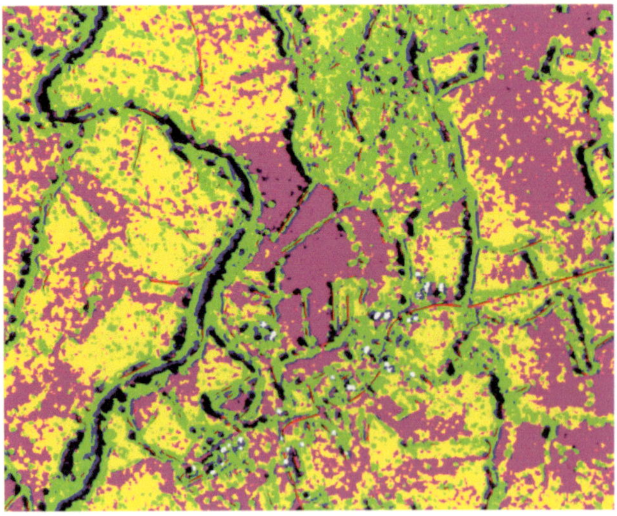

Figure 28. Classification map obtained from the single-channel X-band SAR image.

Class name	SAR		Optical		Fusion	
	PA	UA	PA	UA	PA	UA
Roads	0.47	0.41	0.86	0.28	0.86	0.30
Water	0.28	0.05	0.93	0.90	0.94	0.91
Residential areas	0.08	0.98	0.32	0.17	0.41	0.30
Abandoned land	0.66	0.78	0.74	0.82	0.81	0.89
Fields with vegetation	0.60	0.51	0.74	0.81	0.81	0.80
Trees and shrubs	0.58	0.55	0.88	0.83	0.88	0.86
Total accuracy	0.59		0.76		0.82	

Table 9. Classification results obtained from SAR processing, optical image processing and fusion.

9.5. PolSAR and PolInSAR

In the past recent years, several polarimetric decomposition methods have been developed for various applications. Each application combines the PolSAR or PolInSAR information in order to characterize a certain type of scattering process which was determined through complementary information as experience, physical grounds [67] or by systematic selection and reduction process [68]. In very complex scenes, it is useful to exploit the discriminative power offered by the combination of a great number of these features.

At RMA, we investigated the complementarity of different frequencies and the fusion of PolSAR and PolInSAR data for land-cover classification in mine-covered areas. Several PolSAR and PolInSAR features are extracted in the imaging process, each combining amplitude, phase and correlation information in order to extract specific characteristics of the scene. For land-cover classification, two levels of fusion are applied: feature and decision levels. In the feature-level fusion, a logistic regression (LR) is used for feature selection and for fusing the selected features by optimizing the log-likelihood function. The obtained probability images are then combined using a soft-decision fusion method, the neural network (NN), in order to obtain the final classification results.

9.5.1. Feature extraction

A polarimetric SAR system measures the electro-magnetic signal and its polarization state that are backscattered from the scene. The interaction of the transmitted wave with a scattering object transforms its polarization. Discrimination of different types of mechanisms is possible using the polarimetric signatures that are strongly depended on the scattering process. It has been demonstrated that the inclusion of SAR polarimetry can lead to a significant improvement in the quality of the data analysis in comparison to conventional single-polarization SAR data [69]. In our study, for describing different properties of the land-cover objects and for the fusion processes, a total of 72 different PolSAR and PolInSAR features were extracted. Nine PolSAR features and nine PolInSAR features are presented in **Figures 29** and **30**, respectively.

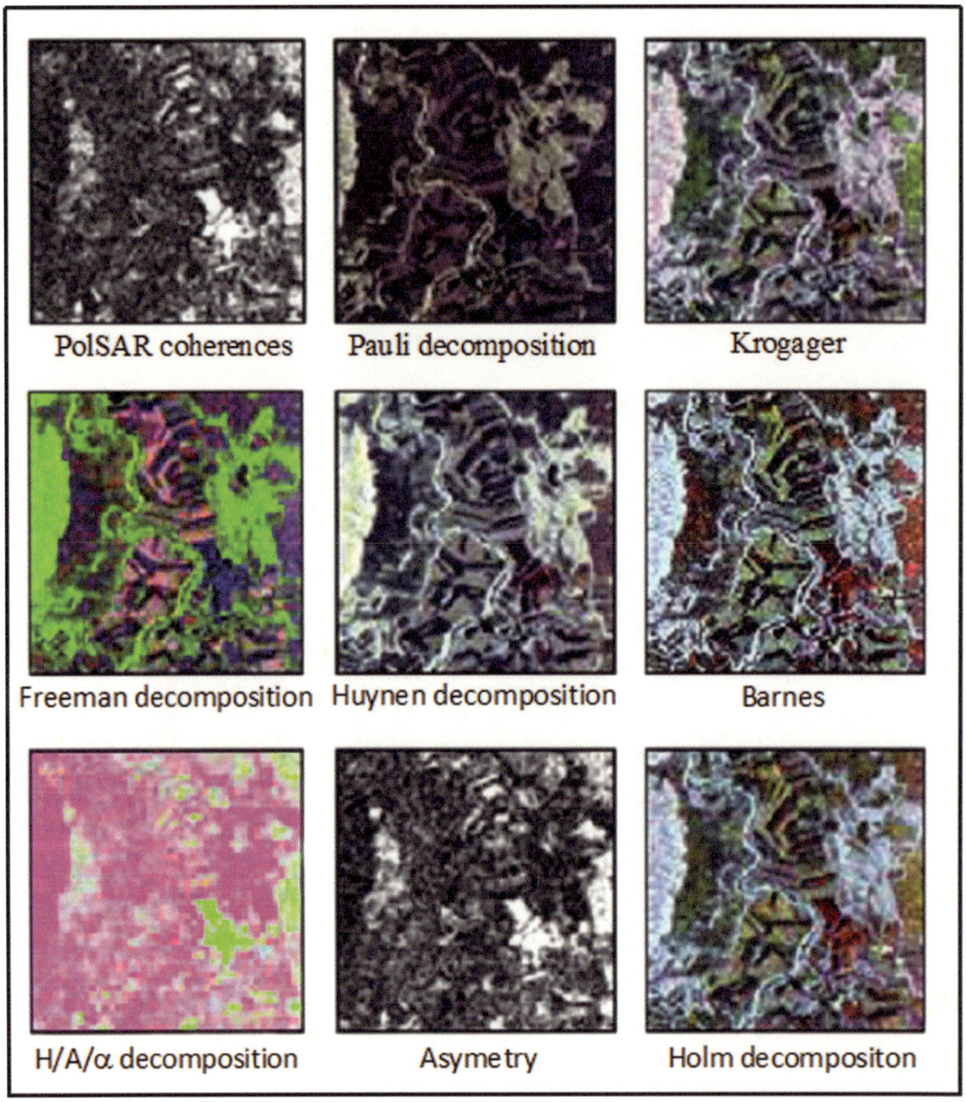

Figure 29. PolSAR features dataset extracted using different decomposition methods and by calculating the PolSAR coherences.

The figures demonstrate the variation and the inherent complementarity between the different features. For a detailed description of the used feature set the authors refer to and the references therein, see Ref. [70].

9.5.2. Fusion methodology

In this study, four independent sources of information: L-band PolSAR, L-band PolInSAR, P-band PolSAR and P-band PolInSAR, are considered. The image processing resulted in the

http://dx.doi.org/10.5772/66691

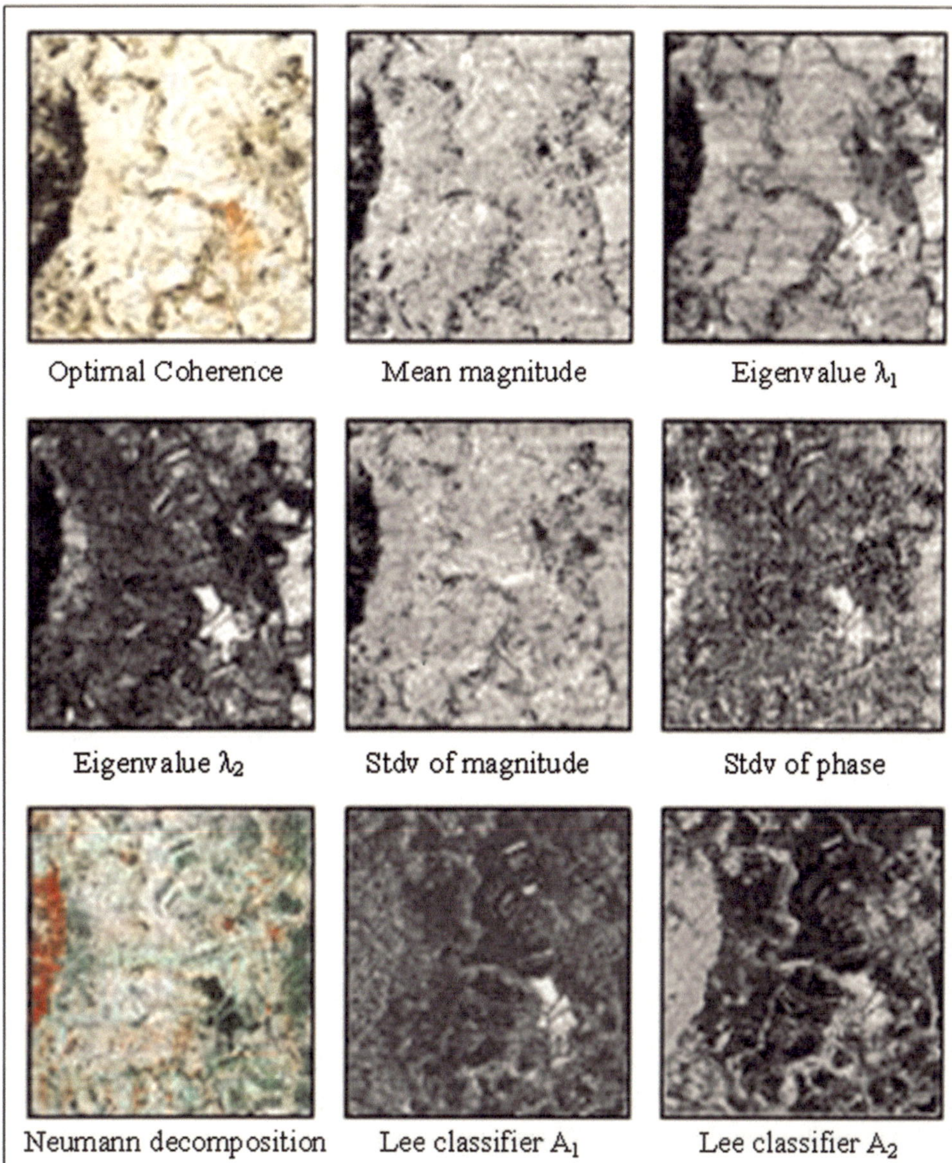

Figure 30. PolInSAR features dataset extracted using different decomposition methods and by calculating the PolInSAR optimal coherences.

extraction of a large features set: 25 L-band PolSAR features, 25 P-band PolSAR features, 13 L-band PolInSAR features and 13 P-band PolInSAR features [70].

The end users identified nine different land-cover classes using a ground-truth survey: roads, residential areas, wheat crop fields, corn crop fields, forests, rivers, bare soils, and abandoned areas (no trees or shrubs). The processing of the classification found two fundamental technical problems. The first consists in identifying the parameters that capture the information from the radar signals and that allow to distinguish the different classes of land cover. The second consists in selecting a technique that fuses these features in an efficient manner that will allow distinguishing the land-cover classes.

Figure 31 presents the selected methodological scheme. The neural network, (NN) which is a non-linear method, is selected as a classifier because linear classifiers do not provide satisfying results in the fusion of data from different sources [71, 72]. However, the NN cannot be applied directly on the original 76 features because the training set should be larger than available. To overcome this shortcoming, we separated the fusion process for two levels. In the feature-level fusion, a logistic regression (LR) [65] is used for combining the different features extracted from the PolSAR and the PolInSAR data. The LR is used as intelligent feature reduction method. The LR is also a linear classifier that is applied separately on each information source and on each of the nine classes. The results are nine probability images for each information source. The obtained probability images are then fused using a multi-layer perceptron neural network with one hidden layer [73, 74] to yield the final classification.

9.5.3. Results

As mentioned before, only the probability images derived by the LR using different combinations of SAR features are used as input to the NN method.

The NN results per class using only L-band or only the P-band E-SAR dataset are, respectively, presented in **Tables 10** and **11**. **Table 10** shows that high UA and PA results using only the L-PolSAR or only the L-PolInSAR dataset were obtained for the classes 'forests', 'wheat fields' and 'abandoned areas'. For the other classes, due to the low PA and UA results and due to the lack of pattern in the results, it is difficult to identify which feature set is more significant, the PolSAR or the PolInSAR. The highest PA (98.87 and 91.10%) and UA (98.71 and 88.38%) results are obtained for the abandoned area class using the L-band PolSAR and PolInSAR probability datasets and the NN classifier.

High UA and PA results are obtained for the 'abandoned area' class by using only the P-PolSAR or only the P-PolInSAR dataset (**Table 11**). For the other classes, lower UA and PA results are obtained using P-PolInSAR probability sets than those obtained using the P-band PolSAR probability sets. Using the NN classifier, similar trend was obtained for the 'abandoned area' class with the highest PA (99.84 and 89.16%) and UA (82.87 and 62.05%) results and using P-band PolSAR and PolInSAR probability datasets, respectively.

The 'pasture' class obtained very low PA and UA results in both L-band and P-band datasets. Analysis of the confusion matrices shows that the 'pasture' class was mainly confused with bare soil. As expected, the signal of the L- and the P-frequencies penetrates the pasture areas that are covered with very short scattered grass, as in bare soil.

A comparison between the results was obtained using single dataset (**Tables 10** and **11**) and the results after fusion (**Table 12**) show that process using the NN significantly improves the results of the land-cover classification.

http://dx.doi.org/10.5772/66691

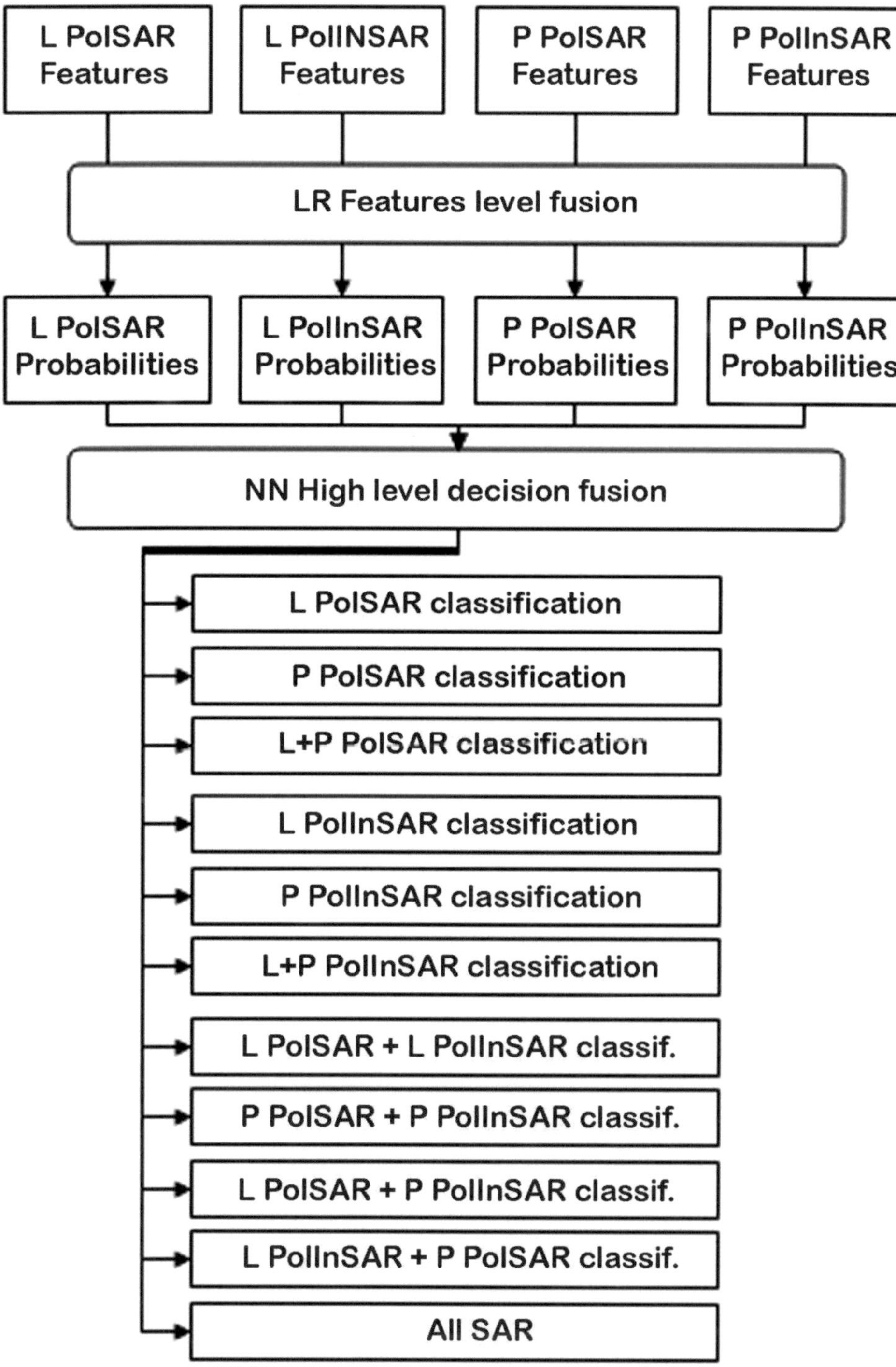

Figure 31. The fusion process scheme.

L SAR/Land cover class	L-PolSAR		L-PolInSAR	
	PA (%)	UA (%)	PA (%)	UA (%)
Residence	31.88	33.55	1.25	4.44
Road	6.28	7.96	24.37	24.25
River	23.92	31.82	13.49	32.89
Forest	88.00	72.33	94.58	71.67
Bare soil	41.00	50.00	67.55	85.60
Abandoned area	98.87	98.71	91.10	88.38
Wheat	87.45	55.18	94.40	77.62
Corn	30.32	16.05	43.15	36.01
Pasture	0.00	0.00	19.43	17.28

Table 10. The NN results per class using E-SAR L-Band datasets.

P SAR/Land cover class	P-PolSAR		P-PolInSAR	
	PA (%)	UA (%)	PA (%)	UA (%)
Residence	95.62	52.85	15.94	17.11
Road	36.93	37.69	14.32	12.98
River	21.94	40.00	7.01	16.18
Forest	83.13	62.66	38.00	65.73
Bare soil	82.76	49.57	39.15	17.97
Abandoned area	99.84	82.27	89.16	62.05
Wheat	51.35	76.88	91.51	40.0
Corn	0.00	0.00	0.0	0.0
Pasture	0.26	5.88	0.0	0.0

Table 11. The NN results per class using E-SAR P-band.

The best results are derived using the 'All-SAR' probability datasets for seven of the nine land-cover classes (**Table 12**). The highest PA and UA results were obtained for all the classes using the fused PolSAR and PolInSAR datasets. These results emphasize again the complementarities of the different SAR frequency datasets and of the PolSAR and PolInSAR information. The results of the 'abandoned area' class are the highest with PA (99.51%) and UA (99.35%). The results for this class have a high importance for the Croatian mine action (i.e., end users) because it considered as suspected area. Moreover, it was demonstrated that using a pixel-wise classification based on optical data or a single SAR dataset, it is difficult to distinguish this class. Applying the LR and multinomial logistic regression (MNLR) methods only on the PolSAR data resulted with lower user accuracy (UA) of 80% and producer

http://dx.doi.org/10.5772/66691

Land cover class	NN classification		
	The best fusion set	PA (%)	UA (%)
Residences	All-SAR	90.31	68.81
Roads	All-SAR	41.84	62.67
River	All-SAR	61.64	81.73
Forest	All-SAR	91.60	88.08
Bare soil	All-SAR	89.32	79.92
Abandoned areas	All-SAR	99.51	99.35
Wheat	All-SAR	99.03	97.53
Corn	L-PolSAR + L-PolInSAR	65.31	66.27
Pastures	L-PolInSAR + P-PolSAR	31.35	29.58

Table 12. The highest PA and UA results per class obtained using the NN classifier.

accuracy (PA) of 66% for the 'abandoned areas' in 'Glinska Poljana' [66], than the results were presented in this study. Using the PolSAR and multi-spectral data and a Dempster-Shafer (DS) method, the 'abandoned areas' are classified with PA of 84% and using a fuzzy method with PA of 89% [25].

After the fusion process, the road and river classes are still confused. However, those classes can be easily distinguished by a photo-interpreter. **Table 13** presents the overall accuracies and the Kappa values obtained using the NN for the land-cover classifications. The results show that the classification accuracy improves for the complete land-cover classification, using fused

LR probability dataset	NN	
	Overall accuracy (%)	Kappa
L-PolSAR	52.13	0.437
P-PolSAR	59.72	0.522
L+P-PolSAR	67.48	0.618
L-PolInSAR	61.04	0.539
P-PolInSAR	37.69	0.522
L+P-PolInSAR	66.16	0.601
L-PolSAR + L-PolInSAR	63.87	0.576
P-PolSAR + P-PolInSAR	65.55	0.596
L-PolSAR + P-PolInSAR	63.22	0.569
L-PolInSAR + P-PolSAR	69.96	0.646
All-SAR	84.00	0.809

Table 13. Overall accuracy for land-cover classification obtained using the NN classifier.

datasets from different SAR frequencies. A kappa of 0.809 and the highest accuracy (84.00%) are obtained using the All-SAR dataset. The results emphasize again the complementarities of the PolSAR and PolInSAR information and of data acquired in different frequencies.

Figure 32 presents the land-cover classification obtained by implementing the All-SAR dataset and using the NN classifier. Many false alarms are present in the class 'road'. Please note that we did not select 'shadow' as class because we did not visualize this phenomenon in the original image scene. However, common artefacts as shadow and layover are also part of the SAR images and future improvement should include this class.

9.6. Conclusion

The results presented here show that a feature-based fusion method for classifying SAR image data is a valuable approach. When only a single-channel SAR image is available, a rough land-cover map can be extracted by combining the available intensity information with spatial

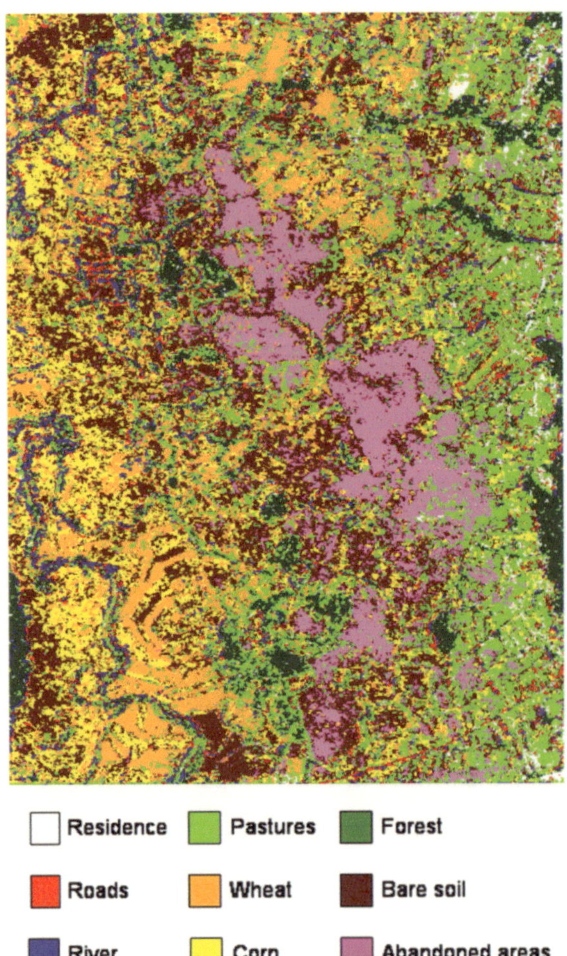

Figure 32. NN land-cover classification of Glinska Poljana obtained using All-SAR dataset.

information such as line detection results and textural features. Fusion with a classification obtained from optical images does provide a significant improvement in classification accuracy with respect to using optical data alone.

Features from different SAR frequencies were found to be complementary and adequate for land-cover classification. It was also found that PolInSAR features are complementary to the PolSAR features and essential for producing an accurate classification of complex scenes composed of very diverse land-cover classes. Fully polarized and interferometric SAR data proved to produce valuable remote-sensing information and can be used to obtain accurate information for areas under danger of mine presence.

The current availability of high-resolution SAR image data from various satellites on one hand and the SAR-processing toolboxes freely available on the other hand makes it easier to apply an approach as described in this chapter. Nevertheless, an efficient exploitation of the full information richness offered by SAR images and its complementarity to information from optical data still requires acquiring a good understanding of the SAR image phenomenology.

10. Conclusion: the future of remote sensing in mine action

The information collected by remote sensing can help mine action in many other ways that survey a suspicious area to decide where to send the clearance teams. In some mined countries, the availability of up-to-date maps can be poor. Remote sensing can be used to create recent maps to help the planning of mine action operations. An analysis of the relief and the land cover can help understand the road network and decide the best route to access a given location. Deciding which areas to clear first is often a difficult decision. Remote sensing can provide input regarding the risk of the presence of mines or the vulnerability of local population. In case of major disasters such as flooding, remote sensing can be used to assess the change in the mine contamination.

For decades, airplanes and satellites were the only vectors to obtain remote-sensing data. Drones offer now new opportunities. With the reduction of size and costs of cameras, they change drastically the way remote sensing can be used. If the autonomy of the smaller drones is still limited, they still allow for fast deployment, which can help reactivity of mine action organizations.

With drones offering a good solution for small-range survey and satellite for large areas, the question now is about the relevance of conventional aircraft. Have they still a place between drones and satellites? U.S. Air Force is currently training more pilots of drones than pilots of fighter aircraft. Will aircraft slowly phase out for remote sensing?

Acknowledgements

The authors of this chapter would like to thank the Belgian Ministry of Defence and the European Union.

Section 2: The aerial campaign was supported by financial aid of the US State Department, through the International Trust Fund for Demining and Mine Victims Assistance (ITF), and kindly shared by HCR Centre for testing, development and training Ltd.

Section 3: The research leading to these information and results has received funding from the European Community's Seventh Framework Programme (FP7/2007–2013) under grant agreement n° 284747 (TIRAMISU). Noveltis and IGEAT, two partners of TIRAMISU, contributed to this crater detection study. Noveltis pre-processed the data (georeferencing and atmospheric correction). IGEAT made the visual interpretation and provided valuable information for the data preparation and results discussion.

Sections 6, 7 and 9: The authors thank all the researchers of the project SMART, co-funded by the European Commission (IST-2000-25044) and involving the following partners: CROMAC, DLR, ENST, IXL, Renaissance ASBL (RMA), RST, TRASYS, ULB-IGEAT and Zeppelin.

Author details

Yann Yvinec*, Nada Milisavljevic, Charles Beumier, Idrissa Mahamadou, Dirk Borghys, Michal Shimoni and Vinciane Lacroix

*Address all correspondence to: yann.yvinec@rma.ac.be

Royal Military Academy, Signal and Image Centre, Brussels, Belgium

References

[1] GICHD. A Guide to Land Release: Technical Methods. Geneva: GICHD; April 2011, ISBN 2-940369-42-9

[2] Van Genderen JL, et al. Airborne Detection of Landmines: A Review of Techniques and Some Practical Results. Disarmament & International Security. Germany; 1998

[3] Maathuis B. RS based detection of minefields. Geocarto International. 2003;**18**:51-60

[4] Yvinec. EU Project of Remote Detection: SMART in a nutshell. In: Proceedings of Robotics and Mechanical Assistance in HuDem and Similar Risky Interventions. Brussels, Belgium; 2004

[5] Bajić et al. Research of the Mine Suspected Area. CROMAC; 2011, ISBN 978-953-99879-7-6

[6] Bajić M. Airborne Detection of UXO, Reconstruction of Former War Zone, EOD FUTURE in the Light of NATO EOD Demonstrations and Trials 2012, A Book of Papers. 2013. pp. 33-41

[7] Krtalic A, Bajić M. Upgrade of the Advanced Intelligence Decision Support System for Mine Action in the TIRAMISU Project. In: 11th IARP Workshop HUDEM, Book of Papers; 25th April 2013, Šibenik, Croatia; 2013. pp. 83-86

[8] Lacroix V, Vanhuysse S. Indicators of mine presence: Focus on trenches. South-Eastern European Journal of Earth Observation and Geomatics. 2014;**3**(2):635-639

[9] Vanhuysse et al. Object-based image analysis for detecting indicators of mine presence to support suspected hazardous area re-delineation. South-Eastern European Journal of Earth Observation and Geomatics. 2014;**3**(2):525-529

[10] Lacroix et al. Feature-extraction using the constrained gradient; ISPRS. Journal of Photogrammetry and Remote Sensing. 1998;**53**:85-94

[11] Hatfield-Consultants. Uxo predictive modeling in MMG LXML, Laos (sepon) mine development area. Technical Report 1791.D6.1, Hatfield consultants, East Lansing, MI; 2014

[12] Lacroix V, Vanhuysse S. Crater detection using CGC—A new circle detection method. ICPRAM. 2015;(1):320-327

[13] Akinlar C, Topal C. Edcircles: A real-time circle detector with a false detection control. Pattern Recognition. 2013;**46**

[14] Haralick RM. Statistical and structural approaches to texture. Proceedings of the IEEE. 1979;**67**:786-804

[15] Cross GR, Jain AK. Markov random field texture models. IEEE Transactions on Pattern Analysis and Machine Intelligence. 1983;**5**:25-39

[16] Bovik AC, Clarke M, Geisler WS. Multichannel texture analysis using localized spatial filters. IEEE Transactions on Pattern Analysis and Machine Intelligence. 1990;**12**:55-73

[17] Randen T, Husoy JH. Filtering for texture classification: A comparative study. IEEE Transactions on Pattern Analysis and Machine Intelligence. 1999;**21**(4):291-310

[18] Neyt X, Acheroy M, Lemahieu I. Directional adaptive image restoration. In: Proceedings of International Workshop on Image and Signal Processing IWISP'96; November 1996

[19] Bezdek JC, Pal SK. Fuzzy models for pattern recognition. IEEE Press. 1992

[20] Gath I, Geva AB. Unsupervised optimal fuzzy clustering. IEEE Transactions on Pattern Analysis and Machine Intelligence. 1989;**11**(7):773-781

[21] Isabelle B. On fuzzy distance and their use in image processing under imprecision. Pattern Recognition. 1999;**32**(11):1873-1895

[22] Yvinec Y, Druyts P, Lacroix V, Ouaghli Y, Idrissa M. Airborne minefield detection: Pilot project RMA/SIC final report. Technical Report. Royal Military Academy/Signal and Image Centre, Avenue de la Renaissance, 30 B-1000 Brussels; 1999.

[23] Axelsson P. DEM generation from laser scanner data using adaptive TIN models. International Archives of Photogrammetry and Remote Sensing. 2000;**XXXIII**:110-117

[24] Keller M, Milisavljevic N, Suess H, Acheroy M. Reduction of mine suspected areas by multisensor airborne measurements: First results. In: Proceedings of SPIE Conference on Detection Technologies for Mines and Minelike Targets. Vol. 4742. Orlando, USA; 2002. pp. 857-871

[25] Bloch I, Milisavljević N, Acheroy M. Multisensor data fusion for spaceborne and airborne reduction of mine suspected areas. International Journal of Advanced Robotic Systems. 2007;**4**(2):173-186

[26] Milisavljevic N, Bloch I. Sensor fusion in anti-personnel mine detection using a two-level belief function model. IEEE Transactions on Systems, Man and Cybernetics—Part C. 2003;**33**(2):269-283

[27] Shimoni M, Van der Meer F, Acheroy M. Thermal imaging spectroscopy: Present technology and future dual use applications. In: Proceedings of the 5th EARSeL Workshop on Imaging Spectroscopy; Bruges, Belgium; April 23-25, 2007.

[28] Liao W-J, Chen D-H, Baertlein BA. Algorithms for detection of surface mines in multispectral IR and visible imagery. Proceedings of the SPIE. 2001;**4394**:310-321

[29] Burt AB. Detection of Spatially Unresolved (Nominally SubPixel) submerged and surface targets using hyper-spectral data [MSc thesis]. Remote Sensing Centre, Naval Postgraduate School Monterey, Canada; 2012

[30] Elbakary MI, Alam MS. Mine detection in multispectral imagery data using constrained energy minimization. Proceedings of the SPIE 6977, Optical Pattern Recognition. 2008; **XIX**:P697707. DOI: 10.1117/12.781735

[31] Baghbidi MZ, Jamshidi K, Nilchi AR, Homayouni S. Improvement of anomaly detection algorithms in hyperspectral images using discrete wavelet transform. Signal & Image Processing: An International Journal. 2011;**2**(4):13-25

[32] McFee JE, Ripley HT. Detection of buried land mines using a CASI hyperspectral imager. In: Proceedings of the SPIE 3079-Detection and Remediation Technologies for Mines and Minelike Targets II. Vol. 738. 1997

[33] Winter EM, Fields DJ, Carter MR, Bennett CL, Lucey PG, Johnson JR, Horton KA, Bowman AP. Assessment of techniques for airborne infrared land mine detection. In: Presented at the Third International Airborne Remote Sensing Conference and Exhibition, July 1997

[34] Winter EM, Miller M, Simi C, Hill A, Williams T, Hampton D, Wood M, Zadnick J, Sviland M. Mine detection experiments using hyperspectral sensors. In: SPIE Proceeding. Vol. 5415; September 2004. pp. 791-798

[35] Zare A, Bolton J, Gader P, Schatten M. Vegetation mapping for landmine detection using long-wave hyperspectral imagery. IEEE Transactions on Geoscience and Remote Sensing. January 2008;46(1):172-178

[36] A. Zare, J. Bolton, P. Gader, and M. Schatten, "Vegetation mapping for landmine detection using long-wave hyperspectral imagery," IEEE Trans. Geosci. Remote Sens., vol. 46, iss. 1, pp. 172–178, 2008.

[37] Tabulo MA. Detecting disturbed earth using hyperspectral imaging. AFDA. 2001;**4**(1)

[38] Gurton KP, Felton M. Remote detection of buried land-mines and IEDs using LWIR polarimetric imaging. Optics Express. 2012;**20**(20)

[39] TELOPS. Infrared Cameras. 2015. Available from: http://www.telops.com/en/

[40] Schaum A, Stocker A. Hyperspectral change detection and supervised matched filtering based on covariance equalization. In: Proceedings of the SPIE. Vol. 5425. 2004. pp. 77-90

[41] Eismann AT, Meola J, Hardie RC. Hyperspectral change detection in the presence of diurnal and seasonal variations. IEEE Transactions on Geoscience and Remote Sensing. 2008;**46**(1):237-249

[42] Eismann MT, Meola J, Stocker AD, Beaven SG, Schaum AP. Airborne hyperspectral detection of small changes. Applied Optics. 2008;**47**:27-45

[43] Eisman MT, Stein DW. Stochastic mixture modelling. In: Chang C-I, editor. Hyperspectral Data Exploitation: Theory and Applications. John Wiley & Sons; 2007. pp. 107-148

[44] Oliver C, Quegan S. Understanding Synthetic Aperture Radar Images. SciTech Publishing; 2004. p. 479

[45] Maître H. Processing of Synthetic Aperture Radar Images. London, UK: ISTE; 2008. DOI: 10.1002/9780470611111.biblio

[46] Richards JA. Remote Sensing with Imaging Radar. Springer; 2009. p. 361. DOI: 10.1007/978-3-642-02020-9

[47] ESA Earth Online. SAR Land Applications Tutorial [Internet]. Available from: https://earth.esa.int/web/guest/training-packages/-/article/sar-land-applications-tutorial [Accessed: 2015-10-30]

[48] Sandia National Laboratories. What Is Synthetic Aperture Radar [Internet]. Available from: http://www.sandia.gov/radar/what_is_sar/index.html [Accessed: 2015-10-30]

[49] Cloude SE. Polarisation: Applications in Remote Sensing. OUP Oxford; 2009. p. 472. DOI: 10.1093/acprof:oso/9780199569731.001.0001

[50] Lee JS, Pottier E. Polarimetric Radar Imaging: From Basics to Applications. Taylor & Francis; 2009. p. 438. ISBN: 9-781-4200-5497-2

[51] Hansen RE. Radar Interferometry: Data Interpretation and Error Analysis [thesis]. TU Delft, The Netherlands; 2001

[52] Papathanassiou K, Cloude SR. Three-stage inversion process for polarimetric SAR interferometry. IEE Proceedings on Radar, Sonar and Navigation. 2003;**150**(3):125-134. DOI: 10.1049/ip-rsn:20030449

[53] Reigber A, Neumann M, Guillaso S, Sauer S, Ferro-Famil L. Evaluating PolInSAR parameter estimation using tomographic imaging results. In: Proceedings of European Radar Conference (EURAD'05); Paris; 6-7 October 2005. pp. 189-192. DOI: 10.1109/EURAD.2005.1605597

[54] Gamba P, Houshmand B. Three-dimensional road network by fusion of polarimetric and interferometric SAR data. In: Proceeding of IGARSS'99; 28 June-2 July 1999; Hamburg, Germany: IEEE; 1999. pp. 302-304

[55] Hong S, Moon WM, Paik HY, Choi GH. Data fusion of multiple polarimetric SAR images using discrete wavelet transform (DWT). In: Proceeding of IGARSS'02; 24-28 June 2002; Toronto, Canada: IEEE; 2002. pp. 3323-3325. DOI: 10.1109/IGARSS.2002.1027170

[56] PolSARPro: The Polarimetric SAR Processing and Educational Toolbox [Internet]. Available from: https://earth.esa.int/web/polsarpro/home [Accessed: 2015-10-29]

[57] ESA Sentinel Online: Sentinel Toolboxes [Internet]. Available from: https://sentinel.esa.int/web/sentinel/toolboxes [Accessed: 2015-10-29]

[58] Canadian Space Agency: Earth Observation Application Development Program (EOADP), Open Source RADARSAT-2 Polarimetric Toolbox [Internet]. Available from: http://www4.asc-csa.gc.ca/AUOT-EOAU/eng/EOADP/Projects/72596.aspx [Accessed: 2015-10-29]

[59] Next ESA SAR Toolbox (NEST) [Internet]. Available from: https://earth.esa.int/web/nest/home [Accessed: 2015-10-29]

[60] Rosenfield G, Fitzpatrick-Lins KA. Coefficient of agreement as a measure of thematic classification accuracy. Photogrammetric Engineering and Remote Sensing. 1986;**52**(2):223-227. DOI: 10.1016/S0034-4257(97)00083-7

[61] Congalton RG, Green K. Assessing the Accuracy of Remotely Sensed Data: Principles and Practices. Lewis Publishers; 1999. p. 630. DOI: 10.1201/978142004856

[62] Haralick RM, Shanmugam K, Dinstein IH. Textural features for image classification. IEEE Transactions on Systems, Man and Cybernetics. 1973;**3**(6):610-621. DOI: 10.1109/TSMC.1973.4309314

[63] Tupin F, Maitre H, Mangin J-F, Nicolas J-M, Pechersky E. Detection of linear features in SAR images: Application to road network extraction. IEEE Transactions on Geoscience and Remote Sensing. 1998;**36**(2):434-453. DOI: 10.1109/36.662728

[64] Pizurica A, Philips W, Lemahieu I, Acheroy M. Despeckling SAR images using wavelets and a new class of adaptive shrinkage estimators (ICIP'01). In: Proceedings of the 8th IEEE International Conference on Image Processing. DOI: 10.1109/ICIP.2001.958467

[65] Hosmer DW, Lemeshow S, Sturdivant RX. Applied Logistic Regression. 3rd ed. Wiley; 2013. p. 528. DOI: 10.1002/9781118548387

[66] Borghys D, Perneel C. A supervised classification of multi-channel high-resolution SAR data. EARSEL eProceedings. 2007;**6**(1):26-27

[67] Hoekman DH, Quirones MJ. Land cover type and biomass classification using AirSAR data for evaluation and monitoring scenarios in Columbian Amazon. IEEE Transactions on Geoscience and Remote Sensing. 2000;**38**(2):685-696. DOI: 10.1109/36.841998

[68] Cumming IG, Van Zyl JJ. Feature utility in polarimetric radar image classification. In: Proceeding IGARSS'89; 10-14 July 1989; Vancouver, Canada: IEEE; 1989

[69] Hellmann M. Polarimetric interferometry and differential interferometry 1 & 2", in radar polarimetry and interferometry. 2004. RTO lecture series SET-081; Paper 6, p. 1-33, Available from: http://ftp.rta.nato.int/public//PubFullText/RTO/EN/RTO-EN-SET-081/EN-SET-081-$$TOC.pdf

[70] Shimoni M, Borghys D, Heremans R, Perneel C, Acheroy M. Fusion of PolSAR and PolInSAR data for land cover classification. International Journal of Applied Earth Observations and Geoinformation. 2009;**11**(3):169-180. DOI: 10.1016/j.jag.2009.01.004

[71] Borghys D, Yvinec Y, Perneel C, Pizurica A, Philips W. Supervised feature-based classification of multi-channel SAR images. Pattern Recognition Letters. 2006;**27**(4):252-258. DOI: 10.1016/j.patrec.2005.08.006

[72] Bloch I, Milisavljevic N, Acheroy M. Multisensor data fusion for spaceborne and airborne reduction of mine suspected areas. International Journal of Advanced Robotic Systems. 2007;**4**(2):173-186. DOI: 10.5772/5698

[73] Lippmann R. An introduction to computing with neural nets. IEEE Signal Processing Magazine. 1987;**4**(1):4-22. DOI: 10.1109/MASSP.1987.1165576

[74] Duda RO, Hart PE, Stork DG. Pattern Classification. Wiley Interscience. 1987. p. 654. DOI: 10.1007/s00357-007-0015-9

Chapter 6

InSAR Coherence and Intensity Changes Detection

Damien Closson and Nada Milisavljevic

Additional information is available at the end of the chapter

http://dx.doi.org/10.5772/65779

Abstract

This research aims at differentiating human-induced effects over the landscape from the natural ones by exploiting a combination of amplitude and phase changes in satellite radar images. At a first step, ERS and Envisat data stacks are processed using COS software developed by the company SARMAP. Various features related to amplitude and phase as well as to their changes are then extracted from images of the same sensor. Combinations of the features extracted from one image, from several images of one sensor as well as from different sensors are performed to derive robust indicators of potential human-related changes. Finally, possibilities of exploiting and integrating other types of information sources such as various reports, maps, historical or agricultural data, etc. in the combination process are analyzed to improve the obtained results. The outcomes are used to evaluate the potential of this method applied to Sentinel-1 images.

Keywords: minefields, suspected hazardous areas, InSAR, coherence

1. Introduction

In postconflict regions, before the starting of demining operations, investigators have to carry out studies to delineate suspected hazardous areas (SHA). Many factors limit the reliability of their assessments: imprecision of the minefields records in military documents; inaccessibility and impassability of the field; lack of accidental records; inability to accurately position on a map the military units, their actions, confrontations, and the minefields' deployment; etc. Inaccuracies and imprecisions become even worse when dealing with unexploded ordnance (UXO). As a consequence, a significant difference exists between the surfaces to be cleared and the actually mined ones. This situation leads to a waste of time, work, and funds while donations are limited.

1.1. Problem statement

How to refine the delineation of SHA and thus reduce significantly the cost of clearing operations? When conflicts end, the economic activities restart for the benefit of the society but landmines and UXO disrupt this process. When activities take place in former battle zones, the daily works that shape the landscape such as plowing fields, pruning trees, grazing activities, and roads maintenance, etc., are becoming surviving "experiments" for both humans and animals. In regions affected by mines during a long period of time, inhabitants have progressively developed mental SHA maps based on their habituation of the danger.

When correctly located, the inventory of accidents and detonations are primary sources of information to delineate SHA. But the intersection of all inhabitants' mental SHA maps should provide another valuable independent source of information to highlight no man's lands that could be minefields.

Mental SHA maps are only partially exploited by the National Mines Action Centers in charge of the surveys because of the great difficulty to collect all the data. Nowadays, social media can improve the collection of pertinent information but only where the Internet is accessible.

Our research is dedicated to find out a way to use the surviving "experiments" for both humans and animals with radar remote sensing techniques. The working hypothesis/approach is based on the fact that in particular circumstances, and for a specific timespan, all the anthropogenic modifications of a given landscape can be accurately mapped from space-borne radar single look complex (SLC) images owing to the exploitation of the coherence information (e.g., [1] and references therein). The technique based on radar coherence images is known as coherent change detection (CCD).

1.2. Preliminary researches and results: setting up of a relationship between suspected hazardous areas boundaries and interferometric coherence

Since the launch of the first two European remote sensing (ERS) radar satellites in 1991 and 1995, the value of interfermetric synthetic aperture radar (InSAR) coherence images computed from the processing of two SAR scenes, have been recognized as a useful information source for environmental studies [2].

In SAR CCD techniques, the phases of two SAR images of the same scene are interfered. The images must be acquired in the same conditions, i.e., the same incidence angle but with a position slightly different (from a few meters to a few hundred meters). The coherence between two SAR images expresses the similarity of the radar reflection between them. Any changes in the complex reflectivity function of the scene are manifested as a decorrelation in the phase of the appropriate pixels between the two images. In this manner, even very subtle changes in the scene from one image to the next can be detected. In other words, in a given pixel, if the reflection and/or dielectric properties have changed between the two acquisitions, the coherence value of that target is reduced and the accuracy over the distance measurement between the antenna and the ground decreases.

This ability to detect and to localize very subtle changes can be used in demining operations to map human and animals activities during a specific time laps, e.g., 35 days for ERS and Envisat satellites. One of the challenges is to classify the observed changes into two main categories: pertinent or not for the delineation of minefields perimeter.

Whatever, the thematic concerned, since 1991; space-borne radar interferometry techniques were successfully applied all around the world. A large number of InSAR results were obtained over cities, and in warm or cold desert areas (e.g, Landers earthquake dislocation pattern; Shiraz glacier displacements field, etc.). This is due to the fact that such environments/landscapes present surface characteristics that match with the most important InSAR condition of application: the preservation of the phase coherence through time.

Pioneering researches conducted in semiarid to hyperarid regions [3–8] showed that radar coherence images were able to bring useful insights for geomorphology, geology, soil moisture, and erosion phenomena. Among them, a few suggested the unique function of coherence imagery as a tool for monitoring environmental impacts of human activities [6, 9–11].

In the 2000s, first experiments were conducted at the Space Centre of Liege, and then extended at the Signal and Image Centre (see **Figure 1** and **Figure 2**) demonstrated the CCD capability for discrimination between topsoil disturbed or not by anthropogenic/animal activities. The areas of interest were centered over the Dead Sea region.

Later on, based on the knowledge gained through projects financed by the Ministry of Defense, other experiments related to Surface to Air Missile sites and military airfields [12, 13] have shown that the CCD technique is able to develop a persistent picture of activities within an area of interest and allows the delineation of the system of fortification, even if hidden by camouflage. In the frame of security applications, CCD is particularly interesting in the surveillance (**Figure 3**) and the detection of intrusions.

CCD technique is less efficient in places where too many human activities occur, or in zones where the substratum is too hard to record detectable surface modifications (such as in urban areas). Seasonal variations can affect significantly the estimated coherence, both in terms of overall average value (weather dependence) and in localized natural effects (e.g., vegetation, steams, and water bodies with seasonal variations). Additional information has to be exploited to carefully analyze and understand the nature of the difference coherence losses.

1.3. First hint to delineate mental suspected hazardous areas

The very first observation of a minefield with CCD was derived from ERS spaceborne radar images processed in the frame of a study dedicated to active faulting in the Dead Sea region [14]. Interpretation of interferograms and coherence images led to the first delineation of a square kilometer-size minefield in the northern part of the terminal lake (**Figure 1**). The preservation of phase coherence through time in a zone that normally should have lost coherence provided the first hints to delineate "mental SHA".

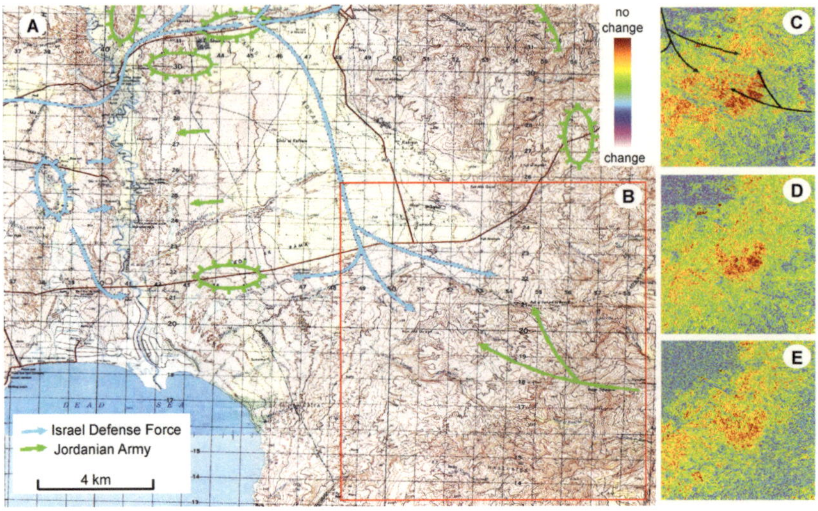

Figure 1. (A) Israel Defense Force and Jordanian army troops' movements during the battle of Al-Karamah 21 March 1968, background 1:50,000 map of Karama, Sheet 3153 IV, series K737 (USA). (B) Location of insets C, D, and E. (C) Interferometric coherence (pixels 20 m), 1995-07-30 and 1995-11-12, temporal baseline = 105 days, perpendicular baseline = 304 m, descending path. Black arrows correspond to Israeli and Jordanian troops' movements. (D) Interferometric coherence, 1995-07-30 and 1996-07-14, temporal baseline = 350 days, perpendicular baseline = 57 m, descending path. (E) Interferometric coherence, 1995-12-16 and 1999-05-29, temporal baseline = 1260 days, perpendicular baseline = 92 m, ascending path. The color palette ranges from change (loss of coherence or surface modifications) to no change.

Figure 1(A) describe some actions of the Al-Karamah battle fought on March 21, 1968, between the Israel Defense Forces and the Jordanian Army. Before and during the battle, minefields were laid and then progressively cleared in the next decades. Systematic demining operations took place only after the Oslo Accords in 1993–1994. Ten years later, at the Space Center of Liege, ERS satellite data acquired from 30/07/1995 to 29/05/1999 were processed to generate differential interferograms and their associated coherence images. Unexpected high coherence values were noticed on the foothills of the Dead Sea basin (**Figure 1B**), Jordanian side. In 2002, during a field survey, rusty barbed wires and warning signs revealed a former battlefield. Farmers and shepherds indicated that the plateau was cleared in 1999 in anticipation of the Pope John Paul II coming in the area (March 20–26, 2000). Indeed, to shorten the trip between Mount Nebo and the Baptism Center next to the Jordan River, the government of Jordan created one new road that had to cross the minefield. Based on these elements, it was deduced that the coherence preservation was a consequence of the absence/presence of grazing activities. Since the Al-Karamah battle, Bedouins had modified their itineraries to protect their flocks and themselves.

In **Figure 1(C)–(E)**, colored coherence images show the presence/absence of perturbation of ground scatterers between two acquisition dates. Coherence images having different time spans (C = 105, D = 350, E = 1260 days) do not display exactly the same amount of changes. Depending on the environmental conditions and land use dynamics, a kind of "decantation process" is necessary to distinguish permanent boundaries of untouched zones (e.g., comparison between **Figure 1C** and **E**).

http://dx.doi.org/10.5772/65779

Based on the previous reasoning, the linear minefield along the Jordanian-Syrian border was mapped with a pair of ERS images acquired 70 days apart (**Figure 2**). The borderline crosses a wide cropped and pastured plateau where coherence is lost rapidly. On the contrary, the no man's land between Syria and Jordan is not affected by human activities and appears highly coherent. Elsewhere over the scene, coherence is preserved on the rocky zones (**Figure 2**, clouds of brown spots). Coherence images do not map minefields but no man's lands in which landmines can be present (**Figure 2B–D**) taking into account the context.

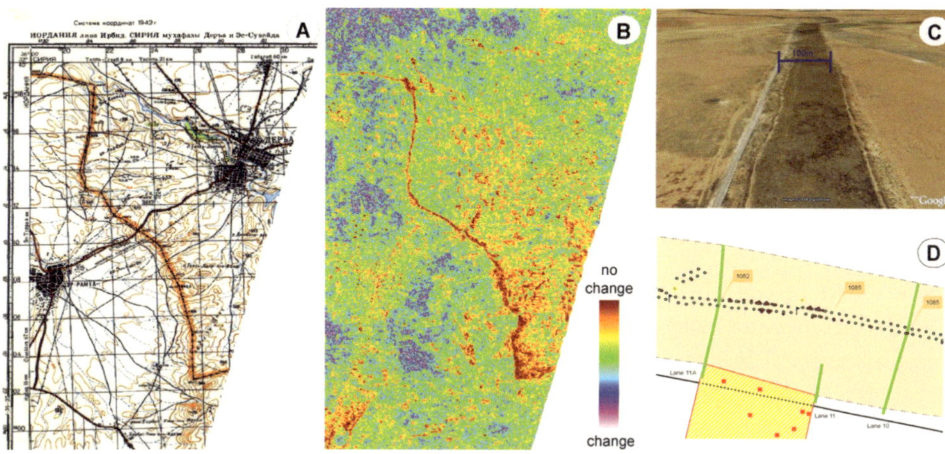

Figure 2. No man's land mapping with ERS coherence image, Jordan-Syria border. (A) 1:100,000 Sheet 9-37-121 (USSR) - graticule 2 km - Irbid, Jordan, is on the left. (B) Interferometric coherence (pixels 20 m), 1995-07-29 and 1995-10-07, temporal baseline = 70 days; perpendicular baseline = 148 m. (C) Birdseye view of the 100 m wide no man's land located over the border (© Google Earth). (D) mines' location into the no man's land (© Norwegian People Aid).

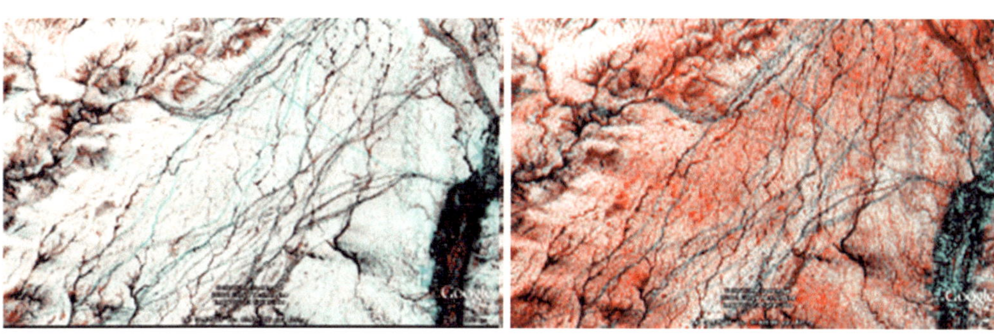

Figure 3. Color composites of a sequence of coherence images derived from Envisat data over an area in Afghanistan. The following coherence combinations are used: 1 – (2004.04.28 – 2004.08.11), 2 – (2004.08.11–2004.11.24), 3 – (2004.11.24–2005.03.09), 4 – (2005.03.09 – 2005.04.13). Left image: 1 – red, 2 – green, 3 – blue. Right image: 4 – red, 2 – green, 3 – blue. Images processed by SARMAP SA.

In 2011, in collaboration with the private company SARMAP, another test was realized in a desert plain, Kandahar region, Afghanistan. The objective concerned the capability of discrimination between ephemeral streams and traces of convoys from a coherence images stack. **Figure 3** compares two color composites created from a sequence of five Envisat ASAR scenes. Natural decorrelation effects can be identified like seasonal and permanent traffic roads as well as seasonal torrents and permanent rivers.

1.4. Purpose of this chapter

Figure 1–3 describe a CCD approach based on a dataset ranging from two to five SLC images. The environmental conditions of applicability are favorable. Considering these preliminary but promising results, a research project untitled SPaceborne Radar INterferometric Techniques for Humanitarian Demining Land Release (SPRINT) [15] was proposed to test and evaluate an approach for SHA area reduction in Croatia based on coherence and intensity images analysis and interpretation. The project was financed by the Belgium Scientific Policy (BELSPO). It received the support of the European Commission TIRAMISU project, and of the private company SARMAP that provided radar images from the European Space Agency and the outputs of the SBAS processing (see below) with their software Sarscape™.

The project idea consisted in processing ERS and Envisat stacked images to generate time series geocoded coherence and intensity data, and then fusing this multitemporal information to map changes in known SHA. It was expected that the types of change and their spatial extension could provide additional inputs in the Decision Support System to reduce the size of the SHA.

In this chapter the main outcomes of this research are presented. The following three scientific questions are addressed as well:

1. Is it useful for SHA area reduction to use InSAR stacking techniques to produce time series of intensity and coherence images of minefields (Gospić, Croatia)?

2. What kind of fusion analysis do we have to perform to extract relevant changes for area reduction?

3. Does it make sense to collect regularly Sentinel-1 radar images over a war zone or a postconflict region to exploit these data with stacking techniques in humanitarian demining?

2. Method

At a first step, C-band ERS and Envisat image stacks are processed using PS-SBAS (permanent scatterer and small baseline) tools developed by the Swiss company SARMAP (Sarscape™ software). Detailed information is provided below. In a next step, unsupervised classification based on the analysis of temporal signatures is performed within each of the stacks of ERS amplitude data, ERS coherence data, Envisat amplitude data, and Envisat coherence data, separately. After that, combination of the obtained results is performed. Validation process is discussed after the presentation of the results.

2.1. Change detection with spaceborne radar

The radar antenna on board satellites transmits a coherent signal that is backscattered by the Earth surface before being detected by the same antenna. A SAR signal is complex. Each pixel of an image holds intensity and phase information. The intensity depends on the ground roughness, the soil moisture, and the incidence angle. The phase information, expressed as an angle, depends on the optical path travelled by the radar wave along its round trip.

Change detection is an application for which SAR is particularly well suited as this type of sensors can consistently produce high-quality well-calibrated imagery with good geo-location accuracy. Two forms of change detection in repeat-pass SAR imagery may be considered, namely coherent and incoherent change detection:

1. Incoherent change detection (ICD) identifies changes in the mean backscatter power of two scenes by comparing sample estimates of the mean backscatter power taken from the repeat-pass image pair. Typically, the sample estimates are obtained by spatially averaging the image pixel intensities (amplitude squared) over local regions in the image pair. The mean backscatter power of a scene is determined by the structural and dielectric properties of the scene, and thus may be used to detect changes in soil or vegetation moisture content or surface roughness.

2. Coherent change detection (CCD) compares the phase of two images for inconsistencies. To detect whether or not a change has occurred, two images are taken of the same scene with a same incident angle, but at different times. These images are then geometrically registered so that the same target pixels in each image align. After the images are registered, they are checked pixel by pixel. Where a change has not occurred between the two images, the pixels remain the same. When the pixels are different, a change has occurred.

Both methods are useful for detecting changes, but they do not measure their magnitude. One of the major issues is defining what "relevant" changes are. By combining the results with ancillary data owing to data fusion techniques, it is possible to improve the detection quality and reliability as well as to differentiate well between changes. Therefore, the SPRINT project was centered over a sequence of activities dedicated:

1. To process and geocode large amount of space-borne radar data, originally, from ERS and Envisat (C-band).

2. To extract anomalies, land-cover types, etc., according to the information available.

3. To fuse, at different levels (feature and/or decision level), various pieces of information coming from one image (in order to improve the quality of the result) as well as change detection results obtained from different sources (images of one sensor or of different sensors).

4. To include existing contextual information, experts opinions, and other knowledge sources in the fusion process and to assess the influence of their inclusion on the final result.

5. To perform qualitative and quantitative assessments of the assets and drawbacks of each radar interferometric techniques as well as of each sensor combination, for dedicated tasks, in specific environment under surveillance, to support future land release actions.

2.2. InSAR processing

SAR repeat pass interferometry is based on two images acquisitions of the same area from slightly displaced orbits of the satellite. The ERS and Envisat repeat pass periods are 35 days but when ERS 1 and 2 were orbiting together (1995–1999), a place was revisited one day apart ("ERS tandem pairs").

During the processing, both images are coregistered ("slave" image over "master" one). In CCD, the phase differences are calculated leading to the so-called interferograms. These interferograms contain information related to local elevation, local displacement, and noise. This noise, or phase decorrelation, records the slight changes in scatterers distribution on the observed Earth surface between the two acquisitions.

There are three "interferometric products" from two SAR scenes. They can be displayed either in radar geometry (slant range) or geocoded:

1. An intensity image that helps to locate points of interest. The radar cross-section of a scatterer is calculated from the ratio of the power density received by the scatterer to the backscattered power density.

2. A phase difference image or interferogram with either only the orbital contribution removed (to show the topographic contribution) or with both orbital and topographic contributions removed (to show ground displacements and phase noise).

3. A coherence image that presents the confidence level of each pixel in the phase difference image.

Coherence (equation 1) refers to a fixed relationship between waves in a beam of electromagnetic radiation. Two wave trains are "coherent" when they are in phase. In SAR interferometry, coherence is used to describe systems that preserve the phase of the received signal. Coherence value can be estimated by means of the "local coherence" of an interferometric SAR image pair. The local coherence estimator is the cross-correlation coefficient of the couple estimated on a small window [16, 17]:

$$\gamma \triangleq \frac{\left|\sum_{i=1}^{L}\sum_{j=1}^{M}\mathrm{Im}_1(x_i,z_j)\,\mathrm{Im}_2^*(x_i,z_j)\right|}{\sqrt{\sum_{i=1}^{L}\sum_{j=1}^{M}\left|\mathrm{Im}_1(x_i,z_j)\right|^2\sum_{i=1}^{L}\sum_{j=1}^{M}\left|\mathrm{Im}_2(x_i,z_j)\right|^2}} \tag{1}$$

where Im_1 is the complex signal of the "master" image and Im^*_2 is the complex signal in the "slave" one. The coherence of a pixel is estimated by the means of kernels with $L \times M$ pixels. In an interferogram, the coherence ranges from 0.0 (total decorrelation, the interferometric phase is pure noise) to 1.0 (phase correlation is preserved). As a statistical value, it cannot provide quantitative measurements of the ground scatterers disturbances. However, a physical interpretation is that it represents the fraction of power scattered by unchanged parts of the scene [18]. The coherence image serves as a measure of the quality of an interferogram and gives information about the surface type (e.g., vegetated versus soil and rock) or shows when a tiny, otherwise invisible change has occurred in the image. Coherence imagery is able to detect centimeter scale changes in the scatterers distribution.

The coherence values are affected by:

1. The local slope (steep slopes oriented toward the satellite line of sight lead to low coherence).

2. The properties of the surface being imaged (e.g., vegetated or water surfaces have low coherence).

3. The time span between the passes in an interferogram. Long lags lead to higher variations in scatterers distribution and hence to lower coherence.

4. The baseline, which is the distance between the satellites positions. The perpendicular component is what matters for coherence purposes. Generally, the optimal baseline length is between 0 and 400 m (large baselines lead to low coherence).

5. The poor coregistration, i.e., the process of lining up two images, a "master" and "slave", covering the same area in a way that they fit exactly on top of each other.

6. The poor resampling leads to low coherence during the InSAR processing [19, 20].

2.3. Advanced InSAR processing

Interferometric stacking techniques emerged in the last decade as methods to obtain very precise measurements of terrain displacements, and in particular of subsidence phenomena. The so-called Persistent Scatterers (PS) [21] and Small BASeline (SBAS) [22] methods can be considered as the two most representative stacking approaches.

In both cases, the exploitation of 20 or more satellite saynthetic aperture radar (SAR) acquisitions obtained from the same satellite sensor with similar geometries on the area of interest allows to measure displacements with an accuracy in the order of few mm/year, and to derive the full location history of " coherent pixels with an accuracy of 1 cm or better for every available date.

The "area-based SBAS-like" tool of Sarscape™ Interferometric Stacking Module is dedicated to compute mm-scale displacements on distributed targets. Geocoded coherence and intensity images are side-products. Therefore, the procedure stopped once the information needed was available.

The first step consisted in the connection graph generation to define the ensemble of pairs to process. These pairs were displayed as connections in a network that links each acquisition to others. Given N acquisitions, the number of maximum theoretical available connections is

$(N^*(N - 1))/2$. The connection graph generation tool permitted to choose the most *a priori* reliable connections.

One acquisition considered as Super Master was automatically chosen among the input acquisitions. The Super Master is the reference image of the whole processing and all the processed slant range pairs were coregistered on this reference geometry.

Then, from these pairs, the "Interferometric Workflow" and the "Refinement and Re-Flattening" tools generated a time series of geocoded coherence and intensity images.

3. Material: training areas and image datasets

In 2008–2010, the Croatian Mine Action Center (CROMAC) successfully tested an Advanced Intelligence-Decision Support System (AI-DSS) over three communities (Gospić, Bilje, and Drniš) to improve SHA circumscription process [23]. In SPRINT, the documented Gospić area was used to test and evaluate a contribution based on the processing of ERS and Envisat satellite radar images time series.

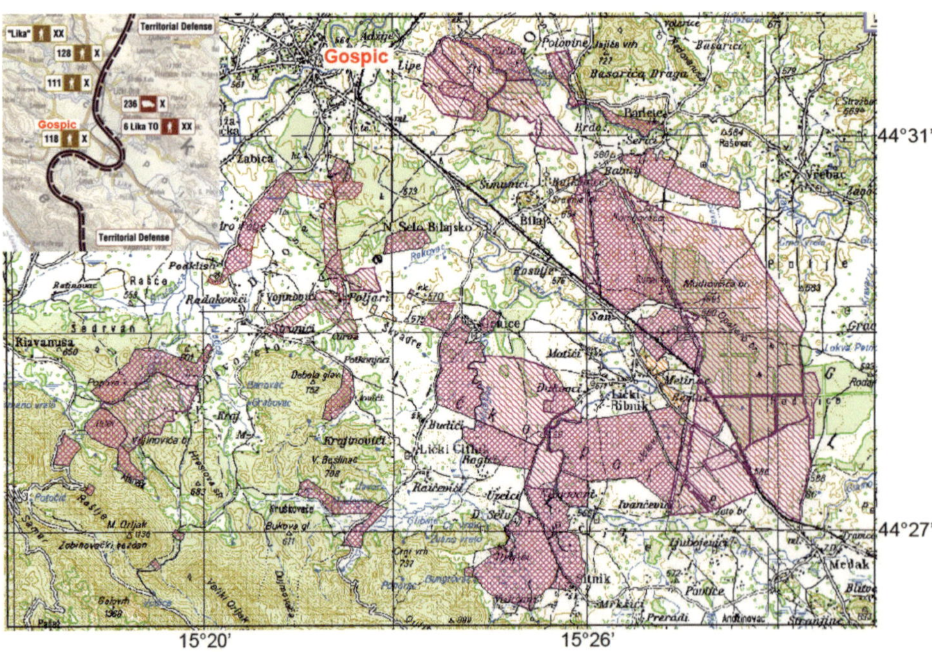

Figure 4. Several battles took place in the Gospić salients from 1991 to 1995 (inset). Minefields are located in a bulge of the front line. Suspected hazardous and confirmed hazardous areas are shown respectively in hatched and crosshatched purple polygons (2009 situation) over a 1:50,000 topographic map, graticule 4 km. Source: Milan Bajic.

The plain of Gospić (**Figure 4**) is characterized by environmental conditions less favorable than those of Jordan or Afghanistan (**Figures 1–3**) because of the forest covering the valley sides (volume decorrelation), the escarpment of the valley sides (shadows; layover), and the climatic conditions (moisture decorrelation).

Apart from these drawbacks, SPRINT demonstrated the possibility to process data and get valuable results for the analysts. The well-documented minefields allowed the study to be carried out in three main steps, without expensive and time consuming field surveys: data acquisition and processing, data fusion, and validation inside a geographical information system.

In the geographical information system ArcGIS™, the geocoded coherence and intensity images (100 ×100 km) have been clipped with confirmed hazardous areas displayed in **Figure 4** (crosshatched). A 300 m wide buffer was also created to compare the results in the minefields with the ones in the immediate vicinity.

Two major datasets were created from radar images (**Table 1**): "minefields" and "buffers". For each one, ERS and Envisat intensity and coherence were computed. SARMAP collected radar images from the archives of the European Space Agency. Sarscape™ Focusing Module generated single look complex images based on a ω-κ frequency domain algorithm. Two stacks of 42 ERS SLC images (1992/08/01–2009/06/02) and of 40 Envisat scenes (2002/11/07–2010/10/21) were the inputs of SBAS processing.

Satellite images track 315 – frame 873 (ascending)	42 ERS intensity	112 ERS coherence	40 Envisat intensity	127 Envisat coherence
26 minefields	1092 subsets	2912 subsets	1040 subsets	3302 subsets
26 buffer zones	1092 subsets	2912 subsets	1040 subsets	3302 subsets
The number of coherence image is higher than the one of the intensity because SBAS process multiple pairs of images				

Table 1. Image dataset processed and analyzed in SPRINT.

Although both radar sensors are using C-band, the interferometric combination of Envisat and ERS data was not performed because the interferometric phase is strongly dependent on the radar frequency. The radar center frequency of Envisat (5.331 GHz) is slightly different compared to the ERS sensor (5.300 GHz). The difference of 31 MHz prevented the simple combination.

4. Results

Figure 5 locates 26 polygons corresponding to the confirmed hazardous areas over a visible satellite image (ArcGIS "world image" layer) to illustrate the land use/land cover. Most of the polygons are located in a plain extending ESE from Gospić. Agricultural parcels, pasture lands, and a few woodlots cover most of the flatten zones. The southern part is hilly and forested.

Figure 5. Location of the confirmed hazardous areas in the plain of Gospić. The minefield in light blue color is used to illustrate the whole process from the intensity and coherence images to the indicators of potential human-related change map.

In terms of land use/land cover dynamics, the comparison between Landsat images acquired from 1993 to 2015 indicated that the landscape is very stable and that most minefields are either covered by woodlots or pasture lands. A very little number presented evidence of agricultural activities.

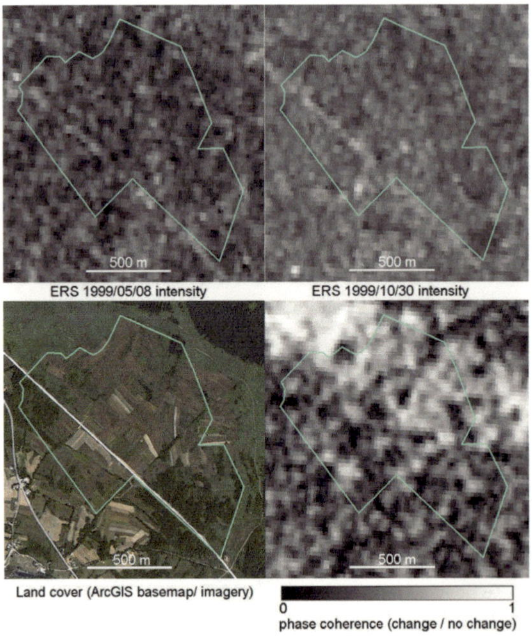

Figure 6. Comparison between intensity and coherence data with the visible information. Radar images have been processed with a pixel size of 25 m.

The wide polygon in **Figure 6** is an illustrative case study. This minefield is almost flat and covered by woods, pastures, and crops. A main road also crosses the area. The landscape displayed in radar image is very different than the one obtained in the visible wavelength because the physical mechanism used to create those scenes is completely different. Radar and visible data are independent, reliable, and complementary sources of information. However, at a first glance, radar data provide no information directly useful in demining operations. In the intensity images the main road crossing the polygon can be distinguished but its width is overestimated. The limits between woods, crops, and pastures cannot be observed. Coherence displayed in gray-scale provides information on the ground scatterers' perturbations between two dates (1999/08/05–1999/10/30). However, radar intensity and phase provide information complementary to the visible ones about the subtle changes of the ground surface.

4.1. Natural decorrelation revealed by ERS tandem pairs

As previously mentioned, slopes, forests, and high moisture zones are able to decorrelate the signal in less than one day. Hence, in those zones, it is clearly impossible to work with radar time series. **Figure 7** illustrates this important element for the approach to delineate SHA perimeters. The inset on the right side shows the land cover of the minefield during the period of war (1991–1995). Green color indicates wooden areas. The main road crossing the zone already existed and appears as a NW-SE linear light grey color feature.

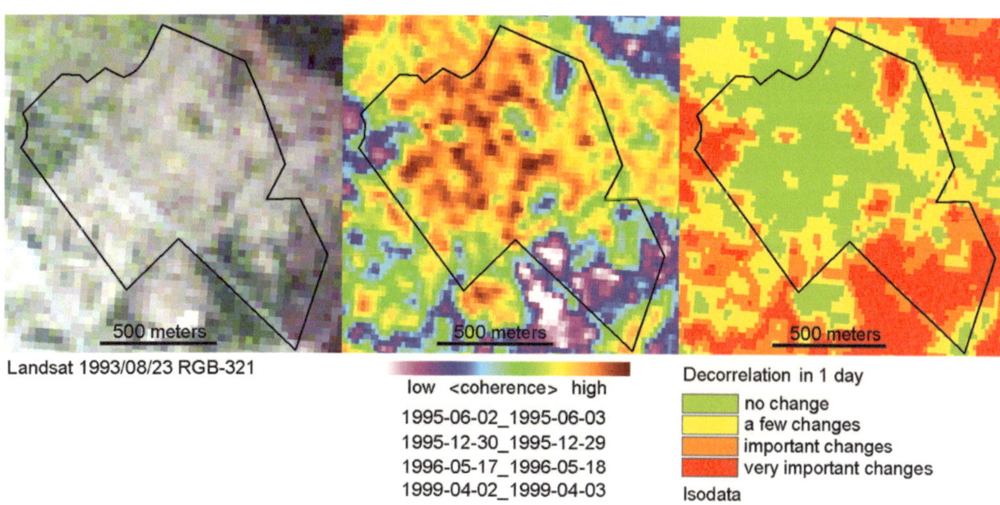

Figure 7. Evaluation of the decorrelation phenomena from tandem pairs (temporal baseline = 1 day). Green and yellow areas allow performing time series analysis. The orange red areas loose very quickly coherence. No information can be retrieved.

The middle inset is the average coherence between the four tandem pairs. The coherence is generally well preserved everywhere in the polygon at the noticeable exception of the wooden parcels. The left inset is a classification of the previous data in four classes (unsupervised). The green color shows where further studies can be conducted while the red color indicates that no information can be retrieved. Orange and yellow colors are areas where studies can be conducted but the environmental conditions are less favorable. In yellow spaces there are still certain levels of confidence that can be obtained.

4.2. Data fusion first level

Elementary pieces of information (intensity and coherence), such as the ones displayed in **Figure 6**, become relevant once exploited through stacks of data. Four data sources have been used (**Table 2**): ERS coherence, Envisat coherence, ERS intensity, and Envisat intensity.

Satellite images track 315 – frame 873 (ascending)	112 ERS coherence	127 Envisat coherence	42 ERS intensity	40 Envisat intensity
Figure 6′ Minefield	112 subsets	127 subsets	42 subsets	40 subsets
Unsupervised classification				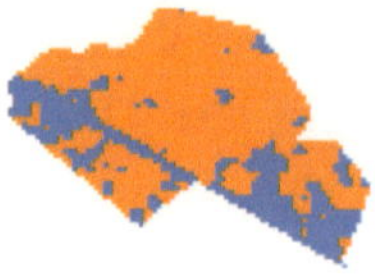

Table 2. Unsupervised classification results.

Data within each of these stacks have been combined and classified using multitemporal signature analysis, where pixels with similar signatures are grouped together. The output is an unsupervised classification of ERS coherence, ERS intensity, Envisat coherence, and Envisat intensity. These classification results are inputs for the second level.

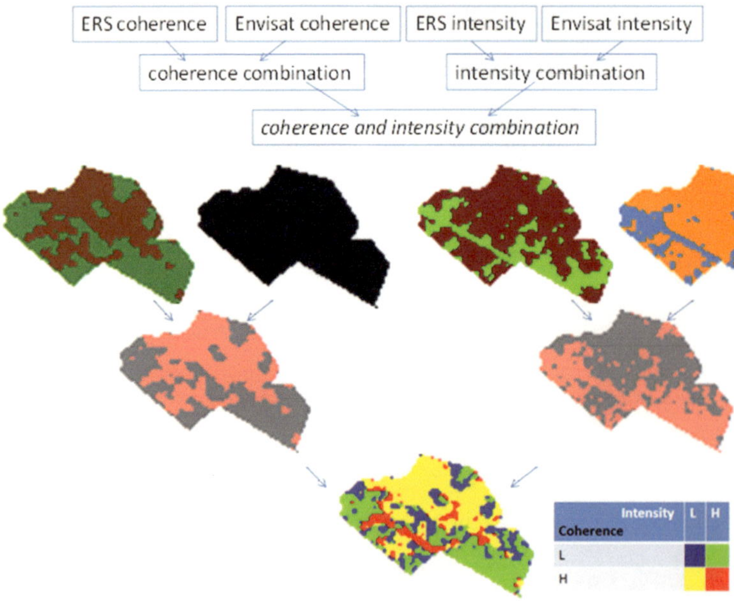

Figure 8. Example of combination, coherence classification results of Envisat and of ERS and intensity classification results of Envisat and of ERS. Then, fusing the two intermediate fusion results to create a map of indicators of potential human-related changes in the landscape. L = low; H = high.

4.3. Data fusion second level

The combination is performed using different strategies:

1. Fusing firstly coherence classification results of Envisat and of ERS, and intensity classification results of Envisat and of ERS. Then, fusing the two intermediate fusion results (**Figure 8**).

2. Fusing first the two ERS results (coherence classification and amplitude classification), and the two Envisat results (coherence classification and amplitude classification), and then fusing the two intermediate results (**Figure 9**).

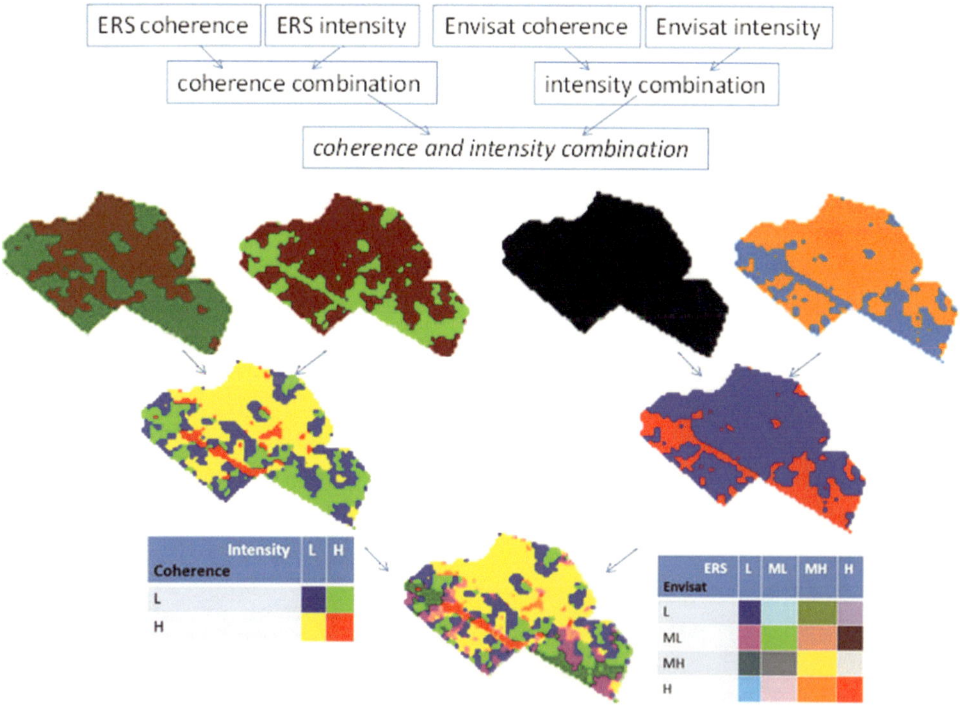

Figure 9. Example of combination, coherence classification results and intensity classification results of Envisat of ERS are combined. Then, fusing the two intermediate fusion results to create a map of indicators of potential human-related changes in the landscape. L = low; ML mean low; MH mean high; H = high.

The two strategies should provide the same final result only if the combination operators are commutative and associative. It is not the case here since different reasoning is applied in function of the data to be fused. The fusion order matters and influences the final result. The two strategies are developed and tested in parallel in order to cover various situations that might occur in reality (only one sensor used; only amplitude or coherence data available, etc.).

5. Discussion

5.1. Performance evaluation

Taking into account that humanitarian demining is a very sensitive matter, it was not easy to validate on the ground, especially, in the case of a demonstration project due to the lack of time, as well as human and material resources. To validate the results, alternative solutions were taken. One of the validation sources provided by CROMAC was photos as well as various indicators of mine presence and mine absence.

– Indicators of mine presence: destroyed houses, abandoned roads and areas, bunker lines, trenches, cover for infantry and artillery, shallow drafts, bridges.

– Indicators of mine absence: safe roads, houses in use, roads in use, and areas in use.

Although these pieces of knowledge were very informative, they were not geocoded, so it was not possible to use them directly for validation.

Due to their sensitive content, CROMAC could not provide geocoded validation data, so this type of validation had to be left to them.

Thus, the only means to validate the results in SPRINT was to compare them with existing (published) aerial data.

Figure 10 shows a superimposition of ERS classification results over aerial photo of the region of interest. Zones with trees are clearly classified as a class labeled by green color. Agricultural zones are well classified in the class labeled yellow. Roads are in red, and the blue color seems to be the zone around forests and trees. There are various possibilities to explain why this class occurred as a separate class in our classification. For example, it could be due to the acquisition mode itself (radar shadows), or due to works in the bordering zones around trees. A further investigation in this direction, involving potentially the terrain observation, would be necessary.

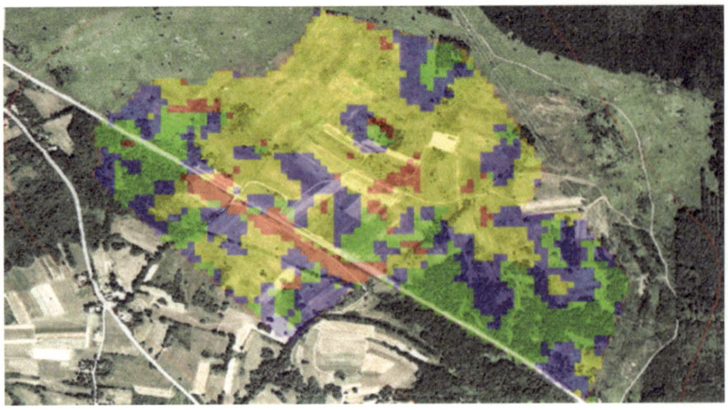

Figure 10. Classification of ERS data over an aerial photograph.

http://dx.doi.org/10.5772/65779

In **Figure 11**, for an easy visual comparison, an aerial image of one of the minefields is given (top left), together with the danger map obtained from CROMAC (top right) and the two classification results at the bottom (ERS left, Envisat right). In the danger map, the level of red indicates the level of danger (more dangerous zones are represented with darker red). Green zones are surfaces in use, while purple corresponds to the roads in use.

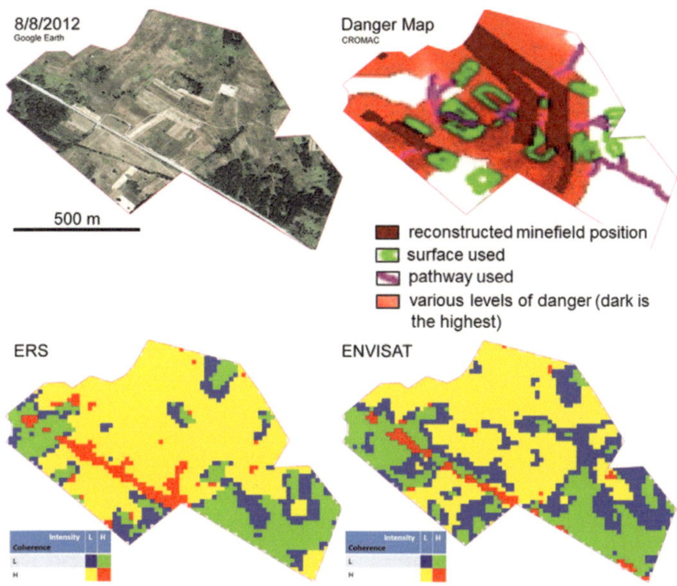

Figure 11. Top: aerial image of one of the regions (left); danger map obtained from CROMAC (right); the color code is: brown = reconstructed minefield position; green = surface used; purple = pathway used; red = various levels of danger (dark is the highest); bottom: ERS classification result (left), Envisat classification result (right).

Based on a visual comparison with the classification results, we can see that fields in use belong to the radar yellow class. Following that logic, we might conclude that the whole yellow classified zone could be suggested to CROMAC as mine-free (in use).

It should be noted that the classification results should be treated as a layer in creating/modifying CROMAC database, and not as a self-sufficient decision-making information, since in creating danger maps, other sources of information should be taken into account as well, that we have not used in our classification, such as: mine laying records, mine accidents, mine incidents, history of battles, etc.

As long as the SPRINT project is concerned, the fact that obtained results are able to distinguish areas in use from forests and other zones and that there is a correspondence with danger maps proves that the method developed in the project is promising.

Taking into account that the terrain itself is not ideal from the point of view of SAR sensors, we might expect even better results for some other areas, where terrain conditions are more beneficial for SAR acquisition (flat zones, not so many trees, etc.—see **Figures 1–3**).

5.2. Research outcomes

When the environmental conditions of applicability are met, to a certain extent, it is possible to apply CCD for the improvement of the SHA delineation (see **Figures 1** and **2**).

CCD techniques present limitations related to the radar interferometric methods. Based on results published in the past 20 years (e.g., [4, 18, 20]), at least nine major factors limit the applicability or the effectiveness:

1. Coherence is a mathematical estimation computed usually with 3 ×3, 5×5, or 7 × 7 windows. The SHA limits are thus fuzzy.

2. Coherence decreases when the perpendicular baseline between acquisitions increases. The use of radar images pairs characterized by short baselines should be favored.

3. Coherence decreases with the angle of local slopes. The interpretation should be confined to flat and undulated surfaces.

4. Coherence is low over vegetation cover (volume scattering) in X-, C-, and L-bands. The approach provides best results in open spaces such as pastured or cropped areas (**Figure 2**).

5. Coherence varies with soil moisture content. Semiarid and arid zones like Iraq or Libya should be preferred.

6. The efficiency of the approach depends on the land-use impact on the ground surface (displacement of the ground scatterers). In desert areas, lot of time can be needed to observe a relevant contrast because of the lack of activities. Coherence remains high most of the time. On the contrary, in busy places, it is possible to detect areas of interest with short temporal baselines (e.g., 35 days for ERS).

7. The ground surfaces must be covered with scatterers able to move when an activity occurs. Hard rock surfaces preserve coherence even when activities occurred. No signal is recorded.

8. Pixel resolution: the smaller the pixel, the longer the coherence is preserved. Indeed, small pixels gather less scatterers than wide pixels. In a small pixel, it is less possible to record an event than in a wider pixel.

9. In time series analysis, the variation of the meteorological conditions through year can affect the efficiency. Regions with relatively stable meteorological conditions will be favored.

Figure 12 presents a qualitative assessment of the approach based on a set of fundamental parameters: human activities, slopes, vegetation cover, and precipitation (climate variable). Up to now, the best results have been obtained in places characterized by human activities modifying the surface the landscape, flat to very gentle slopes, relatively poor vegetation cover, and climate relatively dry to extra dry.

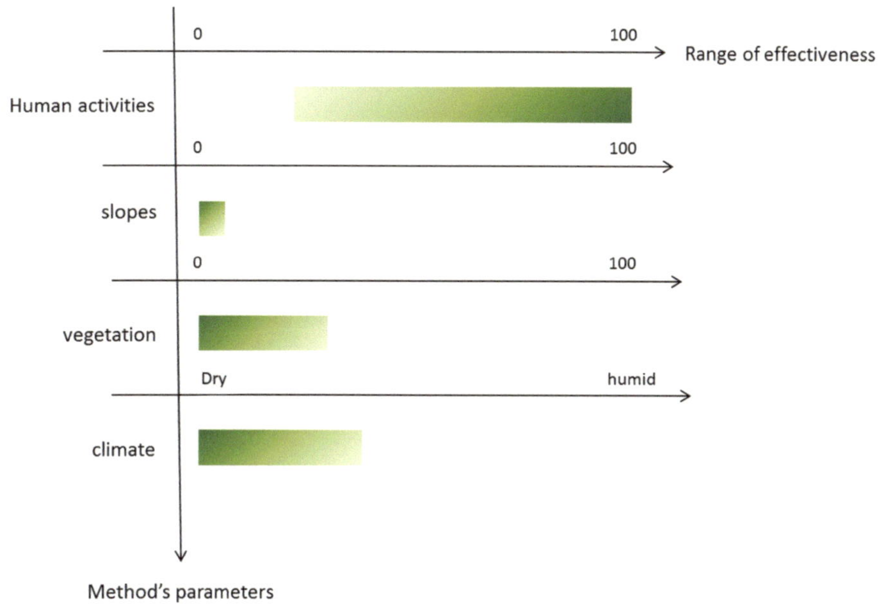

Figure 12. Qualitative assessment of the approach. Effectiveness of the method for C-band datasets. Based on Gospic (Croatia) and Jordan experiments.

6. Conclusion

Is it useful for SHA area reduction to use InSAR stacking techniques to produce time series of intensity and coherence images of minefields (Gospić, Croatia)?

SPRINT demonstrated that valuable information for the mines action centers can be provided through the processing of radar image stacks acquired during and after a conflict. SPRINT used successfully the COS Sarscape™ software and the objectives were achieved. The outputs of the interferometric processing have been successfully manipulated in ArcGIS to generate subset images useful for the fusion analysis step. All outputs, whatever the level in the processing chain, were in a standard format compatible with any types of (open) integrated geospatial systems.

What kind of fusion analysis do we have to perform to extract relevant changes for area reduction?

Unsupervised classification based on the analysis of temporal signatures had been performed within each of the stacks of ERS amplitude data, ERS coherence data, Envisat amplitude data and Envisat coherence data, separately. After that, combination of the obtained results have been performed.

Does it make sense to collect regularly Sentinel-1 radar images over a war zone or a postconflict region to exploit these data with stacking techniques in humanitarian demining?

The answer is positive. Since 2015, a large part of the world is covered by Sentinel-1 SAR (C-band) images. They are for free by the EU-ESA owing to the Copernicus project. The resolution is better than ERS and Envisat (5 vs. 20 m); the revisit time 11 days (and then 6 days when 2 satellites) versus 35 days.

Acknowledgements

This research was funded by the FP7 TIRAMISU project. In particular we wish to thank Dr. Vinciane Lacroix and Prof. Yvan Baudoin for their support. Thanks also are due to CROMAC for information exchanges and the advices of Prof. Milan Bajic. Finally, we thank Paolo Pasquali and Paolo Riccardi of SARMAP for the processing of SAR data.

Author details

Damien Closson* and Nada Milisavljevic

*Address all correspondence to: damien.closson@yahoo.fr

Eurosense Belfotop, Wemmel, Belgium; European Commission, Brussels, Belgium

References

[1] Bouaraba A, Belhadj-Aissa A, Closson D. Man-made change detection using high-resolution Cosmo-SkyMed SAR Interferometry. Arabian Journal for Science and Engineering January 2016; 41(1) 201–208 DOI: 10.1007/s13369-015-1736-4 .

[2] Massonnet D, Feigl K L. Radar interferometry and its application to changes in the Earth's surface. Reviews of Geophysics 1998; 36(4) 441–500.

[3] Ichoku C, Karnieli A, Arkin Y, Chorowicz J, Fleury T, Rudant J P. Exploring the utility potential of SAR interferometric coherence images. International Journal of Remote Sensing 1998; 19(6) 1147–1160.

[4] Liu J G, Capes R, Haynes M, Moore J McM. ERS SAR multi-temporal coherence image as a tool for sand desert study (dune movement, sand encroachment and erosion). In: Conference Proceedings of the 12th International Conference and Workshop on Applied Geologic Remote Sensing, Denver, Colorado, November 1997, 17–19. (Ann Arbor: ERIM), I-478–485.

[5] Lee H, Liu J G. Spatial decorrelation due to the topography in the interferometric SAR coherence image. In: Conference Proceedings of the International Geoscience and

Remote Sensing Symposium (IGARSS'99), Hamburg, Germany, 28 June–3 July 1999 (Piscataway, NJ: IEEE), 1, 485–487.

[6] Liu J G, Black A, Lee H, Hanaizumi H, Moore J McM. Land surface change detection in a desert area in Algeria using multi-temporal ERS SAR coherence images. International Journal of Remote Sensing 2001; 22(13) 2463–2477.

[7] Manspeizer N, Karnieli A, Arkin Y, Chorowicz J. Analyzing potential cliff erosivity from ERS-SAR satellite imagery. International Journal of Remote Sensing 2001; 22(5) 807–817.

[8] Wegmuller U, Strozzi T, Farr T, Werner C L. Arid land surface characterization with repeat-pass SAR interferometry. IEEE Transactions on Geoscience and Remote Sensing 2000; 38(2) 776–781.

[9] Usai S, Klees R. SAR interferometry on a very long time scale: A study of the interferometric characteristics of man-made features. IEEE Transactions on Geoscience and Remote Sensing 1999; 37(4) 2118–2123.

[10] Usai S. An analysis of the interferometric characteristics of anthropogenic features. IEEE Transactions on Geoscience and Remote Sensing 2000; 38(3) 1491–1497.

[11] Weydahl D J. Analysis of ERS SAR coherence images acquired over vegetated areas and urban features. International Journal of Remote Sensing 2001; 22(14) 2811–2830.

[12] Milisavljević N, Closson D, Bloch I. Detecting human-induced scene changes using coherent change detection in SAR images. In: Proceedings of ISPRS Commission VII Symposium, vol. XXXVIII-7B, pp. 389–394, Vienna, Austria; 2010.

[13] Milisavljević N, Closson D, Bloch I. Detecting potential human activities using coherent change detection. In: Proceedings of the 2nd International Conference on Image Processing Theory Tools and Applications (IPTA), pp. 482–485, Paris, France; 2010.

[14] Closson D. Cooccurrence between the geo-hazards induced by the dead sea level lowering and the geological setting—Lisan Peninsula, Lynch Strait, Ghor Al Haditha, Jordan, University of Liege; 2005.

[15] Milisavljević N, Closson D. Spaceborne Radar Interferometric Techniques for Humanitarian Demining and Land Release. Final Report (unpublished), Belgian Science Policy (Belspo). STEREO II, project SR/20/165 – 30 SPRINT; 41 pages; 2013.

[16] Derauw D. Phase unwrapping using coherence measurements. In: Proceedings of SPIE 2584, Synthetic Aperture Radar and Passive Microwave Sensing, Paris, 21 November, 1995; DOI: 10.1117/12.227141

[17] Seymour M S, Cumming I G. Maximum likelihood estimation for SAR interferometry. Geoscience and Remote Sensing Symposium, Pasadena, California, 1994. IGARSS '94. Surface and Atmospheric Remote Sensing: Technologies, Data Analysis and Interpretation. DOI:10.1109/IGARSS.1994.399711

[18] Bouaraba A. Coherent change detection using high resolution SAR images. PhD thesis, Ecole Militaire Polytechnique/Ecole Royale Militaire; 2014.

[19] Rosen PA, Hensley S, Joughin IR, Li F, Madsen SN, et al. Synthetic aperture radar interferometry. Proceedings of the IEEE 2000; 88(3) 333–382.

[20] Zebker H A, Villasenor J. Decorrelation in interferometric radar echoes. IEEE Transactions on Geoscience and Remote Sensing 1992; 30(5) 950–959.

[21] Ferretti A, Prati C, Rocca F. Permanent scatterers in SAR interferometry. IEEE Transactions on Geoscience and Remote Sensing 2001; 39(1) 8–20.

[22] Berardino P, Fornaro G, Lanari R, Sansosti E. A new algorithm for surface deformation monitoring based on small baseline differential SAR interferograms. IEEE Transactions on Geoscience and Remote Sensing 2002; 40(11) 2375–2383.

[23] Bajić M, Turšič R. Operations with advanced intelligence decision support system for mine suspected area assessment in Croatia and Bosnia and Herzegovina. UNMAS/GICHD Mine Action Technology Workshop; 2010. http://www.gichd.org/fileadmin/GICHD/what-we-do/events/Technology-Workshop-2010/C-6Sept2010-SMART-TechWS.pdf

Chapter 7

PARADIS: Information Management for Mine Action

Vinciane Lacroix

Additional information is available at the end of the chapter

http://dx.doi.org/10.5772/65781

"Science is built up of facts, as a house is with stones. But a collection of facts is no more a science than a heap of stones is a house"

Jules Henri Poincaré

Abstract

The use of information management can increase mine action safety and efficiency. Indeed, mine action includes the collection of a large amount of data coming from different sources and data needed for various processing flows. An information management system enables the manager to monitor and visualise critical variables, to get an overview of the situation, to produce customised reports and to perform geospatial analysis in order to make the best decisions. For the field workers, the system offers secure data collection. A Prototype for Assisting Rational Activities in Demining using Images from Satellites (PARADIS) designed and developed between 1999 and 2007, was an early example of how such a tool may improve the planning of humanitarian demining campaigns. The prototype was designed for the SEDEE-DOVO, the Mine Clearance Service [explosive ordnance disposal (EOD) Battalion] of the Belgian Army. This tool involves a software package working from the country scale to the field scale, software embedded in a geographical information system (GIS). The system was built on two user interfaces, one for the desk office, the other, mobile, for the field. The proof-of-concept of the tool aimed at encouraging similar ideas to help both mine action centres and the various actors involved in mine action in the realisation and coordination of day-to-day activities.

Keywords: information management, GIS, remote sensing, user interface

1. Introduction

1.1. Motivation and overview

Which areas are to be cleared in priority? This question is a key issue in humanitarian demining and, so is the planning of demining campaigns. The benefits of a demining campaign are not assessed by a mere computation of the number of mines removed but rather by the socio-economic impact of the clearing. Before 2011, International Mine Action Standards (IMAS) referred to as "Landmine Impact Survey" (LIS); involved teams that worked with the local population to evaluate how landmines and unexploded ordnance (UXO) were affecting their daily lives. Due to their inefficiency and high cost–roughly one–half of the areas suspected by LIS were not mine nor UXO-affected [1]–the term got a bad connotation, so the process was changed but the aim of measuring the impact remained. In order to estimate the socio-economic benefits of mine action, a huge amount of data need to be collected and processed. These data should be efficiently managed in order to guarantee data integrity and to avoid data loss–in the past, demining teams were sent to already cleared areas because of data loss. The data forms to be filled and their format should be explicit enough to be easily understood by the different types of users and should be compliant with tools that enable analysts to make pertinent decisions on their basis. Most of these data are geospatial. As demining campaigns generally take place in developing countries, and recent and accurate maps of the contaminated region are often missing, we suggested to use satellite data instead.

At the beginning of the project, there was a lack of tools to follow the course of campaign, browsing through the identified scales of work, from global (country) scale to local (field) scale, adapted to the users' specific needs. These requirements were the development guidelines for the Prototype for Assisting Rational Activities in Demining using Images from Satellites (PARADIS) system.

PARADIS was developed in three phases. During the initial phase (1999–2001), it was developed in close collaboration with ULB-IGEAT, focusing on the analysis of the needs in spatial information and their relevant scales, and on the design of a geographical information system (GIS) interface; results were published in Refs. [2, 3].

During the second phase (2002–2006), the interface was split into a "planning interface" and a "field interface"; the first was storing all data in a central database GeoDb (safe data storage) while the second was running on a mobile device, a Personal Digital Assistant (PDA), for secure data collection as illustrated on **Figure 1**. The efforts of the second phase were dedicated to the development of GIS tools for the two interfaces and constantly adapting them to the needs of the SEDEE-DOVO, the mine clearance service of the Belgian Army. The results were published in Refs. [4–7].

Finally the last phase (2007) aimed at adapting the tool to the needs of the mine action activities of APOPO, a non-governmental organisation that researches, develops and implements detection rats technology for humanitarian purposes, in particular for mine action.

http://dx.doi.org/10.5772/65781

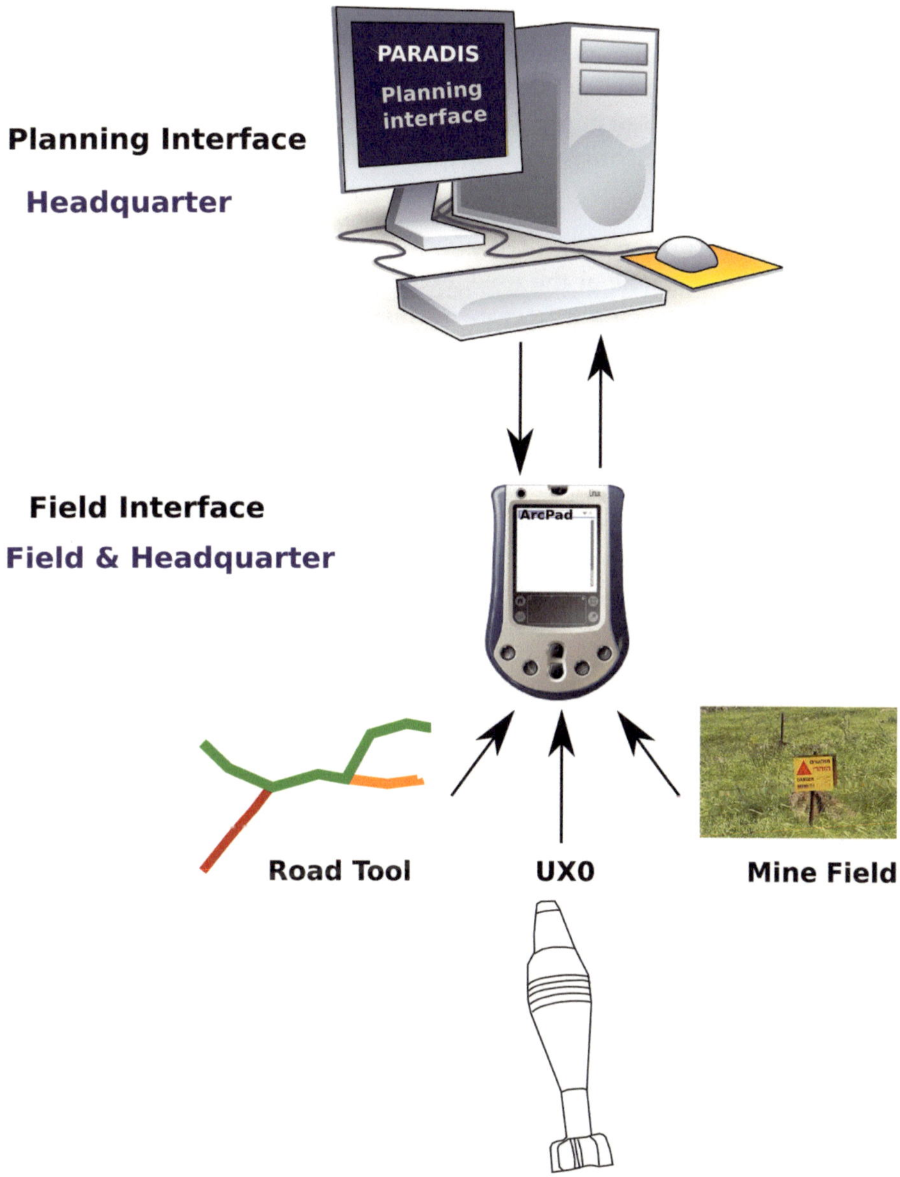

Figure 1. PARADIS "planning" and "field" interfaces.

1.2. PARADIS early dream

The early dream was to build an intelligent system based on a blackboard architecture linked to a GIS. This system would enable the user to view satellite images and extract information at various scales at run time, thanks to image processing routines called by the blackboard. On the finer scale, we were expecting to detect minefield indicators.

The early dream was soon followed by an early disappointment: after the first mission in Mozambique, it turned out that no minefield indicator could be found, except for a minefield near Songo, which could be identified by its specific texture on SPOT images. However, Songo minefield was an atypical minefield everybody knew about, and its texture could not even be considered as a representative of a category of minefield.

The following projects (SMART, TIRAMISU, see Chapter 5) showed that the requested resolution to find an indicator for presence or absence of mine required finer resolution than the satellite data available at the time, except if relief is used, but that would have required multi-view satellite data. Therefore the image processing work was limited to the extraction of objects of interest for the deminers such as roads, rivers, land cover, buildings, etc., objects that were in general not available on the existing maps, or at least, not with a sufficient precision. These techniques, enhanced during the following research projects, are detailed in Chapter 5. We also realised that an intelligent system that would perform semi-automatic image analysis on the spot was not what the users needed. Indeed, a mission should be prepared in advance and, on the other hand, additional tools for reporting were asked for. Moreover, it appeared that a specialised partner would be needed in operational mode: the work of choosing the appropriate satellite, consulting the archives, selecting the appropriate season, programming the satellite, elaborating a budget, interpreting the images, etc., were all tasks performed by the PARADIS team during the project, tasks that should not be assigned to deminers nor to mine action managers. This led us to the elaboration of a method that gives these tasks to a specialised team. The security service section imagery (SGR-IM), a specialised unit of the Belgian Ministry of Defence, played that role in PARADIS. More than 10 years later, the TIRAMISU project is still advocating the same philosophy [8].

2. GIS and remote sensing for mine action

From a review of existing GIS tools and remote sensing for humanitarian demining made in 2000 [2] including (1) the information management system for mine action (IMSMA), initiated at the Geneva International Centre for Humanitarian Demining, (2) the project developed at James Madison University, (3) minedemon, elaborated at ITC, (4) the digital mine documentation system prototype, undertaken by the Business for Industry Concerns (IABG), (5) "FOCUS HD", a system proposed by Landair International Ltd., (6) the mapping information system designed by UK's Defence evaluation and research agency and (7) some studies using remote sensing data in order to detect mine fields [9], we concluded the following:.

- The benefit of airborne or satellite images to help the detection of hidden mines and general mine fields had not yet been demonstrated.
- None of these tools was designed keeping in mind the tasks assigned to deminers during a mission.

http://dx.doi.org/10.5772/65781

- IMSMA, despite its limited spatial capability at the time, was becoming the UN standard for collecting and managing minefield data; during those early days, the tool was mainly dedicated to information centralisation.

Therefore, we decided to build a tool based on IMSMA and ArcView in order to plan the humanitarian demining campaign, following the tasks of deminers during a mission. PARADIS and IMSMA would exchange data using the emerging standard protocol mine action XML (maXML).

At the time of PARADIS, "mine clearance standards" implied a sequential Level 1, Level 2 and Level 3 process (see Chapter 1, **Table 1**) while a SEDEE-DOVO humanitarian demining mission involved:

- The gathering of general data at mission announcement.
- A Field Survey to get approximate mine field locations.
- The planning of demining campaigns.
- In-field data collection devoted to minefields and UXOs, in order to better locate them, understand the contamination and prepare clearance.
- The clearance of the minefields and the removal of UXOs.
- A quality evaluation of the work performed.

Tasks to perform	**Global scale (1:1,000,000–1:250,000)**	**Region scale (1:250,000–1:50,000)**	**Field scale (1:10,000–1:1500)**	**Advancement scale (1:500)**
(1) Collecting data needed for general survey: climate, relief, roads, inhabited areas, etc.	Mission announcement			
(2) Collecting information not available at mission announcement: locations of minefields, accidents, populations, etc.			General survey	
(3) Using collected data to plan demining activities: priorities, personnel, material, etc.		Planning		
(4) Clearing minefields and disposing of UXOs.			Technical survey	
(5) Evaluating how the work has been done. Ex: sampling the cleared zone with another type of detector.	Quality evaluation			
	Planning interface		Field interface	

Table 1. PARADIS was based on the tasks performed during a campaign.

3. PARADIS design

Mine action activities can be sub-divided according to the scale of the geographical areas where they take place and according to the scale of the data needed.

In the first phase of PARADIS, we made a distinction between "world scale", "country scale", "region scale" and "field scale". Desks activities of gathering general information, statistics collection or production, international co-ordination between programs, would take place at the "headquarter". Headquarters can be located in the contaminated countries or abroad. In PARADIS, the headquarter was located in Brussels, where SGR-IM is working. Planning and organising campaigns, collecting mine information, disseminating this information to task force units, and instructing mine awareness classes rather took place at the "mine action centre" which is located in contaminated countries. Finally, field data collection and demining take place among the communities and in the Suspected Hazardous Areas (SHA: see vocabulary in Chapter 1); these tasks are performed by mobile teams.

In the second phase of PARADIS development, we realised that a finer scale was necessary to follow clearance on the field; we called it the "advancement scale". Besides, given the different activities involved, one at the desk, the other in the field, we built two interfaces; one on the PC and the other on the mobile device, a PDA. The first one was called "the planning interface", the second one, "the field interface", as displayed in **Figure 1**.

An analysis of the data user needs was performed for all the tasks involved and led to a "table of needs" partially published in Refs. [1, 2]. **Table 1** could be considered as a very short summary of this table: it relates the tasks performed during a campaign to the scale of the data needed to perform the task.

3.1. The planning interface

The planning interface works at the global scale (~1:1,000,000) and regional scale (1:250,000–1:50,000). It provides a global overview of the demining activities to the planners (i.e. managers), which enables them to set priorities on different zones. At this scale, the information such as administrative limits, airport location, mine clearance centres and their responsibility areas, inhabited areas and local names, refugee camps, roads, relief, hydrographic networks, climate zones, and soil types, should be available as separate layers, so that the user can combine them at his/her best convenience. The geographical information is made of topographic maps, maps of refugees, or of any free geographical data available such as the digital chart of the world (DCW), or OpenStreetMap data (not available during the project lifetime).

The "region scale" (1:250,000–1:50,000) may contain satellite images (Landsat MSS and Landsat-TM, TERRA-ASTER, SPOT, ERS and RADARSAT), topographic maps, OpenStreetMap data and information from field surveys. Based on these more precise data, the user should complete the global scale data with, on the one hand, data obtained from the image interpretation and, on the other hand, data coming from the IMSMA database or from the field survey. The first set includes up-to-date roads and bridges (according to data capture time), village extensions, land cover, all obtained thanks to the image interpretation. The second set includes hospitals, military buildings, accident localisation, campaign schedules, and mine-field locations. At this scale, thanks to specific tools, priorities, regional constraints and logistic facilities can be combined in order to plan campaigns and organise teams. In particular, the planning interface offers a team management and reporting tools. Team management involves managing team location, logistics and tasking. Statistics can also be derived. For example, a tool is proposed to prevent lack of gas for roving team–a roving team is a team going from village to village to interview local populations about the location of minefields and UXO. Another tool is designed to compute how much explosive must be made available in order to answer the demands of the explosive ordnance disposal (EOD) teams. The reporting tool enables mine action staff to automatically extract data and introduces them into reports according to standards (OTAN standards were used), avoiding errors and reducing time devoted to that task. As illustrative examples, **Figures 2** and **3** show the monitoring of a minefield clearance.

3.2. The field interface

The field scale (1:10,000–1:1500) devoted to the field work and its management at the mine action centre, displays aerial photos, very high resolution satellite images (IKONOS) and

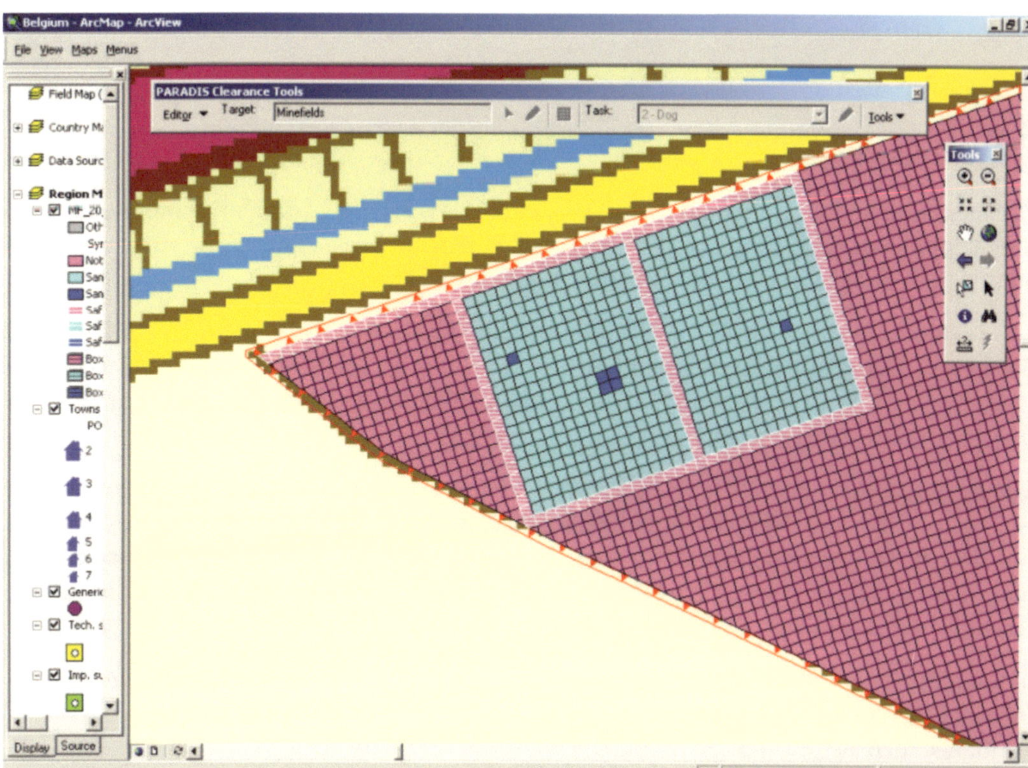

Figure 2. Mapping a square meter grid element on a minefield and following the status of each element as reported by the PDA users.

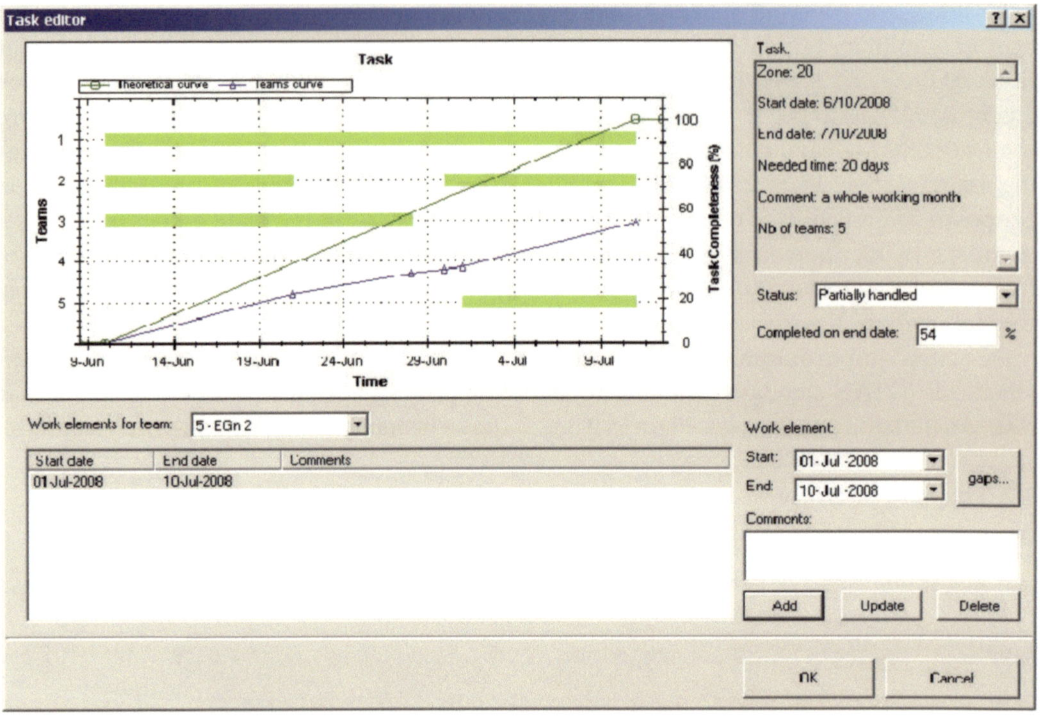

Figure 3. Monitoring the work on a certain zone depending on the allocated resources and surface area; estimating the time needed to complete the clearance of the zone.

sketches. Today, this scale could include QuickBird, WorldView2 data or photography from remotely-piloted aerial systems (RPAS) as well. Highly accurate maps of the suspected areas and cleared mine fields could be available as overlays.

Several tools were developed for this interface: the Road Map tool, the Minefield Grid tool, the Layers tool, the GPS measurement tool, the shifting map tool and the Ammunition Reporting tool. Most of them are detailed in Refs. [5–7], and some of them are summarised hereafter.

The *Road Map tool* enables field operator to enter a detailed description of a road. This is useful during missions, such as ammunition disposal task, prevention patrol or roving. The Road Map tool can also be useful to the planners (i.e. managers) too, as they may want to assign a specific itinerary to the team of deminers they are sending to the field.

In the planning interface, the planner selects the roads and exports the resulting 'road map' to the PDA. On the field, each road segment may be edited; it is given a name and several attributes, such as:

- a practicability state which depends on the season (dry/wet season) associated with a colour code (unknown: black/not practicable: red/poor: orange/good: green),

- a type (tarred road, track, etc.) and
- a form describing UXO present in the vicinity of the road segment.

As one goes along the road, he/she may want to edit the status of the road (e.g. from poor to good); this action is possible in the field interface which allows the edition of each road segment. If some UXO is reported by local population or found by the roving team, a description of the ammunition could be entered in the road form, as an attribute to the corresponding segment. Generally, one may want to enter precise location of the ammunition found; specific ammunition reporting tool was developed for that purpose. Back at the central office, the team import these data into the central GeoDb. A regional road map is then obtained by grouping the road maps of the different teams.

The Minefield Grid tool gives an overview of the work performed on a minefield. Once delineated, the minefield is overlaid with a grid whose parameters (orientation, cell size, etc.), are defined by the user. On the field, each cell corresponds to the working area assigned to a unique deminer; it is typically one square meter. Then, daily, the work performed by the deminers is associated with each cell they worked on: the ammunition found, the status of the cells (not cleared, being cleared, cleared) are stored. When a reasonable part of the grid is completed, the tool can produce an estimate time for achieving the complete clearance of the minefield, given an estimated number of deminers and their equipment; it assumes the same working method for the remaining work and the same contamination for the areas that are not yet cleared. Conversely, the tool can provide the number of deminers (equipped with metal detectors) needed in order to finish the complete clearance of the minefield before a given deadline. A screenshot of the tool is displayed in **Figure 4**.

The *Layers tool* enables the production of maps corresponding to the users' needs. Each layer contains data of similar type: towns are grouped into one layer, roads into another one, etc. The layer can be edited using an "edit layer" box, which is adapted to the type of objects contained in the layer. The user can compose his/her own map by selecting one or more layers from a list of possible pre-defined layers (e.g. minefields, hospitals, towns, UXO locations, etc.); an appropriate symbology is automatically assigned to these layers.

The *GeoNote tool* enables users to associate some geographical position with heterogeneous information, such as text, voice recordings, photos, videos, etc. For example, if the field operators interview some local inhabitants to know the position of some UXO, they may create a GeoNote at the location of the interview and tie several files to the point: the audio files of the interview, image files containing sketches showing the position of the ammunitions, textual files including the information of the estimated size of the UXO.

Adaptation for *APOPO needs* led us to design a specific tool to visualise the true and false detection of each rat, for comparing the rat performance with respect to field parameters, such as temperature, humidity, type of soil and to add mine risk education information.

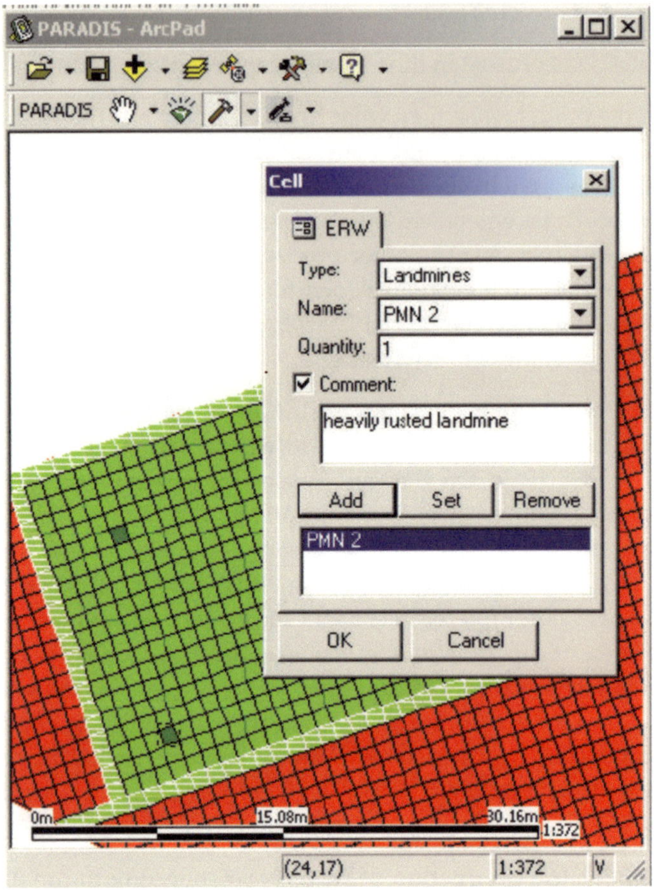

Figure 4. The minefield grid tool.

4. Organisation of a campaign

The proposed method involves a demining organisation, an image interpretation team and, optionally, an image processing team. The latter may collaborate with the interpretation team in order to accelerate the data production.

At the mission announcement, the first task is the collection/production of geo-information on the areas of interest. The demining team thus contacts the interpretation team. Given the geographical zones, this team collects maps and all relevant information; they identify the most appropriate sensors and the best season for acquiring satellite data; they order them, and, upon reception, georeference them; they also combine the various image bands in colour composite images in order to highlight specific terrain features. They may send the georeferenced images to the image processing team for an automatic analysis as the extraction of information from satellite images and scanned maps could be very tedious. Alternatively, they may proceed by a semi-automatic visual interpretation themselves. The aim of such a process is to

http://dx.doi.org/10.5772/65781

produce the vectorial overlays made of the road and hydrographic networks, inhabited and cultivated areas, infrastructures, etc., i.e., all visible relevant information for the demining team. Meanwhile, the deminer team introduces the field survey into the IMSMA database. All data are then introduced in the prototype described in the "PARADIS interface".

In the PARADIS project, the demining organisation was the SEDEE-DOVO while the interpretation task was initially (i.e. during the designing phase) assigned to the scientific partners of the project: the "Institut de Gestion de l'Environnement et d'Aménagement du Territoire" (IGEAT) and the Signal and Image centre of the Royal Military Academy (SIC). During the assessment phase, in order to demonstrate the feasibility of the method in a routine process, an operational partner had to be found inside the Belgian Defence itself. Thus, in this second phase, the interpretation team was the section imagery (SGR-IM), of The General Intelligence and Security Service of the Belgian Defence. Any well-chosen subcontractor could play this role in the case of non-Belgian missions.

4.1. Test sites

As co-lateral data over a test site in Mozambique (Tete province) were already available from the "airborne mine detection: pilot project" [9], this site was chosen as the first site. Another test site in Laos, where the Belgian deminers were active, was selected in order to show the adaptability of the method in a different context. All missions involved data collection, data interpretation and ground survey, as well as collaboration with demining organisations working in the concerned areas: Norwegian People's Aid (NPA) in Mozambique and UXO Lao in Laos. For the second phase, the two interfaces were tested in Afghanistan by the SEDEE DOVO during their EOD missions in Kabul.

4.2. Mozambique

Test sites located in the province of Tete close to Mameme and Songo were selected. Three missions took place. The first one aimed at obtaining all the missing information, establishing contacts with NPA, agreeing with them on a working method and testing the use of high-resolution images for demining activities. During this first mission, the team also checked the relevance of the image interpretation. The aim of the second mission was to present the final results to NPA and to confirm the use of very high-resolution data. The mission took place in the second development phase in order to test the "planning" and the "field" interfaces, and their associated tools.

LANDSAT MSS images (resolution 80 m), LANDSAT TM (30 m), SPOT multispectral (20 m), panchromatic (10 m) and RADARSAT (13 m) have been purchased and processed for the first mission. IKONOS panchromatic data (1 m) were acquired for the second one.

Information about roads, railways, villages and crops were extracted from high-resolution satellite images (resolution between 10 and 30 m). Sequence of images enabled to observe an increase of inhabited areas, of crops and even of tracks in the Songo area. Only roads and rivers were visible on the RADARSAT image; such type of data should be used for mapping very cloudy areas, as RADARSAT is able to get data despite the presence of clouds. Finally, the resolution of LANDSAT MSS images was too low for our purpose.

4.3. Laos

Test sites located in the Champasak province were chosen. SPOT multispectral data (resolution 20 m), SPOT panchromatic (resolution 10 m) and IKONOS panchromatic data (resolution 1 m) were made available for the project. The SPOT data were purchased by Belgium's "Secrétariat pour la Coopération au Développement" while UXO LAO purchased the IKONOS data, as they wanted to assess the use of high-resolution images for mine action activities.

4.4. Afghanistan

During the second phase of PARADIS, a mission took place in Afghanistan in order to validate the various tools and the PDA interface. The reporting tool was adapted in order to be compatible with NATO forms.

5. PARADIS assessment

We assessed the method described above and the tools provided in the PARADIS interface. The assessment included an analysis of the most appropriate remote sensing data for deriving the information listed in the table of needs which was also assessed for its completeness and appropriateness. We validated the image interpretation and the use of image processing tools in order to accelerate the information layers production.

5.1. Method

In order to assess the method at the end of the project, we simulated a campaign in Laos (see Section 4.3.). SGR-IM looked in the archives and found SPOT and IKONOS data covering the zone of interest. Cartographical 1:50,000 maps were used to georeference the imagery with a high accuracy, providing a location at about 80 m (SPOT) and 50 m (IKONOS) respectively. Scientists applied image processing tools to these data and sent results back to the photo analysts. The latter interpreted the images focusing on the legends set up by the team. They delivered products, such as space maps, overlays, documentation, according to the agreed time schedule. Information from Level One and Level Two Surveys for which IMSMA forms were used, were also introduced in the interface. The end-users considered that all products were up to their expectation.

5.2. Use of remote sensing data

Our scientist performed a visual interpretation of the satellite data collected for the first mission in Mozambique at our head quarter (i.e. at RMA). In this case, despite a few errors, the visual interpretation was helpful for non-experts. Because of these potential interpretation errors, interpreted layers should always be checked on the field.

For the mission in Laos, we purchased an IKONOS panchromatic image, which offered a better spatial resolution. Therefore, punctual and linear elements, such as houses, trees, roads, rivers, etc., were well identified by the photo-interpreter. On the other hand, due to the complexity of the land use in that region (e.g. the mixture of coffee plants and bushes) the land cover interpretation seemed more difficult to use. In fact, the land cover interpretation was not even necessary: deminers and villagers could read the raw image as it was a map or a familiar representation of their environment; indeed, deminers could draw on the image a recently cleared area, and conversely, without any hesitation, they could delineate on the field an area delineated on the image. The photo interpretation was even considered as disturbing as the interpreted layers could hide the reality of the image; they definitely preferred the raw image.

Therefore, in the case of a limited budget, one should prefer purchasing panchromatic images (with a spatial resolution of 1 m) rather than multi-spectral data (resolution 4 m). Indeed, apart from the case of some interesting and big enough objects that would have a very different spectral signature (i.e. colour) compared to their neighbourhood, an increase of spatial resolution is, in general, more useful than an increase in spectral resolution.

The IKONOS panchromatic images can also be used by the survey and the roving teams as a background map instead of schematic maps. For example, these teams could overlay UXO location or the roving itinerary. Today, google map offers this kind of background data, but the date of capture should be checked, as the information could be outdated in case of rapid changes or of natural disasters.

SPOT images on the other hand have a lower spatial resolution so that a non-specialist has more difficulties to interpret them. Nevertheless, a well-chosen colour composition combining the different spectral bands, can be helpful as it highlights some specificities of the landscape. Relating artificial colours to the natural scenes, however, requires some training. Thus, a visual interpretation is definitely helpful for the deminers. In our missions in Laos and Mozambique, deminers and villagers used the visual interpretation as a map: they could show their position and use it to follow some paths, while guiding us.

In order to use any type of satellite images as reference maps to locate mine fields or UXO, they should be properly geo-referenced. For this task, if topographic maps are available and are precise enough, the user has to identify specific points in the image and find their geographical coordinates on the map. Otherwise, the inverse pairing should take place: a set of ground control points should be collected on the field with a GPS, and their corresponding point should be identified on the image. The set of matched pairs (i.e. points on the image and point either on a map or located by GPS) is used for the geometric correction.

5.3. Tools

Some core features of the field interface were too difficult to use; for example, locating a square meter within a wide grid on the small screen of the PDA was tough because of the insufficient accuracy of the measuring device available at the time of the project.

In the case of the customisation for APOPO, the use of expensive equipment (one PDA per rat was planned) led to an overall poor cost-effectiveness.

6. Conclusions and perspective

In the first phase, PARADIS demonstrated the benefits of using remote sensing data in mine action. Very high-resolution satellite images (resolution 1 m) are useful for the work in minefields since they do not require an interpretation. High-resolution images (resolution of 10–30 m) are useful as regional maps for planning the teams' work. However, they require interpretation by an expert. The automatic image processing tools for image interpretation were at that time not efficient enough. Visual interpretation and semi-automated tools seemed more adequate. The study of image sequence in high-resolution images also provides relevant information and can be used as a substitute for topographic maps. In order to make the best of these remote sensing data, an efficient operational method involving a demining team and an image interpretation team was proposed and assessed, thanks to field missions in Mozambique and Laos.

In the second phase, the benefit of splitting PARADIS in two interfaces, namely a "planning" and a mobile "field" interface was demonstrated. Specific tools (such as the Minefield Grid tool, automatically integrating GPS measurements and shifting the scanned maps) showed improvements on the deminers' work both for the everyday job and for the planning process at the office. The secure data collection, data sharing, the easy and fast reporting through customised forms, and the various scales of analysis of a situation were seen as important valuable features of the system.

The last phase showed the adaptability of the system to fit other specific user needs, such as APOPO. However, as the tool was developed by a research lab, only limited maintenance could be proposed to the end-user; the prototype was rather a proof of concept.

Places for improvements have been identified.

As far as the field interface tools are concerned, the precise positioning using multi-satellite navigation systems like GPS, Glonass and Galileo would solve the problem of imprecise location [10]. Besides, the PDA used as a mobile device was considered as too expensive. Today smartphones and tablets would make a better field interface platform. The Grid tool should also take into account the required safety distances between deminers in order to derive appropriate statistics.

With respect to symbology, PARADIS made use of its own set of symbols; today the set of recommended symbols exists [11] and could be used as default.

More recent projects can be seen as improvements of PARADIS design. In TIRAMISU [12, 13], the tools developed for "advanced general survey" (equivalent to GMAA; see vocabulary in Chapter 1) correspond to the country scale proposed in PARADIS, while the tools for non-technical survey (NTS) correspond to the region scale, field scale or advancement scale (RPAS tools). Today all scale levels can be fed with more recent remote sensing data and included in more powerful GIS [12–14]. T-IMS [15], the field data collection tool developed in TIRAMISU has several features similar to the field tool proposed in PARADIS. In TIRAMISU, data were stored in dedicated repository in the TIRAMISU repository service [10], which can also be seen as a further development to the PARADIS GeoDB.

The trends of splitting the work in two specific interfaces is confirmed nowadays and the mobile phone devices [16] or field tablets [15] are advantageously replacing the PDA for the field interface. IMSMA has been in constant evolution in order to catch up with these innovations [17]. Other mine action centres, such as CMAC and CROMAC have their own information management system integrating these new concepts and tools [18].

Acknowledgements

PARADIS involved at RMA: D. Borghys, S. Delhay, M. Idrissa, V. Lacroix, M. Shimoni and Y. Ouaghli. At SEDEE-DOVO: M. Lambrechts, V. Muylkens, G. De Decker and E. Henin. At ULB-IGEAT: (first phase) E. Wolff and R. Fricke. At SGR-IM: (first phase) A. Vechiato. PARADIS (first phase) was funded by the TELSAT programme of the Scientific, Technical and Cultural Affairs of the Prime Minister's Service (Belgian State). Belgian Defence financed the other phases.

Author details

Vinciane Lacroix

Address all correspondence to: Vinciane.Lacroix@elec.rma.ac.be

Royal Military Academy (RMA) Belgium

References

[1] R. Gasser, "The surprisingly constant cost of landmine impact survey", The Journal of ERW and Mine Action, Issue 15.1, 2011.

[2] V. Lacroix, M. Shimoni, E. Wolff and M. Acheroy, "A geographical information system for humanitarian demining", ISPRS2000 GeoInformation for All, Amsterdam, 2000, paper 738.

[3] V. Lacroix, E. Wolff and M. Acheroy, "Paradis: A prototype for assisting rational activities in humanitarian demining using images from satellites", Journal for Mine Action, Issue 6.1, 2001.

[4] S. Delhay, V. Lacroix and M. Idrissa, "Paradis, a GIS tool for the management of humanitarian demining campaigns", EUDEM2-SCOT-2003 International Conference on Requirements and Technologies for the Detection, Removal and Neutralization of Landmines and UXO, September 2003, Brussels, Belgium.

[5] S. Delhay, V. Lacroix and M. Idrissa, "Paradis: GIS tools for humanitarian demining", ISCRAM 2005 Information Systems for Crisis Response and Management Conference", April 2005, Brussels, Belgium.

[6] S. Delhay, V. Lacroix and M. Idrissa, "A GIS tool for humanitarian demining", Proceedings of AGIT Conference, July 2006, Salzburg, Austria.

[7] S. Delhay, V. Lacroix and Y. Yvinec, "Examples of the use of remote sensing and GIS for humanitarian demining", 6th International Conference on Earth Observation and Geoinformation Sciences in Support of Africa's Development, 30 October–2 November 2006, Cairo, Egypt.

[8] TIRAMISU Geospatial tools for Demining [Internet]. 2016. Available from: http://www.geospatial.fp7-tiramisu.eu. [Accessed: 2016-06-22].

[9] J. van Genderen and B. Maathuis, Airborne Detection of Landmines: A Review of the Techniques and Some Practical Results, Disarmament and International Security, Regensburg Germany, 1998.

[10] D. Peeters, D. Schmidt, G. Waizmann and J. Riesch, "Centralised data management in the demining world: The Tiramisu Repository Service", International Workshop on Humanitarian Demining, Biograd (Croatia), April 2015. Available from: http://www.fp7-tiramisu.eu/publications [Accessed: 2016-06-27].

[11] Technical Note 07.11/01-2016 "Land Release Symbology" [Internet]. 2016. Available from: http://www.mineactionstandards.org/news/detail/article/1457344361-land-release-symbology-and-more/. [Accessed: 2016-06-23].

[12] Y. Yvinec, V. Lacroix and Y. Baudoin, "TIRAMISU final technology demonstration at SEDEE-DOVO", The Journal of Conventional Weapons Destruction, Issue 20.1, 58-60, 2016.

[13] S. Vanhuysse1, V. Lacroix and E. Wolff, "Geospatial analysis of the conflict landscape for supporting non-technical survey", International Workshop on Humanitarian Demining, Biograd (Croatia), April 2015. Available from http://www.fp7-tiramisu.eu/publications [Accessed: 2016-06-27].

[14] Kruijff et al., "Space assets for demining assistance". Acta Astronautica, Vol. 83, 2013, 239-259.

[15] T. Vikstrôm and S. Kallin, "TIRAMISU information management tool-T-IMS", International Workshop on Humanitarian Demining, Biograd (Croatia), April 2015. Available from http://www.fp7-tiramisu.eu/publications [Accessed: 2016-06-27].

[16] E. Lajoie, "Going mobile: Information sharing and changing face of digital data collection", Journal of ERW and Mine Action, Vol. 17, Issue 2, Article 14, 2015. Available from: http://commons.lib.jmu.edu/cisr-journal/vol17/iss2/14. [Accessed: 2016-06-27].

[17] Welcome to IMSMA Wiki, [Internet]. 2016. Available from: http://mwiki.gichd.org/IM/Main_Page [Accessed: 2016-06-27].

[18] C. Matic, D. Laura, R. Tursic and A. Krtalic, "Analytical assessment for the process of collecting additional data on a suspected hazardous area in humanitarian demining". [Internet]. 2014. Available from: http://www.fp7-tiramisu.eu/publications [Accessed: 2016-06-27].

Chapter 8

Assessing the Performance of Personal Protective Equipment[1]

Georgios Kechagiadakis and Marc Pirlot

Additional information is available at the end of the chapter

http://dx.doi.org/10.5772/65778

Abstract

Lightweight protective structures and materials such as the personal protective equipment (PPE) for explosive ordnance disposal (EOD) personnel are frequently under investigation globally. Their mechanical response to impulsive loads such as blast and ballistic impacts is critical for establishing the spectrum of their performance against various types of threats. This chapter presents a novel testing technique that incorporates three near-simultaneous impacts in one shot in order to acquire deeper understanding of the dynamic interactions that take place during an explosion. A numerical model of an aramid fabric is developed to examine the parameters that influence the ballistic performance under multiple impacts. Fragment cluster impacts with dense dispersion have increased probability to perforate the target material. Heterogeneous, non-isotropic materials, like most of the ballistic grade protective materials, distribute the energy of the impacts in the form of stress wave streams causing the material to behave differently depending on the formation of the impacting fragments. Experimental work with aramid fabrics against single and triple impacts with the fragment-simulating projectile (FSP, 1.102 g) indicates that the ballistic limit in triple impacts is considerably lower that the ballistic limit in single impacts. The actual ballistic performance against multiple fragment impacts is severely underestimated with the classical single-impact methodologies.

Keywords: PPE, fragmentation protection, ballistic test, multiple impacts, triple impacts

[1] "Sometimes I would carry around 100 with me in a sack…"
Aki Ra, Deminer. He cleared 50000 mines
Source: Time magazine

1. Introduction

The deminers are the backbone of mine action. They consider the incidence of accidents during demining operations as low, only once per 25 work years. Most of them have lost family and friends from mine accidents, and in some cases, they have experienced the terrible face of war. An accident is something that happens. But when it happens, the life of the deminer depends on the personal protective equipment, he wears.

The consequences of an accident go beyond the injury of a deminer. Their occurrences affect the morale of the whole demining operation, maiming the motivation of the endeavour. Personal protective equipment serves this way a double purpose. The obligation of the scientific community is to provide with efficient tools to the manufacturers to tune the performance of their products, and clear guidelines to the end users for selecting them.

2. The threat

The physics of anti-personnel (AP) mine explosions may be examined from a science point of view or from a functional point of view. These two perspectives are complementary. The scientific approach aims at describing the details of the fundamental processes that come into play when an AP mine is actuated. This might involve chemistry, physics, material science, etc. The functional approach aims more at the end state, trying to describe the terminal effects of the mine against people and equipment.

All mine explosions share the same fundamental elements: a chemical reaction transforms an explosive material into a hot, high-pressure gas. This gas then performs work, in a thermodynamics sense, on its surroundings. This could be fragmenting the casing of the mine, pushing soil or water, or driving a shock in neighbouring air. It could also be damaging an object or injuring a person nearby. Chemistry and physics provide mathematical equations to describe and quantify these processes. Waves are one of the fundamental quantities in nature. They appear in almost every branch of physics, being associated with light, heat, electromagnetism, etc. In the case of the explosions, the transmission of mechanical waves through a deformable or elastic medium is the most fundamental phenomenon. Such waves originate in the displacement of some portion of an elastic medium from its normal position, causing it to oscillate about an equilibrium position. Because of the elastic properties of the medium, the disturbance is transmitted from one layer to the next. This disturbance, or wave, consequently progresses through the medium. Energy can be transmitted over considerable distances by wave motion. Thus, waves might be construed as being one of nature's mechanisms to transmit and propagate information to materials. Therein lays their importance in describing the physics of mine explosions. This includes the propagation of a detonation wave through the explosive and the subsequent behaviour of the explosive by-products with their surroundings.

When an explosive device detonates, a blast wave is generated, along with multiple accelerating fragments, propagating away to all directions. The blast over-pressure, in extreme

http://dx.doi.org/10.5772/65778

conditions, when it interacts with an individual, may cause amputations and shearing of body parts. It may also cause damage to the ears, which can lead to permanent hearing loss, and due to rapid acceleration to the head, a range of concussive injuries may occur, which can be lethal. The fragmentation of the casing of an explosive device (primary fragmentations), together with dirt, pebbles and debris from the ground (secondary fragmentations), are accelerated by the violently expanding gases of the explosive to supersonic velocities. These fragments, if no adequate ballistic protection is worn, can cause perforations to critical parts of the body resulting to even death.

2.1. Fragmentation mines

Despite significant variations in geometry, mass and materials, all fragmentation mines share the same basic elements and the same working principle. There are four main elements to a fragmentation mine: a main explosive charge, a fragmentation case, a detonator and some actuation (fuse) system. Triggering the fuse usually releases a spring-loaded striker that hits the detonator. The latter is simply a small receptacle that contains a highly sensitive explosive that can be ignited by a simple mechanical hit. This first explosion ramps up to ignite the high explosive that makes up the main charge of the mine. It should be noted that a high explosive is usually much less sensitive than the explosive used in the detonator. This makes it safer to handle and transport the munitions. Detonators are much more sensitive, requiring special packaging and careful handling.

The explosion of the detonator causes the main charge to detonate. As the shock front propagates through the high explosive, it transforms the solid (or liquid) explosive into detonation products. Given that the maximum physical dimension of the main charge for the average fragmentation mine is of the order of 100–200 mm, and assuming an average shock front velocity of 7 km/sec, it takes approximately 15–30 μs to transform the solid explosive into detonation products. In the thermodynamics sense, this mass of hot, high-pressure gas then performs work on its surroundings. For a fragmentation mine, this means breaking up the metallic case into a large number of small fragments and propelling these fragments away from the centre of the explosion at high velocity.

The aim of the fragmentation mines is to cause as many casualties as possible to the opponent, therefore they are extremely dangerous. Current deminer PPE are not able to protect from fragmentation mines.

2.2. Blast mines

An average blast anti-personnel mine contains about 110 g of explosive. The maximum reported charge mass is 300 g and the minimum is 28 g. Blast mines are typically classified in terms of small (less than 50 g explosive), medium (50–100 g of explosive) and large (greater than 100 g of explosive). The most common explosive used in blast mines is Trinitrotoluene (TNT), with Composition B, RDX and other explosives being less common. Most blast mines are concealed in ground.

The vast majority of blast mines are pressure actuated. Two scenarios prevail against this class of mines. In the first scenario, the victim steps on the mine and detonates it as per the intent of the design. In the second scenario, the victim is conducting a mine clearance drill and is in a low-down position when the mine goes off. The main difference between these scenarios is the stand-off distance between the mine and the victim.

A blast mine is a simple device that usually consists of a plastic container filled with the main explosive charge, a detonator and a mechanical actuation (fuse) system. Most detonators rely on stab initiation or a friction sensitive compound to start the combustion process that quickly transitions to a detonation. When a downward force of sufficient magnitude (often less than 10 kg) is applied to the fuse, the mine detonates. It takes between 5 and 10 μs to transform the explosive charge into detonation products. As the detonation wave reaches the mine case, the shock is transmitted to the case, then to the soil and finally, to the sole of footwear or to air. Similar to fragmentation mines, the case breaks up due to the combined effect of the transmitted stress wave and the extreme pressure from the detonation products. However, since the ignition temperature of plastics is much lower than for metals, the case materials also start to combust and continue to do so as they are ejected by the expanding gas. One can surmise that a blast mine relies on three main mechanisms to impart energy to an object nearby. First, it transmits a high-amplitude stress wave into the soil, footwear and air. Secondly, the detonation products expand and perform work on the immediate environment. Thirdly, part of the explosive energy imparts motion to the soil that is then ejected at relatively low velocity (< 1000 m/s) and strikes the victim.

Blast mines are typically used to inflict severe injuries to the enemy unit, causing it to stand in place to attend to the wound. Because of the severity of the injuries, it inflicts a major hit to the morale of the survivors. Current PPE are designed to protect the life of the deminer from all known AP blast mines and reduce the severity of possible injuries.

2.3. Sub-munitions

The term 'sub-munitions' generically describes any item of ordnance carried in or ejected by a dispenser. Anything from a scatterable mine to a remotely delivered radio jammer might be described as a sub-munition, the enormous diversity making it difficult to define their characteristics and effects. A wide variety of launch or delivery system is used, including missiles, rockets, projectiles and mortars. Sub-munitions warheads for these types of land service ammunitions are normally called 'carriers' while bombs containing sub-munitions are generally known as 'cluster bombs'. Some air-delivered sub-munitions are dispensed from canisters that remain attached to the aircraft; examples include the British JP-233, German MW-1 and Russian KMG-U.

Once the dispenser is fired, launched or dropped, opening is normally determined by a time delay or proximity fuse. The sub-munitions are normally dispensed in one of three ways: base ejection, nose ejection or case rupture. Base ejection is most common in projectiles, but is also used in other carriers and some cluster bombs. In both, nose ejection and base ejection, the fuse usually initiates a small propellant charge that opens the carrier and pushes the sub-munitions out. Case rupture, used in some rocket and missile warheads, sometimes involves the use of small linear cutting charges to split the casing and may also use a propellant charge to eject the sub-munitions. Most (but not all) sub-munitions incorporate a separate arming mechanism (such as wind-driven vanes), which functions, after ejection, as the sub-munition falls. The majority use some form of stabilisation (normally, the fins, a streamer or a chute) to bring them into a nose-down attitude but some are designed to spin in the air stream and use this movement for arming. Since sub-munitions disperse after ejection, the density of the impact footprint is dependent on the speed and altitude at which the dispenser opens. Most sub-munitions are designed to detonate on impact but some (such as scatterable mines) are victim operated or incorporate delays.

3. Current methodologies

Current practices for assessing the performance of PPE have been proven significant for providing basic certified measurements to the humanitarian demining community but they lack physical reasoning. This has led to designs of PPE that over-estimate (expensive and non-ergonomic), or under-estimate the threats deminers are dealing with. The problem of designing PPE is the compromise between adequate protection and comfortability. A better protection is required against various types of threats (splinters, debris, fragments, blast, etc.) but the overall weight and the thermal impact have to remain acceptable for the demining personnel.

Currently, there are mainly two standardised techniques used by the manufacturers to test and certify their products according to the International Mine Action Standards (IMAS) provided by the United Nations Mine Actions Service (UNMAS). Multiple single ballistic impact tests are used to assess the V50, which is the velocity at which there is 50% probability of perforation, in order to test the impact resistance of the PPE against multiple fragment impacts usually generated by an anti-personnel blast-mine explosion event, according to the North Atlantic Treaty Organization (NATO) STANAG 2920. Blast tests are also used to test the structural integrity of the PPE against the blast loads of the explosion. These techniques have helped the humanitarian demining community to understand the necessity of standardised performance and the implementation of validating tools in order to provide the men on the field with certified PPE. However, the above-mentioned techniques have certain limitations. The single impact tests are based on the hypothesis that the effect of a simultaneous multiple-impact event can be simulated with single-impact events, something that is contradictory to the observations on the field. For example, police and armed forces have reported armour system failure when multiple shots impact within certain space and time intervals, called the "double-tap" technique. Also, the blast tests are not that enlightening if expensive measurement devices are not implemented as well, such as biofidelic "crash test" mannequins that cost a few hundreds of thousands of euros. The last point of this critique is that the assumption of analyzing two concurrent phenomena separately, viz. the blast wave over-pressure and the impacts of the fragments, would lead to an accurate and detailed description of the explosion effects, is already a farfetched hypothesis.

4. The triple-impact technique

In 2013, a novel technique for testing protective materials against fragment impacts was presented. Three projectile impacts within close space and time proximity were incorporated in order to simulate the effects of multiple fragment impacts on protective structures. These fragments are typically generated when an explosive device detonates, and they are lethal if no sufficient protection is worn.

4.1. State of the art

The announced ballistic performance of a protective material is highly dependent on the methodology that is followed to assess it [1]. A few attempts have been made to examine the ballistic response of a protective material under near-simultaneous multi-site impacts [2–5]. Qian et al. investigated the terminal effects of fragment cluster impacts on a metallic plate [2,3]. Norris developed a five steel cube launcher for hazard assessments for non-nuclear munitions [3]. Deka et al. developed an air gun triple launcher for conducting near-simultaneous triple-impact tests against S2-glass/epoxy composite laminates [4]. Kechagiadakis et al. examined the response of fragmentation vests against dual impacts from 9 mm to investigate the effect of proximity [6].

After one impact occurs, the material transfers the kinetic energy of the projectile away from the point of impact. The path on which the energy travels depends on the micro- and macro-mechanical structural complexity of the material. Most of the ballistic grade materials are highly complex, heterogeneous and anisotropic. This complexity affects the way the internal stress waves manifest and propagate. Orthotropic materials polarise the three basic (pure) stress waves, the longitudinal and the two shear, on their two main axis of symmetry (for plates) causing the creation of high energy cross-like stress streams[7]. Dynamic interactions of multi-site impacts generate multiple stress wave patterns that superimpose elevating the stress-strain energy density of the material in levels that may decrease its ballistic performance.

Moreover, a target plate after being impacted is beginning to deflect towards the direction of the velocity vector of the impactor till it reaches a maximum. After that point, the transverse wave that caused the deflection is reflected initiating a backwards motion of the plate that begins to oscillate generating Lamb waves. Second and third projectile impacts on an oscillating plate may result to a different terminal effect depending on the phase of this oscillation. A first consequence would be a variant relative velocity the second or third impactor would have with the target plate moving towards or away from the impactor. A second consequence would be the inclination of the deflected plate causing an oblique impact. This effect is more apparent to softer target materials, such as textiles, than metals. For this reason, the present article is focused on the response of aramid fabrics against multiple small size medium velocity impactors.

http://dx.doi.org/10.5772/65778

5. Parameters of the technique

The triple-impact technique is based on the idea that the interactions between multiple impacts influence the ballistic performance of the target material. The complex phenomena that occur during the impact process result in variant responses. These responses are probabilistic, so an enormous number of tests would be required for a sufficiently accurate statistical analysis. Alternatively, with the use of numerical simulations, it is possible to acquire adequate understanding of such a phenomenon. To realise these simulations, a typical aramid fabric is modelled using a homogenised orthotropic elastic shell plate to exhibit the stress wave propagating patterns and three fragment-simulating projectile (FSP) 5.56 mm simple elastic solids are chosen as projectiles [6]. Different configurations of projectile formations with respect to the material's axes of elastic symmetry are modelled in order to acquire a more thorough understanding of the variations in the impact interactions that may be encountered in real ballistic triple impact tests.

The parameters that influence these responses have non-constant weighting factors. There are two major categories of parameters. The ones that have been examined thoroughly in classic ballistics and the ones more related to the interactions of multiple impacts.

Classic ballistics parameters: velocity, projectile shape, impacting materials, obliquity of impact and boundary conditions.

Interactions parameters: projectile formation (dispersion between impacts, rotation of projectile formation, rotation of projectiles and time intervals between impacts) and material directionality (material anisotropy, material heterogeneity and damage propagation). These parameters in most cases are inter-coupling. This means that one parameter influences the weighting factor of another parameter and thus may not be investigated separately.

5.1. Material anisotropy vs. projectile formation

5.1.1. Initial velocity vs. residual velocity

The influence of the projectile formation on the fabric's ballistic response is examined for four cases of triple impacts. The synergistic effects of stress waves and the contact stress concentrations differ from case to case. This differentiation adds further probabilistic properties to the phenomena of terminal ballistics in the cases of multiple impacts. The distance between the projectiles is set to 22.5 mm (**Figure 1**).

5.1.2. Two on-axis impacts and one off-axis impact (case 1)

In this configuration, two of the projectile impacts are aligned with one of the main axis of elastic symmetry of the material. Due to the orthotropic symmetry of the material, the waves exhibit pure longitudinal and shear modes on the main material axes of symmetry and hence they propagate faster. These two impacts interact with each other more intensely than with the third impact because of the constructive/destructive interference their wave stress streams invoke. In this set of simulations, the two on-axis impacting projectiles present generally higher

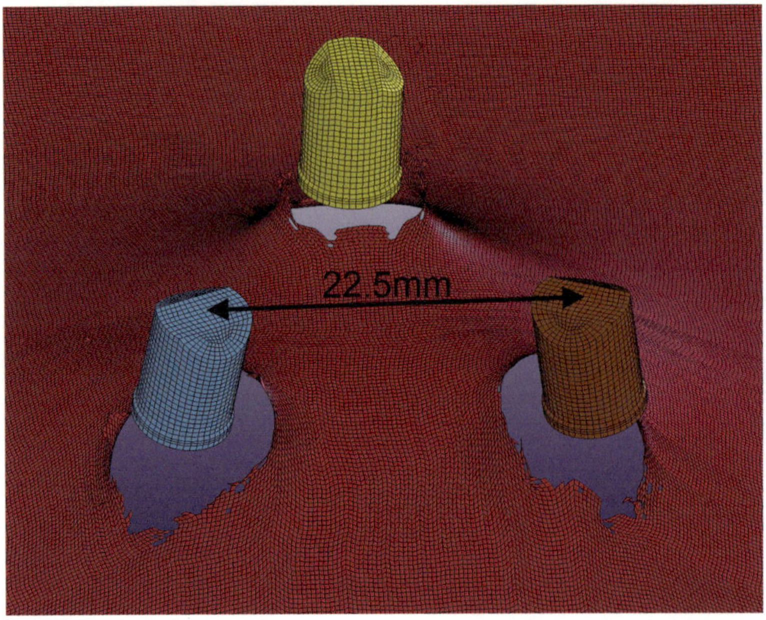

Figure 1. A picture of the parts participating in the simulations.

residual velocities than the third off-axis. Characteristic examples of this phenomenon are the simulations at 420 m/s in which the off-axis impacting projectiles are either blocked or have significantly lower residual velocities (51 m/s).

5.1.3. 15° rotated triple formation (case 2)

Given the orthotropic elasticity of the fabric and the equilateral formation of the three projectiles, there are infinite different rotational configurations that may result in significant deviations of the apparent ballistic performance. However, the rotational ranges in which these configurations may appear in a triple ballistic test are only between 0° and 15°. Beyond the 15°, the line that connects two of the three impact points gradually converges with the second axis of elastic symmetry of the fabric. In this case, the interactions of the three impacts exhibit less dependency on the degree of anisotropy of the fabric.

5.1.4. 90° rotated FSPs (Cases 3 and 4)

The FSPs are projectiles with low length/diameter ratio, resulting in unstable flight after they leave the barrel. Usually, such a problem is minimised by inducing rotational movement on the projectiles, balancing them gyroscopically. While in single-impact tests, this rotational movement induces indifferent ballistic results, in the triple impact case, small deviations may appear. Because of the non-axisymmetrically shaped tip of the FSP, the contact forces that manifest during an impact are not constant along its nose-tip surface. Two areas of stress concentrations appear at the 90° edges of the FSPs tips.

http://dx.doi.org/10.5772/65778

Random rotational orientation of the FSPs result in random stress concentration points on the target and random stress waves source points. Cases 1 and 2 are re-examined for 90° rotated projectiles (**Figure 2**). The resultant residual projectile velocities opposed to the initial velocities for all four cases can be found on (**Figure 3**).

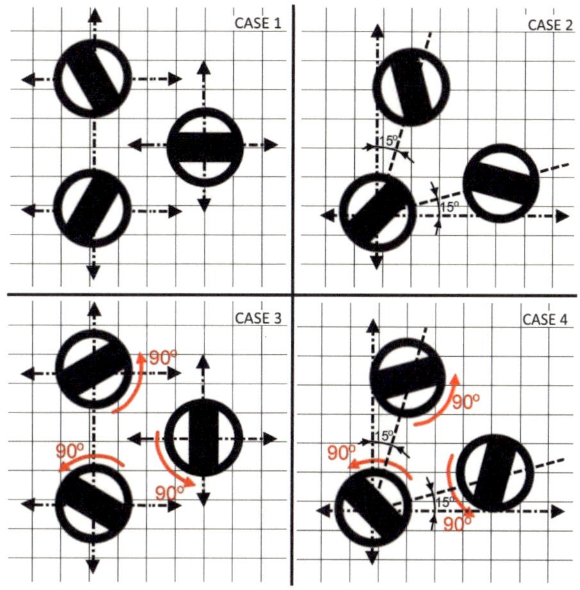

Figure 2. Different triple-impact configurations that were examined. The "dash-dot" axes represent the two axes of elastic symmetry of the material. The black angular increment represents the rotation of the projectile formation around its centre and the red rotational arrows correspond to the rotation of the projectiles around their axial direction.

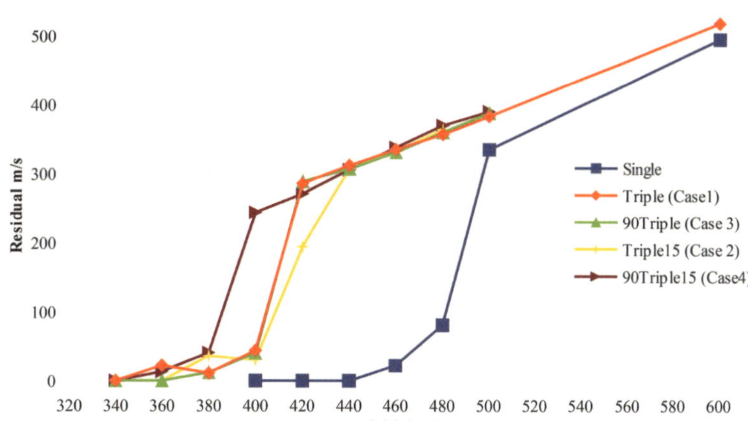

Figure 3. Initial versus residual velocity of single impact and different configurations of triple-impact simulations. The maximum velocities are presented.

5.2. Distinct periods of interactions

The distances between the projectiles, and between the projectiles impact points and the plate's edges were taken into account for calculating the times of arrival of the quasi-longitudinal incident and reflected stress waves. Three distinct periods are observed. In the first period, the projectile has perforated the plate before the stress waves from the other two impacting projectiles reach the point of its impact. The third period is characterised by the influence of the reflected stress waves bouncing at the edges of the plate. The second period is defined between the arrival of the incident and reflected stress waves. In the first period, the projectiles perforate the plate without the influence of multiple-impact interactions due to high initial velocities, over 125% V50. Most cases of multiple plate perforations occur in the second period during which the constructive/destructive interactions coupled with the contact forces and the deflection of the plate have increased probability of failure initiation. In the third, the projectiles have low residual velocities. In **Table 2**, the times of block and initiation of perforation, after contact of the projectiles, are presented.

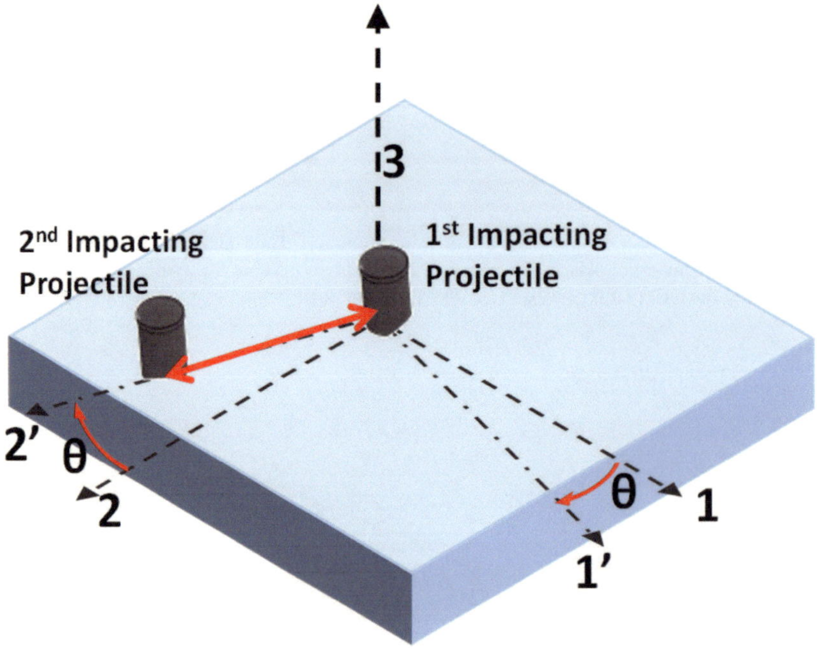

Figure 4. An orthotropic plate being impacted by two projectiles. The axes 1, 2 and 3 represent the axes of material symmetry. The axis that connects impact points of the two impacting projectiles deviates from the axis 2 by θ°.

The times or arrival of stress waves were calculated using the composite material mechanics.

Supposedly, there is an orthotropic plate with axes of material symmetry represented by three dashed arrows 1, 2 and 3 (**Figure 4**). A projectile impacts the plate (at its centre for illustration purposes). Then, another projectile impacts at a second impact point. The line that is defined by the two impact points defines the path on which the interactions, from and between the two impacts, will occur (marked with red). This line, however, is having an angle in relation to one of the material's axes of symmetry, in this case the second axis.

Then, the stiffness matrix of the plate is [8]

$$\begin{Bmatrix} \sigma_1 \\ \sigma_2 \\ \sigma_3 \\ \sigma_4 \\ \sigma_5 \\ \sigma_6 \end{Bmatrix} = \begin{bmatrix} C_{11} & C_{12} & C_{13} & & & \\ C_{12} & C_{22} & C_{23} & & & \\ C_{13} & C_{23} & C_{33} & & & \\ & & & C_{44} & & \\ & & & & C_{55} & \\ & & & & & C_{66} \end{bmatrix} \begin{Bmatrix} \varepsilon_1 \\ \varepsilon_2 \\ \varepsilon_3 \\ \varepsilon_4 \\ \varepsilon_5 \\ \varepsilon_6 \end{Bmatrix} \tag{1}$$

The material response in directions other than the "on-axis" directions can be calculated by rotating the stiffness matrix, in this case by θ°.

$$\begin{Bmatrix} \sigma'_1 \\ \sigma'_2 \\ \sigma'_3 \\ \sigma'_4 \\ \sigma'_5 \\ \sigma'_6 \end{Bmatrix} = \begin{bmatrix} C'_{11} & C'_{12} & C'_{13} & & & C'_{16} \\ C'_{12} & C'_{22} & C'_{23} & & & C'_{26} \\ C'_{13} & C'_{23} & C'_{33} & & & C'_{36} \\ & & & C'_{44} & C'_{45} & \\ & & & C'_{45} & C'_{55} & \\ C'_{16} & C'_{26} & C'_{36} & & & C'_{66} \end{bmatrix} \begin{Bmatrix} \varepsilon'_1 \\ \varepsilon'_2 \\ \varepsilon'_3 \\ \varepsilon'_4 \\ \varepsilon'_5 \\ \varepsilon'_6 \end{Bmatrix} \tag{2}$$

This it is possible to calculate the longitudinal stress wave velocities with the formula:

$$V_2 = \left(\frac{C'_{22}}{\rho} \right)^{\frac{1}{2}} \tag{3}$$

where

$$C'_{22} = C_{11} n^4 + 2m^2 n^2 (C_{12} + 2C_{66}) + C_{22} m^4 \tag{4}$$

$m = \cos(\vartheta)$, $n = \sin(\vartheta)$.

It is possible to draw a curve of the stress wave velocities as a function of degrees (material anisotropy)(**Figure 5**).

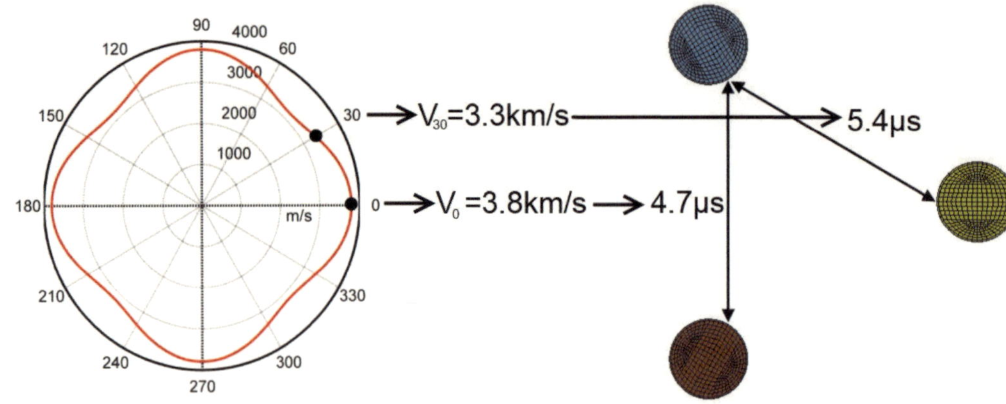

Figure 5. Quasi-longitudinal stress wave velocity as a function of degrees linked to the distances between the projectiles. The times of arrival of the incident waves are calculated.

On the results from the simulations (cases 1-4) are presented (**Table 1, Table 2**).

Velocity (m/s)													
Initial	**Single**	**Triple**			**90 Triple**			**Triple 15**			**90 Triple 15**		
340		0	0	0	0	0	0				0	0	0
360		22	0	0	0	0	0	0	0	0	13	0	0
380		11	11	11	12	0	0	36	0	0	41	22	0
400	0	44	34	0	40	29	0	30	10	0	244	32	0
420	0	286	279	51	289	287	0	194	187	32	271	266	80
440	0	312	311	303	307	299	299	307	298	298	306	302	292
460	22	335	328	326	331	330	320	332	331	331	338	326	323
480	81	357	354	354	360	360	351	364	360	354	370	352	351
500	335	383	379	376	388	386	381				390	389	376
600	495	518	517	498									

Table 1. Initial and residual velocities of the projectiles in single- and triple-impact simulations.

http://dx.doi.org/10.5772/65778

Initial velocity	Time of block/perforation (μs)												
	Single	Triple			90 Triple			Triple 15			90 Triple 15		
340		*124*	*126*	*126*	*121*	*123*	*136*				*123*	*123*	*123*
360		*98*	*124*	*125*	*138*	*145*	*188*	*123*	*125*	*125*	*102*	*138*	*172*
380		*97*	*99*	*103*	*98*	*152*	*180*	*97*	*125*	*125*	*97*	*99*	*148*
400	101	*95*	*96*	*128*	*97*	*98*	*122*	*94*	102	*123*	12	*100*	*138*
420	101	9	9	*96*	9	10	*191*	32	34	*94*	9	11	*84*
440	101	9	9	9	9	9	10	9	9	10	9	10	10
460	88	9	9	9	8	8	9	8	8	8	8	8	8
480	76	9	8	8	7	7	8	7	8	8	7	7	8
500	14	7	8	8	7	7	8				6	7	7
600	6	**3**	**4**	6									

Table 2. Time of block (underlined) of the projectiles and time of initiation of erosion of the fabric in single- and triple-impact simulations. With bold are perforations in **Period 1**, with Italic in *Period 3* and normal in Period 2.

5.3. Effect of heterogeneity and anisotropy

Most protective materials apart from their directional composition (anisotropy) have a degree of heterogeneity. This discontinuity within their structure exacerbates the anisotropic distribution of stress waves. As the stress waves propagate inside the material, the internal boundaries between different material segments determine whether the propagation continues or reflects. At the interface between "internal" boundaries within the bulk of the structure, part of the stress waves is transmitted and the rest is reflected depending on the material impedance ratios. The heterogeneity and the anisotropy determine the distribution of the energy of an impact, thus the damage propagating patterns and the interactions between impacts. In the example of the aramid fabric, the stress waves travel within the filaments. Because of their long and thin shape they force the stress waves though the longitudinal direction. The filaments are grouped to yarns. The yarns interchange to form a woven structure. At the contact points between filaments and yarns stress waves are partially reflected and transmitted.

5.4. Effect of dispersion in triple impacts

Dispersion between impacts is an important factor when analysing interactions of multiple impacts. Simulations at three different impact distances were performed in order to define the magnitude of the influence of the dispersion in the ballistic performance of the fabric. The projectile formation was identical to the first case in the triple impacts with two of the projectiles impacting on one of the axes of elastic symmetry of the plate. Simulations with over 420 m/s projectiles initial velocity have shown a decrease of the residual velocity of the projectiles with increasing dispersion. Also, the influence of the orthotropic stress wave polarisation is more intense with the increase of the dispersion. The off-axis impacts have considerably lower "ballistic limit", a phenomenon more apparent in higher ranges of dispersion. The stress wave streams become narrower with increasing radius. The residual projectile velocities as a relation to the dispersion between the impacts are presented in the (**Figure 6**, **Figure 7**) and on the (**Table 3**, **Table 4**).

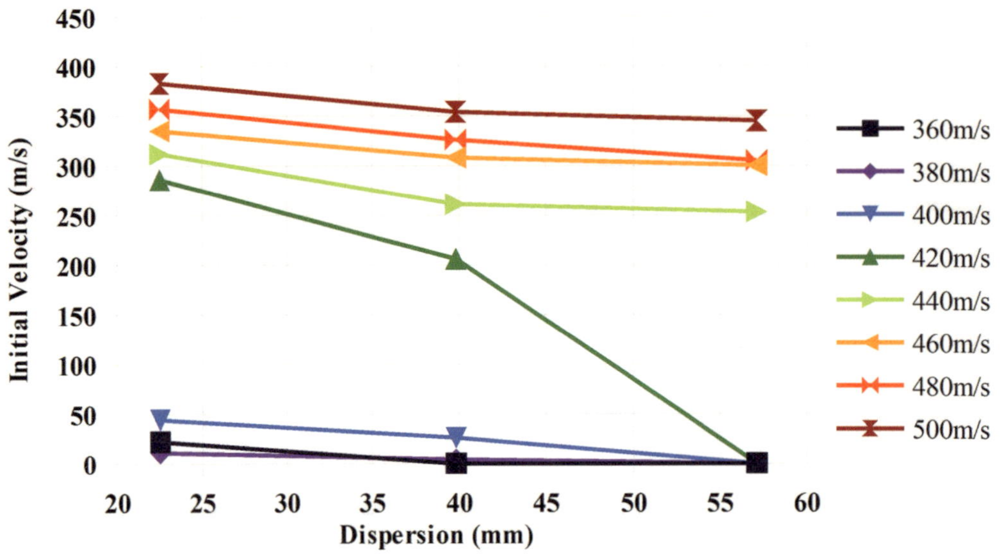

Figure 6. Maximum residual velocity of projectiles in triple impacts as a function of distance between impacts.

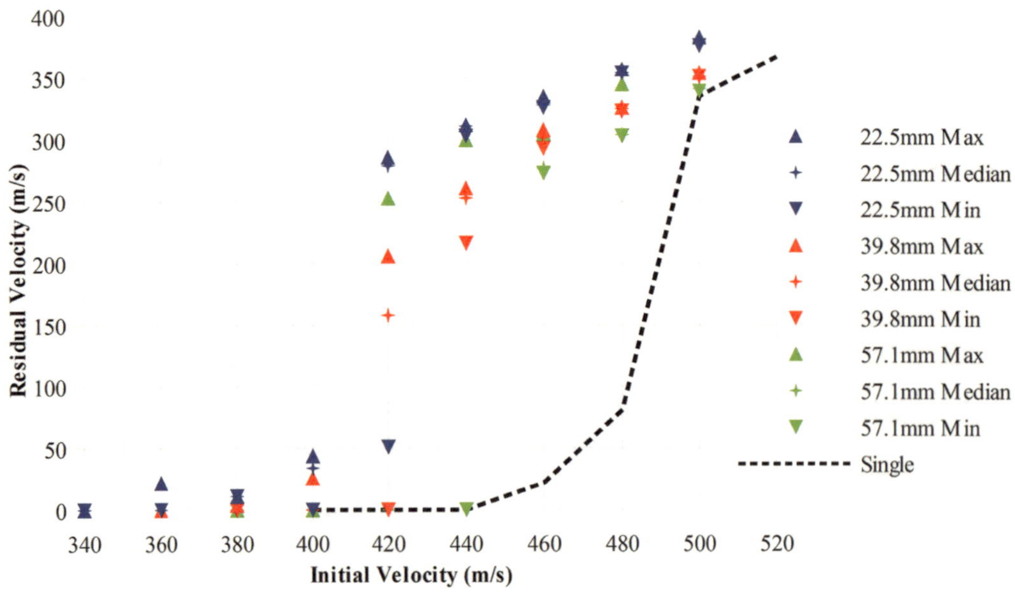

Figure 7. Residual vs initial velocities for three ranges of dispersion.

http://dx.doi.org/10.5772/65778

Velocity (m/s)									
Initial	**22.5 mm**			**39.8 mm**			**57.1 mm**		
340	0	0	0						
360	22	0	0	0	0	0	0	0	0
380	11	11	11	4	4	0	0	0	0
400	44	34	0	26	0	0	0	0	0
420	286	279	51	206	158	0	0	0	0
440	312	311	303	261	253	216	253	253	0
460	335	328	326	308	302	293	300	277	273
480	357	354	354	326	325	323	305	304	303
500	383	379	376	354	351	351	345	341	339

Table 3. Velocities of triple impacts with different radial dispersions.

Initial velocity	Time (μs)								
	22.5 mm			**39.8 mm**			**57.1 mm**		
340	*124*	*126*	*126*						
360	*98*	*124*	*125*	*86*	*89*	*89*	*83*	*83*	*84*
380	*97*	*99*	*103*	*80*	*82*	*87*	*82*	*82*	*84*
400	*95*	*96*	*128*	*75*	*86*	*92*	*81*	*82*	*83*
420	9	9	*96*	24	28	*90*	*80*	*81*	*82*
440	9	9	9	18	19	25	22	22	92
460	9	9	9	14	15	15	16	20	21
480	9	8	8	13	13	14	18	18	18
500	7	8	8	12	12	13	14	15	16

Table 4. Time of block (underlined) of the projectiles and time of initiation of erosion of the fabric in triple-impact simulations with three levels of projectile dispersion. With Italic in *Period 3* and with normal in Period 2.

5.5. Effect of time interval in multiple impacts

The influence of time and space intervals on the interactions between multiple impacts is direct. As described, the increasing dispersion between impacts results in a gradual convergence with single impacts. The same applies with time. Very long durations between two impacts result in less interactions. The response of the material tends to be indifferent from single impacts. However, a distinction between simple single impacts from single subsequent impacts has to be made. An impact in many cases may cause permanent changes in the target material. Plastic deformation, crack formations and structure distortion are the examples of residual damage from an impact. If the same material is subjected to a subsequent impact (at a different impact point), this residual damage will influence the terminal ballistic response.

The extent of the influence of cumulative damage effects and dynamic interactions in the initiation of perforation in multiple impacts is examined in this section. Four cases of fabrics with pre-defined damage formations are investigated under single and dual ballistic impacts.

The kinetic energy absorption of the fabric is analysed in all four cases and is compared with the single and triple impacts with no pre-defined damaged plates.

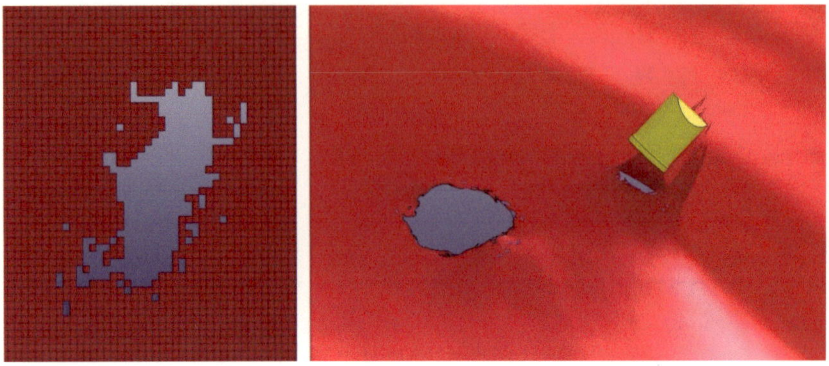

Figure 8. Pre-defined damage pattern, before and during single impact.

The damage patterns that were induced in the plate were taken by the eroded shell elements in the triple-impact simulations at 440 m/s. A small script in Matlab© was written in order to filter and document the eroded elements from the message files from the triple-impact simulations at 440 m/s so to be deleted in the key files. The calculations for the absorbed kinetic energy are based on the difference between the initial/residual velocities of the projectiles. The shapes of the damaged fabrics are presented in the **Figures 8** and **9**. The results are presented in the (**Figure 10**) and on the (**Table 5**).

5.5.1. Fabric with one hole under single impact (cases 5 and 6)

Two simulations with one projectile impacting at two different points of a plate with pre-defined damage were performed. Results show a reduced ballistic performance compared to single-impact simulations on intact plate. More specifically, the energy absorption of the plate was reduced by 4.58–9.37% compared to single-impact simulations and by 37.32–40.46% compared to triple-impact simulations with no pre-defined damage. These results prove the hypothesis that the reduced ballistic performance of the plate is due to the dynamic interactions of multiple impacts. Cumulative damage on the plate during the multiple impacts plays less important role in the initiation or continuation of perforation.

5.5.2. Fabric with one hole under dual impact (case 7)

In this simulation, the cumulative damage effects together with the dynamic interactions of two simultaneous impacts are examined. A pre-defined single perforated plate under dual ballistic impact was performed. The results show elevated energy absorption of the fabric 93.07% more than the single impact with intact plate and 26.83% more than the triple-impact simulations with intact plate. In this simulation, the fabric instead of perforating at the two impact points, the combined impulse from the two projectiles increases the surface tension resulting in shear tearing of the pre-defined hole. Dynamic interactions of the two simultaneous impacts seem to increase the energy absorption of the damaged fabric by invoking more extensive damage patterns.

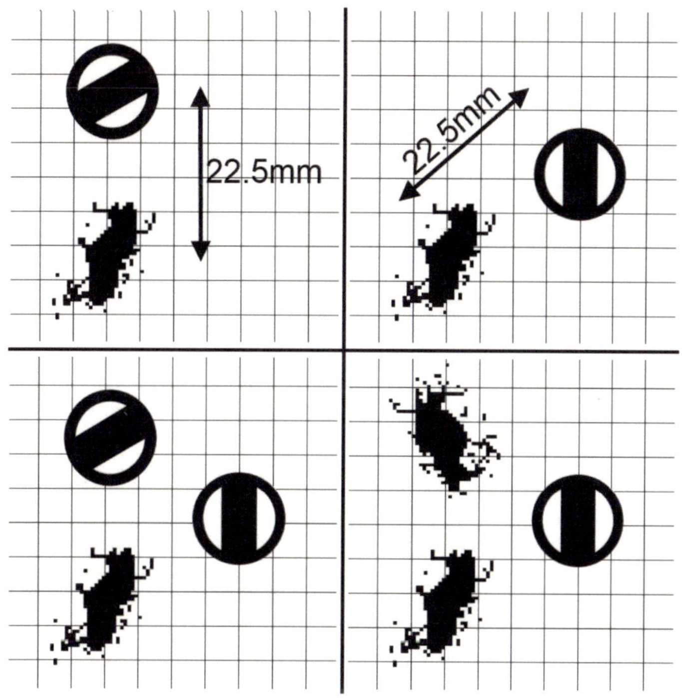

Figure 9. Configurations of simulations with pre-perforated plates.

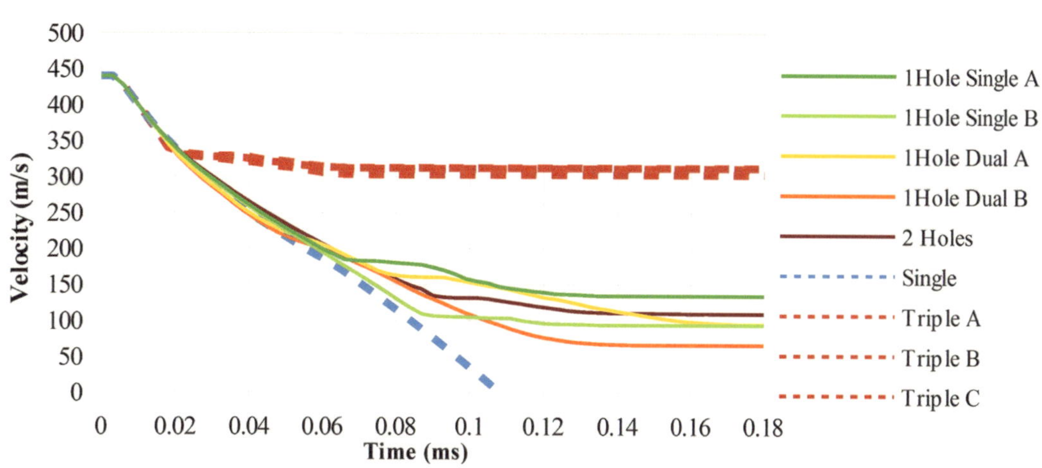

Figure 10. Velocity of projectiles during impact, for single and dual impacts on damaged fabrics and comparison with single and triple impacts on non-damaged fabrics.

Parameter	Effect	Penetrability	Interactions
Velocity	In higher velocities, multiple impacts converge with single impacts. The material responses locally due to inertia. Perforation sequence is fast, thus less interactions are exhibited between multiple impacts.	Increases	Decrease
Projectile sharpness	Sharp impactors induce more localized material responses.	Increases	Decrease
Material hardness	Hard materials respond locally.	Depends on other material properties	Decrease
Obliquity	The impact energy is more distributed.	Decreases	Increase
Boundary conditions	More restricted material response.	Increases	Decrease
Dispersion	In off-axis impacts, the interactions decrease severely with increasing distance. On-axis impacts are less affected.	Decreases	Decrease
Projectile rotation	Flat non-axisymmetric impactors enhance the influence of projectile rotation. The stress concentration points are shifted. Also, anisotropy and heterogeneity exacerbate its influence.	Increases	Depends on anisotropy/ heterogeneity
Formation rotation	Isotropic/homogeneous materials behave indifferently when the formation of the impacting projectiles rotates. Materials with higher degree of anisotropy/ heterogeneity alter their response if the projectile formation rotates.	Depends on Anisotropy/ Heterogeneity	Depends on Anisotropy/ Heterogeneity
Time interval	The time interval between impacts affects other parameters as well, such as the obliquity and the relative impacting velocity. Its effect on penetrability depends greatly on the oscillation phase of the material.	Depends on Phase and/or oscillation phase	Increase generally though some phases are more synergic
Anisotropy	Defines the distribution of stress waves. The material response depends strongly on the formation of the impacting projectiles.	Depends on angle with respect of material axes $f(\theta)$	Depend on angle with respect of material axes $f(\theta)$
Heterogeneity	Defines the distribution of stress waves. The material response depends strongly on the formation of the impacting projectiles.	Depends on coordinates $f(x,y,z)$	Generally decreases Depends on coordinates $f(x,y,z)$

Table 5. Map of the parameters involved in triple impacts.

5.5.3. Fabric with two holes under single impact (case 8)

In this simulation, one projectile impacts on a plate perforated on two points. The same Matlab script was used as with the single-hole simulations without filtering the higher number ID elements. The ballistic performance was in the same levels as with the simulations with a single hole and a single impact with 6.24% absorbed kinetic energy compared to single-impact simulations with no pre-defined damage and 38.41% compared to triple-impact simulations without pre-defined damage.

5.6. Experimental validation of the triple impact technique

A triple launcher was developed for conducting multiple-impact tests on protective structures and materials. The system consists of a 12.7 mm barrel (0.50 in), three 5.56 mm barrels and a custom made diffuser that connects the 12.7 mm barrel to the three 5.56 mm barrels. The whole assembly is fitted on a laboratory universal receiver (**Figure 11**). The triple launcher is capable of shooting three 1.102 g NATO fragment-simulating projectiles (FSP) simultaneously within a range of velocities between 390 and 580 m/s depending on the amount of charge mass that is inserted. The three 5.56 mm barrels are arranged in an equilateral triangular formation.

The charging protocol for the triple launcher is the following:

1. The front piece of the triple launcher, which holds the three 5.56 mm barrels, is removed in order to have access to the chambers of the 5.56 mm barrels.

2. The three 1.102 g FSP are loaded in the three 5.56 mm barrels.

3. The front piece is reattached to the rest of the assembly.

4. The 12.7 mm barrel is then loaded with a cartridge containing the desired amount of combustion powder.

5. The weapon system is armed and ready to shoot.

After ignition, the combustion gases are directed from the 12.7 mm barrel to the diffuser. The diffuser redirects the gases through the 5.56 mm barrels accelerating the three 1.102 g FSP.

The triple-impact technique was evaluated experimentally against 15 layers of plain weave Kevlar® K29. The fabrics were fully clamped on a 100 x 100 mm clamping device and they were tested against single and triple impacts of 1.102 g FSP (**Figure 12**). The single-impact tests were conducted with the use of a standard single barrel of 5.7 mm and the triple-impact tests with the triple launcher. In total, 10 fabrics were tested against single shots and eight against triple shots. The velocities were measured using a Photron high-speed camera that was positioned laterally, filming perpendicular to the projectiles flight. A second high-speed camera was also used to monitor the status of the material during impact and to assist evaluating which of the shots was perforating (in the case of triple impacts).

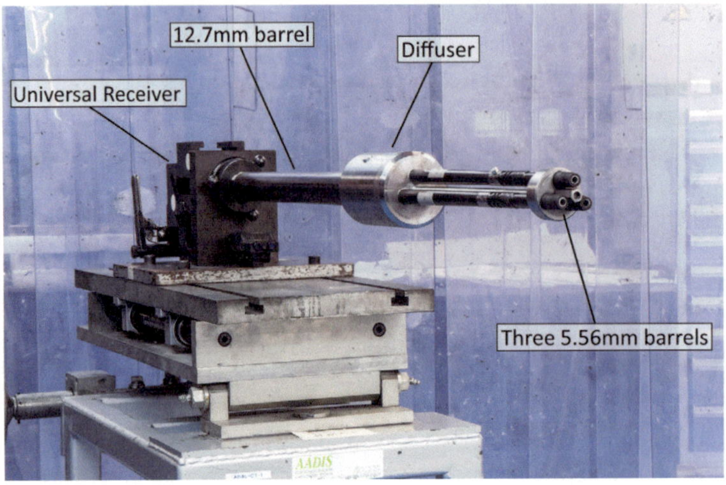

Figure 11. The triple launcher.

Figure 12. Clamped 15 layers of Kevlar® K29.

The experimental results for the single-impact tests were processed using the Probit calculation method and the results were fitted with the normal cumulative distribution. For the triple-impact tests, the results were calculated as an average velocity value per triple shot. The results from the triple-impact tests were fitted with the Weibull cumulative distribution using the method of least squares. The ballistic limit of the aramid fabric in single-impact tests was 455 m/s while based on the triple-impact tests, the ballistic limit dropped to 426 m/s (**Figure 13**). The terminal ballistic results can be found on the (**Table 6**). Seven consequent instances of a triple impact test, with three FSP being ejected towards an aramid specimen can be found on the (**Figure 14**).

Test	Velocity/average velocity (m/s)	P/NP	V50 (m/s)
Single	433	NP	
Single	438	NP	
Single	449	NP	
Single	451	P	
Single	453	NP	**455**
Single	459	P	
Single	465	P	
Single	465	NP	
Single	466	P	
Single	467	P	
Triple	349	NP	
Triple	419	NP	
Triple	421	3P	
Triple	428	NP	**426**
Triple	430	1P	
Triple	433	NP	
Triple	434	3P	
Triple	435	2P	

Table 6. Results from tests single versus triple impacts.

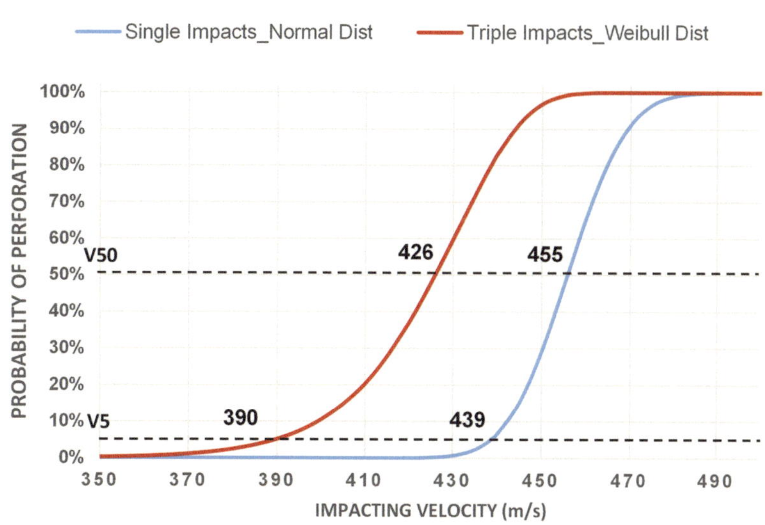

Figure 13. Probability of perforation as a function of impacting velocity for single and triple impacts.

The probability density function that is produced reveals the true value of the triple-impact technique. The V50 is a statistical value that has not direct practical use. The velocity at which perforation is improbable, i.e. 5%, is more usefull for the end users and the manufacturers. Due to the unpredictable nature of the interactions between multiple impacts, the actual difference of the ballistic performance may be much greater.

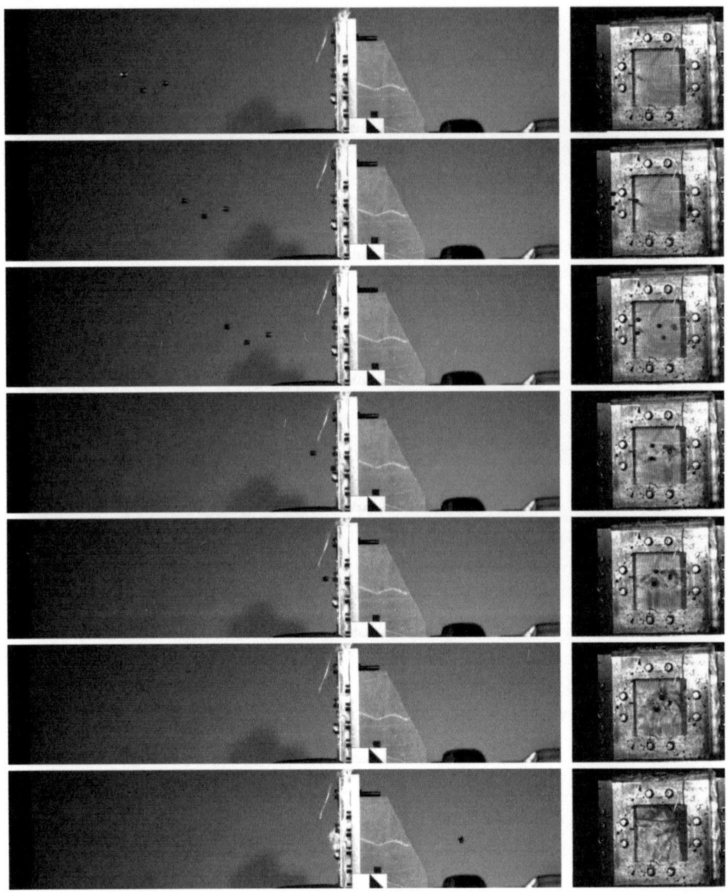

Figure 14. A triple impact test that resulted in single perforation.

6. Conclusions

- Numerical simulations and experimental results have shown a significant decrease of the ballistic performance of the aramid fabric under triple impacts compared to single impacts. The constructive/destructive interference of the stress wave patterns of the three impacts increase the stress-strain energy density of the material closer to its strength.

- Different configurations of triple impacts, such as the orientations of the projectile formation, the rotation of the projectiles (in the case they are not axisymmetric) and their radial dispersion, and modify the material's apparent ballistic performance.
- With the increase of the initial velocity of the triple impacts, their residual velocity converges with the residual velocity of the single impacts.
- The on-axis impacts due to generation of pure mode longitudinal stress waves interact more intensely and exhibit higher residual velocities.
- With increasing distance between the projectiles, the triple impact interactions decrease; thus the penetrability tends to converge with the one in single impacts.
- With increasing distance between the projectiles, the influence of paired on-axis impacts becomes more obvious. This happens because of the stress wave streams that become narrower in higher distances. Non-pure (quasi-longitudinal) waves become evanescent.
- The ballistic performance in simultaneous multiple impacts cannot be assessed with subsequent impacts.
- The existence of damage on the target material does not imply necessarily reduction of the ballistic performance.

Acknowledgements

The research leading to these results has received funding from the European Community's Seventh Framework Programme (FP7/2007–2013) under grant agreement number 284747 (TIRAMISU project).

Author details

Georgios Kechagiadakis* and Marc Pirlot

*Address all correspondence to: georgios.kechagiadakis@rma.ac.be

Royal Military Academy, Brussels, Belgium

References

[1] Roey J.V, Rabet L, Reck B, Bousu F, Imad A. Experimental investigation of the failure of fabric armour during ballistic impact. Proceedings of the 2nd International

Conference of Engineering against Fracture (ICEAF II); June 2011; Mykonos, Greece; p.22—24.

[2] Qian L, Qu M, Feng G. Study on terminal effects of dense fragment cluster impact on armor plate. Part I: Analytical model. International Journal of Impact Engineering; 31; 2005; 755–767.

[3] Qian L, Qu M. Study on terminal effects of dense fragment cluster impact on armor plate. Part II: Numerical simulations. International Journal of Impact Engineering; 31; 2005; 769–780.

[4] Norris FE. Description of the five steel cube launcher used in the multiple fragment impact test. Naval Air Warfare Center Weapons Division China Lake, CA, No. NAWC-WPNS-TP-8062. June 1993.

[5] Deka L.J, Bartus S.D, Vaidya U.K. Multi-site impact response of S2-glass/epoxy composite laminates. Composites Science and Technology; 69; 2009; 725–735.

[6] Kechagiadakis G, Pirlot M, Doucet A. The material response of PPE against Dual Impacts: The Double Tap. Proceedings of the 12th International Symposium "Mine Action 2015"; 27–30 April 2015; Biograd, Croatia; ISSN 1849-3718; p. 67–71.

[7] Nayfeh A.H. Wave propagation in layered anisotropic media with applications to composites. North-Holland Series in Applied Mathematics and Mechanics; 39; 1995; 31–115.

[8] Jones R.M. Mechanics of Composite Materials. Technology and Engineering; CRC Press; 1 July 1998; p. 55–66 ISBN 9781560327127.

Chapter 9

Unmanned Ground and Aerial Robots Supporting Mine Action Activities

Yvan Baudoin, Daniela Doroftei, Geert de Cubber,
Jean-Claude Habumuremyi, Haris Balta and
Ioan Doroftei

Additional information is available at the end of the chapter

http://dx.doi.org/10.5772/65783

'Technology developed on the basis of real needs, in a participatory way together with people who exf cost-legged robots, as illustrated pressed these needs, contributes significantly to their human development by enhancing their knowledge and creativity'

—E.E. Cepolina (Snail-Aid, Italy)

Abstract

During the Humanitarian-demining actions, teleoperation of sensors or multi-sensor heads can enhance-detection process by allowing more precise scanning, which is useful for the optimization of the signal processing algorithms. This chapter summarizes the technologies and experiences developed during 16 years through national and/or European-funded projects, illustrated by some contributions of our own laboratory, located at the Royal Military Academy of Brussels, focusing on the detection of unexploded devices and the implementation of mobile robotics systems on minefields.

Keywords: robotics, demining, navigation, sensors, image processing

1. Introduction

The mobile robotics systems are beginning in applications related to - security and the environmental surveillance: prevention of disasters and intervention during disasters with all possible kinds of missions ensuring the security.

The general objective of the International Advanced Robotics Program (IARP) [1] is, as foreseen by its status, 'to encourage development of advanced robotic systems that can dispense with human work for difficult activities in harsh, demanding or dangerous environments, and to contribute to the revitalization and growth of the world'.

Through a first book devoted to the 'Emerging Robotics and Sensors Technologies for Humanitarian Demining and Risky Interventions' [2], the IARP working group HUDEM (Robotics Assistance to Mine-clearing) summarized some important results of R&D activities focusing on the robotics and associates technologies applied to the solution of the Humanitarian demining.

The challenges posed by the use of robotics systems have been examined and partially solved in the FP7-TIRAMISU project. The following chapters focus on the contribution of the Royal Military Academy (RMA) to these challenges.

2. Robotics systems developed at RMA

2.1. 1997–2002

Under the HUDEM'97, project funded by the Belgian Ministry of Defence, with the close cooperation of our partners of the European Network CLAWAR [4], we focused on the design and development of small low-cost-legged robots [10], as illustrated in **Figure 3**.

Wheels have the advantages of engineering simplicity, low friction on a smooth surface and they enable the robot to move at a high speed. Laying down a track for wheels to run on is a way of extending the use of wheels to soft and rough ground. But wheels and tracks have a main weakness. They have poor performance in an unstructured environment when faced with a vertical step or a discontinuous surface. More than half of Earth's land area precludes wheeled and tracked vehicles. In many of these natural terrains, legs are well suited.

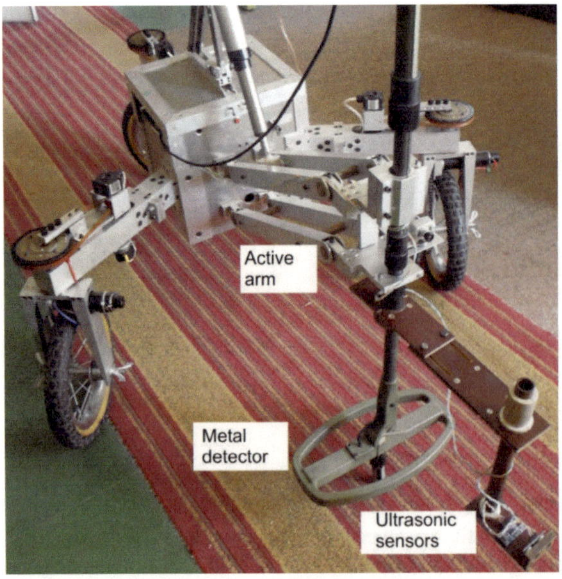

Figure 1. Multi-legged robot, as first prototype (AMRU 5).

http://dx.doi.org/10.5772/65783

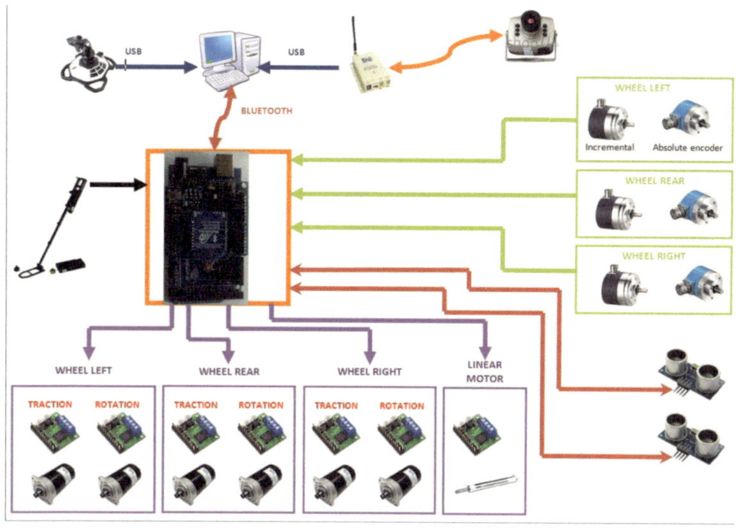

Figure 2. The three-wheeled TRIDEM robot.

Some reasons explain the choice of walking robots [6, 7, 8, 9]:

The last six-legged robot we designed [5] could be equipped with a chemical sensor and followed up with a series of electro-pneumatic platforms, among which a sliding robot could be equipped with several tools.

Several partners of the IARP and CLAWAR Working Groups HUDEM developed similar concepts, but no one could satisfy the basic requirements of cost-effectiveness. The control of such machines is not obvious (**Figure 1**).

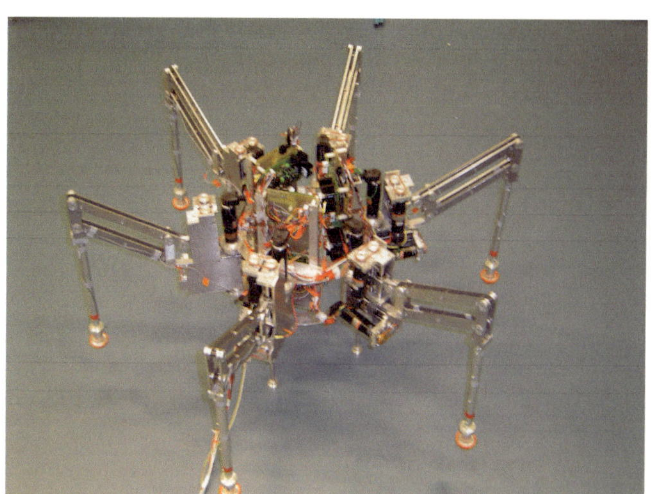

Figure 3. Robot control hardware.

2.2. 2002–2006

From the previous experience and the cost-effectiveness requirements, it quickly appeared that a basic constraint on the design of a robot was the modularity and the conviviality of the human-machine interface (HMI) in order to ease the interpretation of the operator.

Our next design focused on a three-wheeled lightweight teleoperated platform (TRIDEM) that was refined under the project FP7-TIRAMISU [11, 14,] (**Figure 2**).

The last prototype of TRIDEM is controlled by a wireless joystick. An overall view of the robot control hardware is shown in **Figure 3**. A microcontroller placed on the robot is responsible for the robot control. The joystick is connected to a computer, via a USB port. The commands of the human operator are sent to the robot through this joystick and wireless connection between computer and microcontroller. Information concerning robot movements, the presence of an obstacle, the presence of a mine, and so on, are sent from the microcontroller to the computer. All these parameters may be visualized by a human operator, thanks to a graphical user interface (see **Figure 4**). A wireless video camera, mounted on the robot arm, sends images to the same graphical interface, so that a human operator is informed about the work environment of the robot.

In most robotic applications, in Humanitarian demining in particular, the robot has to be operated by an inexperienced user. Therefore, a simple and intuitive interface is required especially when the robot has many motors and degrees of freedom. We developed such an interface for TRIDEM (**Figures 3** and **4**).

2.3. 2006–2010

Although multi-legged or similar multi-wheeled robots offer promising solutions and despite the current maturity of such platforms, the preference was given to the conversion of existing commercial platforms. Exchange of information was then pursued through the organization of annual IARP workshops, some of them located in countries confronted with mined areas:

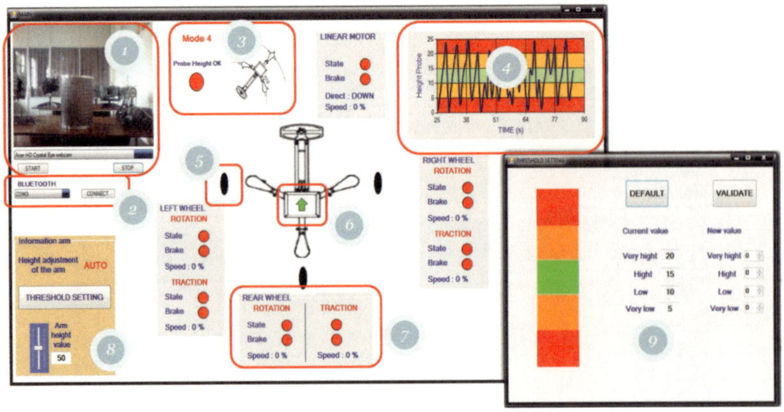

Figure 4. Graphical user interface (HMI).

Kosovo, Egypt, Tunisia, Croatia or in countries pursuing national funded R&D related to the Humanitarian demining: Austria, Japan and Belgium (**Figure 5**).

At the RMA, we then focused on the adaptation of two commercial platforms: the ROBOSOFT, renamed ROBUDEM, laterally equipped with a three-dimensional (3D) scanning carrier of a VALLON metal detector (**Figure 5**) and progressively adapted for an autonomous behaviour-based navigation control [12], finally tested in the context of the FP6 project VIEW-FINDER focusing on the Robotics assistance of security services.

A detailed study of the state-of-the-art has been entrusted to Daniela Doroftei [13].

An optimal combination of sensors (location of the robot, detection of explosives and vision sensory package) was analysed and tested under the project FP7-TIRAMISU. A partial combination has been developed at the RMA, focused on the use of a limited number of sensors: a behaviour-based architecture for mine detection (project RSTD MB07 [13]) and a behaviour-based navigation for search and inspection interventions (project VIEW-Finder [15]). **Figure 6** summarizes the control architecture.

2.4. 2011–2015

Under the FP7-TIRAMISU project [20], another commercial mobile robotic platform (the tEO-Dor) was used for the integration with a metal detector array developed by VALLON (the MCMD) (**Figure 7**).

An important drawback of the standard EOD tEODor platform is that it did not feature any autonomous capabilities. To overcome such a constraint, the platform was upgraded by RMA

Figure 5. The ROBUDEM.

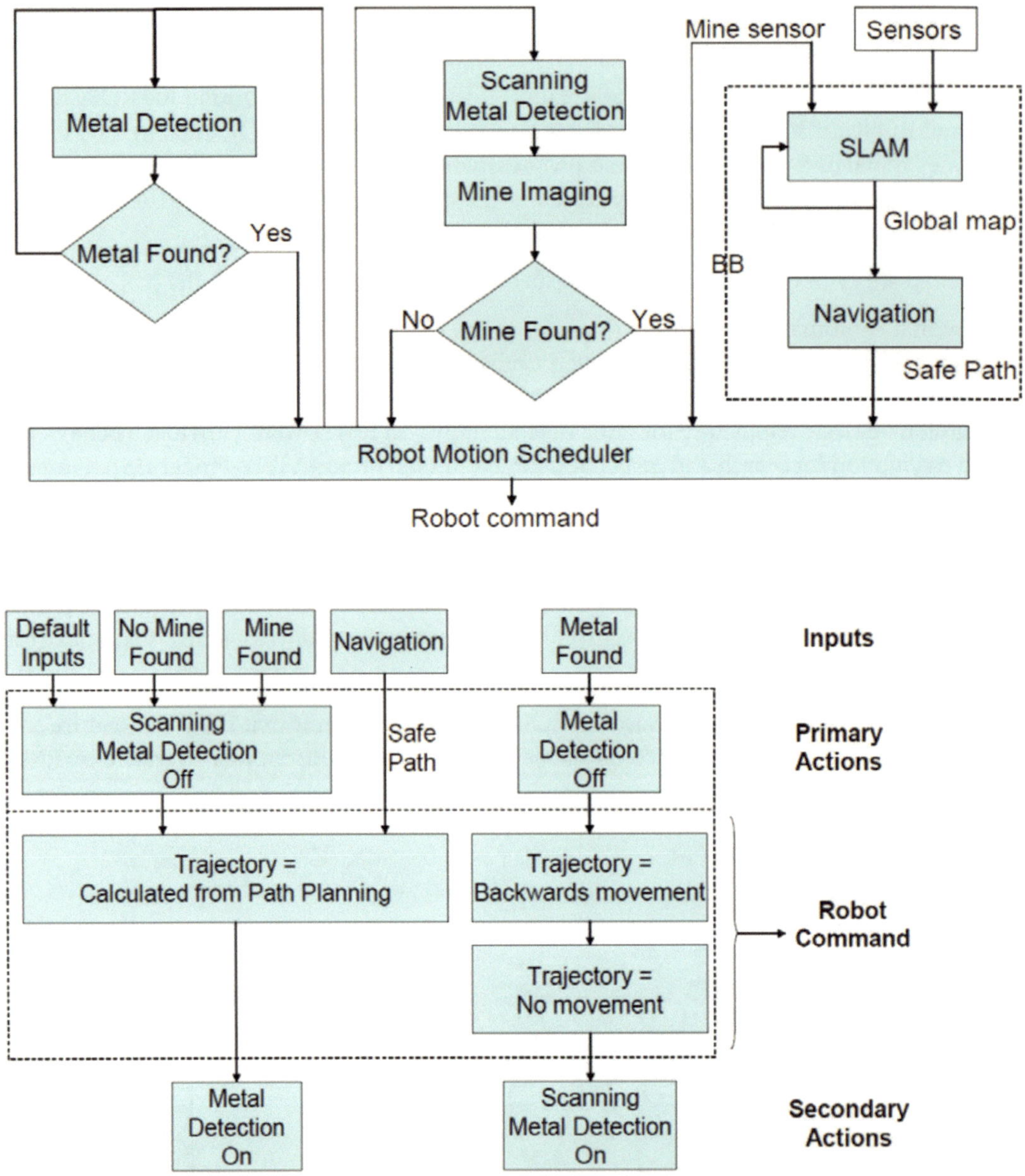

Figure 6. General control architecture of the ROBUDEM.

with necessary electronics, sensors, computing power, motor control units and power sources in order to be able to execute remote-controlled and semi-autonomous tasks [18].

A geo-positioning and communication device, the TCP box developed by DIALOGIS/ PROTIME, partners of TIRAMISU, was integrated with the system for geo-referencing the data.

The validation tests of the system were conducted on a dummy minefield at the SEDEE-DOVO (Demining Service of the Belgian Defence) and led to the desired Technical Readiness Level 7.

Figure 7. The teoDOR equipped with the MCMD and a TCP box positioning system.

Furthermore, a similar platform was adapted under the FP7-ICARUS project (2012–2016—Robotics Assistance for Search and Rescue Operations), including vision capabilities enabling an assessment of terrain traversability and thereby allowing a semi-autonomous navigation facing the traversability issues [16, 17, 19, 21].

3. Remotely piloted aircraft system (RPAS) deployment for mine action

The deployment of the remotely piloted aircraft system (RPAS) was extensively used within the TIRAMISU project [20] for the support of different scenarios (**Figure 8**).

Figure 8. Overview of the complete RPAS.

Over the 3 years (2012–2015), the RPAS was deployed in test areas as well as in real missions. The next sections shortly introduce some of the missions done with the RPAS.

4. RPAS survey of honeybees

Survey of honeybees with an RPAS was a cooperation we did in 2014 together with the Croatian partners CTRO and the University of Zagreb within the TIRAMISU project. We used the RPAS in order to make the survey of honeybees by monitoring their activities and to analyse their ability to detect buried mines. The initial first proposal was prepared by Milan Bajic (CTRO) and Haris Balta (RMA) and field mission was done in the period of end of July beginning of August 2014 at the CTRO test centre in Cerovac, Croatia.

The scenario was that the conditioned (trained) honeybees fly over the area where landmines (real, with explosives, not the dummy landmines) were placed, or over area where remains of the explosive exist after the use of mechanical machines. The images shot from a RPAS (hovering or flying very slow) provided a time sequence of images of one area. Processing and analysis of time sequence enables assessment of bee's density in space and time.

Therefore, we used the RPAS in hover-mode-flying altitude around 25 m with the high-resolution digital camera and collecting sequences of 60–85 images per each section of the minefield. Collected images were processed and analysed offline by CTRO and the University of Zagre.

5. RPAS use for hazardous suspected area

In 2015, we prepared a campaign for operational validation of RPAS [3, 22–24]. This work was conducted together with CTRO and ULB, Belgium, and partially with the support of the University of Zagreb. We used the RPAS in order to perform oblique flights and near infrared

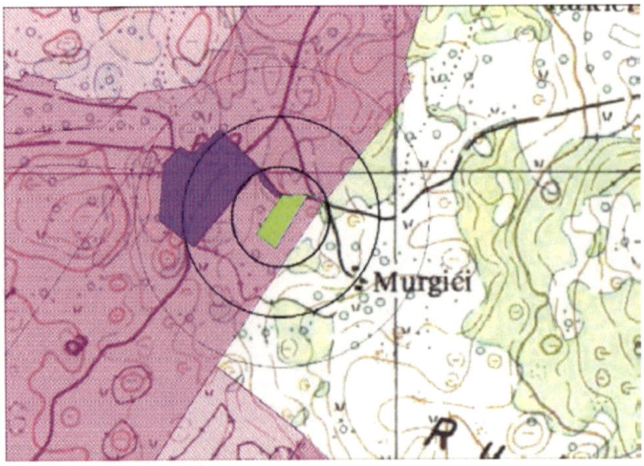

Figure 9. HSA operational area Murgici, Croatia.

http://dx.doi.org/10.5772/65783

(NIR)-mapping flights of the suspected hazardous area (SHA) and the minefields. The initial first proposal was prepared by Milan Bajic (CTRO) and Haris Balta (RMA) and A field mission was done in April 2015 in the Region Murgici, Croatia (44°38'53.28″ N 15°28'10.55″ E.). **Figure 9** shows the SHA (cross-hatched in pink colour).

6. Oblique survey of the indicators of mine presence

Some of the existing tools for the detection of mine presence indicators are limited by nadir images shooting. On the plane terrain, it is acceptable, but on hilly terrain or in mountains this is very limiting. When the RPAS is used on terrains where objects of interest (indicators of mine presence) exist, the oblique images can provide very valuable data of the mine-suspected area. We used the RPAS on our second mission and collected data for these activities (**Figure 10**).

General overview flight above the SHA. Flying altitude is around 120 m. Detection of a possible anomaly is shown in **Figure 11**.

Vertical inspection flight above the anomaly is to confirm it as shown in **Figure 12**.

Figure 10. Sequence of NiR images over the SHA.

Figure 11. Flight above the SHA with a recognition of an anomaly.

Figure 12. Vertical flight above the anomaly with zoom.

7. RPAS deployment in post-flood mine action with TIRAMISU end-user Bosnia and Herzegovina Mine Action Centre (BHMAC), 2014

RPAS was deployed in post-flood mine action activities in Bosnia and Herzegovina in the period of May to June 2014 [3]. We used the RPAS to assist the team from the Bosnia and Herzegovina Mine Action Centre (BHMAC) in detecting the locations of Explosive Remnants of War (ERWs) and performing damage assessment, mapping and aerial inspection. The ERWs were displaced as a result of landslides caused by the floods.

Detected mine, re-located due to the landslides, is shown in **Figure 13**.

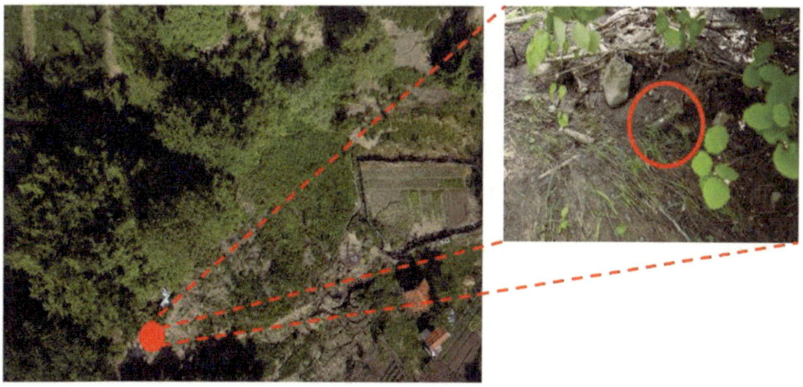

Figure 13. Detected mine in the region: Sarajevo-Vogosca, Bosnia and Herzegovina, 4 June 2014.

Figure 14. Sequence of oblique images around the SHA.

First post-processed results of the regions Zavidovici- Dolac and Olovo, in central Bosnia and Herzegovina, with geo-referenced aerial images, were produced (as shown in **Figure 14**). The RPAS was used for providing 3D-maps, orthophotos and digital terrain models of the environment to analyse the effects of the landslides on mines and ERW. This result has been used by Bosnia and Herzegovina Mine Action Centre (BHMAC) for the localization of displaced ERW, damage assessment and documentation purposes.

8. Conclusions

The development of a Robotics System for demining operations has to take several constraints into account: a high level of protection against the environmental conditions (dust, humidity, temperature, etc.), protection and resistance against vibration, mechanical shocks and instability factors, a sufficient autonomy and reliable communications between the mobile platform and the operator.

In this chapter, some of the most relevant aspects of both technical and environmental aspects have been underlined. Other robotics systems have been co-developed with the partners of the FP7-TIRAMISU project. We invite the reader to consult [20].

Acknowledgements

These activities were partially funded by the Belgian Ministry of Defence (HUDEM'97, 1997–2002) (BEMAT and MB07, 2003–2008), by the European Commission (CLAWAR 1998–2002) (FP6 VIEW-FINDER Project Contract 045541 2007–2010), (FP7 TIRAMISU Contract 284747 2012-2015), and (FP7 ICARUS Contract 285417 2012-2016).

Author details

Yvan Baudoin[1]*, Daniela Doroftei[1], Geert de Cubber[1], Jean-Claude Habumuremyi[1], Haris Balta[1] and Ioan Doroftei[2]

*Address all correspondence to: yvan.baudoin@skynet.be

1 Royal Military Academy, Department of Mechanical Engineering, Unmanned Vehicle Centre, Brussels, Belgium

2 Department of Mechanical Engineering, Technical University GS Asachi, IASI, Romania

References

[1] IARP. www.iarp-robotics.org *(accessed on 2016-06-21).*

[2] Baudoin Y., Habib M.K. Using Robots in Hazardous Environments. Landmine detection, demining and other applications. Woodhead Publishing in Mechanical Engineering, 2011. ISBN 978-1-84569-786-0

[3] Haris B., Geert De C., Yvan B., Daniela D. UAS deployment and data processing during the Balkans flooding with the support to Mine Action"8th International Advanced Robotics Programme (IARP) Workshop on Robotics for Risky Environment (RISE), 28-29 January 2015, Lisbon, Portugal.

[4] CLAWAR. www.clawar.org *(accessed on 2016-06-22).* Clawar Task 9 report, 1999.

[5] Habumureyi, J.C. Adaptive neuro-fuzzy control for a walking robot with 6 pantograph-based legs. PhD Thesis, Free University of Brussels, 2004.

[6] Gonzalez de Santos P., Garcia E., Estremera J., Armada M.A. SILO-6: design and configuration of a legged robot for humanitarian demining. International Symposium Clawar, Brussels, 1998.

[7] Berns K., Kepplin V. Lauron-II a general purpose walking machine for rough terrain', International Symposium Hudem'97, Royal Military Academy, Brussels, June 1999.

[8] Nonami K. Development of autonomous mine detection six-legged walking robot for Humanitarian demining, 5th IARP Workshop, Tokyo, Japan, June 2005.

[9] Randall M.J., Pipe A.G. An intelligent control architecture and its application to walking robots. Proceedings of International Workshop on Advanced Robotics and Intelligent Machines, Salford, UK, 1997, ISSN 1363-2698.

[10] Baudoin Y., Dorofteï I. Hierarchical control of a Hexapod walking robot. International Journal of Robotica & Management 14-1, 2009, ISSN 1453-2069.

[11] Doroftei I., Baudoin Y. Using mobile robots for a clean and safe environment-a difficult challenge, in Proceedings of 2012 International Conference and Exposition on Electrical and Power Engineering (EPE), Iasi, Romania, 2012, pp. 41-46.

[12] De Cubber G., Berrabah, S.A., Doroftei D., Baudoin Y., Sahli H. Combining dense structure from motion and visual SLAM in a behavior-based robot control architecture. International Journal of Advanced Robotics Systems, vol. 7, issue 1, 2010.

[13] Doroftei D. Behavior based navigation techniques. Internal report MoD Study MB07, October 2006 (http://mecatron.rma.ac.be/Research/Behavior-based_Robot_Control.html).

[14] Doroftei D., Baudoin Y. Development of a semi-autonomous demining vehicle. Proceedings of the 7th IARP WS HUDEM'2008, The American University in Cairo (AUC), Cairo, Egypt, 28-30 March 2008.

[15] De Cubber G., Dorofeti D., Marton G. Development of a visually guided mobile robot for environmental observation as an aid for outdoor crisis management operations, Proceedings of the IARP Workshop on Environmental Maintenance & Protection, July 22-23, 2008, Baden-Baden, Germany.

[16] De Cubber G., Doroftei D. Human victim detection and stereo-based terrain traversability analysis for behavior-based robot navigation, handbook using robots in hazardous environments. Landmine detection, demining and other applications. Woodhead Publishing in Mechanical Engineering, 2011, pp. 476-496. ISBN 978-1-84569-786-0.

[17] De Cubber G. Variational methods for dense depth reconstruction from monocular and binocular video sequences, PhD Thesis. VUB-RMA, 2010 (http://mecatron.rma.ac.be/Publications.html).

[18] Balta H., Wolfmayr H., Braunstein J., Baudoin Y. Integrated Mobile Robot System for Landmine Detection. Proceeding of 12th IARP WS HUDEM 2014, Zadar, 25 April 2014 (http://mecatron.rma.ac.be/pub/2014/IARP-6-BALTA.pdf).

[19] De Cubber G., Balta H. Terrain traversability analysis using full-scale 3D processing. in Proceeding of 8th IARP Workshop on Robotics for Risky Environments, Lisbon, January 28, 2015.

[20] www.fp7-tiramisu.eu (accessed on 2016-06-17).

[21] www.fp7-icarus.eu (2016).

[22] De Cubber G., Balta H., Doroftei D., Baudoin Y., 2014. UAS deployment and data processing during the Balkans flooding. 12th IEEE International Symposium on Safety, Security, and Rescue Robotics 27-30 Oktobar 2014, Hokkaido, Japan.

[23] Mechanical Demining Equipment Catalogue 2010, GICHD, Geneva, January 2010. ISBN 2-940369-33-X, http://www.gichd.org/fileadmin/GICHD-resources/rec-documents/MDE-Catalogue-2010.pdf, accessed on 2016-06-16

[24] A Study of Mechanical Application in Demining, GICHD, Geneva, May 2004. ISBN 2-88487-023-7, http://www.gichd.org/fileadmin/GICHD-resources/rec-documents/Mechanical_study_complete.pdf, accessed on 2016-06-16.

Chapter 10

Testing and Evaluating Results of Research in Mine Action

Yann Yvinec

Additional information is available at the end of the chapter

http://dx.doi.org/10.5772/65780

Abstract

This chapter summarizes the experience of the Royal Military Academy in testing and evaluating new tools for mine action. It first underscores the importance of testing and evaluating new methods in general and in mine action in particular. Some recommendations are given to help the design of test protocols: defining carefully the objectives of the test and what is to be measured, the importance of blind and double-blind tests, choosing between realism and statistical relevance, the importance of how to display the results, etc. These recommendations are illustrated by real-life examples, mainly from test and evaluation of detectors of mines in which RMA has been involved. A test protocol is detailed. It is the one that RMA designed and used to evaluate a detector that was proven to be useless and that led to the criminal conviction of its designer in the United Kingdom. Sources of available test protocols and test reports are also listed.

Keywords: test and evaluation, test protocol, standardisation, test design

1. Problem statement

Evaluating a solution for a problem against other, state-of-the-art solutions to the same problem is an integrant part of research. In mine action, however, this is even more important because the mine action community is both anxious to use new methods to help their work and reluctant to trust new tools. Some activities of mine action, for instance the detection of mines or other explosives, may be dangerous and changing a well-know detector for a new and possibly unknown detector is not something all mine action practitioners will do very easily. One way to promote the adoption of new tools or methods is to evaluate them in a way that would convince potential users of the added value of the proposed solution.

A trial is usually an activity you perform in order to answer a given question; does my method achieve the expected results? Is it better than other, existing methods? What are its limitations? The trial must be designed to provide answers to the question.

Therefore, designing correctly a test protocol is of paramount importance.

2. Test protocols and why bothering about them

A test protocol describes how the trial must be performed. It serves several purposes.

When correctly designed it ensures that the objectives of the test will be achieved. This is essential in order not to waste the time, money and resources required for the trial.

During the trial itself the protocol indicates clearly the steps to carry out. When the trial is carried out to decide which of two options is better and some participants of the tests differ on their favourite options, it is important that they all agree on the test protocol before the test begins in order to prevent arguments after the results are known. It is usually considered as a good practice to select the criteria to decide the outcome of a test (success or failure for instance) before the start of a trial.

When included into a test report, the protocol allows readers to assess the conclusions of the report. Conclusions of test reports can be debated just because there are some concerns about the protocols that have been used. How to analyse the results is an important part of a test protocol. The authors of a study published in 1990 considered that they had empirically proven the reality of dowsing, the ability to discover underground water with instruments such a wooden stick or a pendulum. A different statistically analysis of the same data showed that, on the contrary, no such effect could be proven [1]. The whole controversy focused on the way to analyse the data.

The current controversy over the lack of reproducibility of research published in very different scientific domains is intense [2–5]. The explanations often given to questions some published results are the use of too small samples in the experimentation and a misunderstanding of statistics in the analysis of the results.

3. Some key points to keep in mind when designing a test protocol

3.1. Introduction

A good trial, that is, a trial that provides trustworthy and meaningful information, requires sufficient allotted time, sufficient funding, adequate personnel and resources. Before starting a thorough test, it could be useful to have a first, rough evaluation of the equipment to test in order to reject right away equipment that is obviously below the expected performance. This first evaluation could be a basic check when the equipment is based on new or uncertain scientific principles. Many new ideas have been proposed to detect landmines: radar, acoustic systems, thermal systems, rats, bees, etc. Some long-range detectors of explosives have been proposed and proved to be fakes. One such system was sold to many clients, although the working principles were clearly scientifically unsound. This led eventually to the conviction of the author to a prison-sentenced of 10 years [6].

Besides, a scientific evaluation of the principle, some basic tests could be performed in order to check if the equipment works in ideal conditions. Once such an entry-test is successful a full evaluation of the equipment with varying and controlled, conditions could be carried out.

A last test could be an evaluation of the equipment in operational conditions.

When devising a test protocol there are a few issues to consider.

3.2. What is the test for?

In 2007 the mine action community was surprised by the results of a test report of a comparison of the performance of metal detectors [7]. The test showed that most metal detectors had a probability of detection significantly lower than 100%. This result conflicted with the daily experience of regular users of metal detectors. The reason of these apparent strange results was due to the very specific objectives of the test as defined by the authors. The goal of the tests was not to evaluate the field performance of metal detectors but to compare them. Since most metal detectors were expected to have good performance in normal field conditions, the test was designed with extreme conditions in terms, for example, of depths at which the target mines were buried, so that the detectors could be sometimes at the limit of their detection capability. In these extreme conditions metal detectors would have performance below 100% and it was expected that the decrease would be different from one detector to the other, thus allowing a better comparison. If the metal detectors performed so poorly in these tests, it was because of the extreme conditions of the tests, not necessarily because of intrinsic poor performance. The purpose of a test had a direct impact on the test protocol, therefore on the results and hence on the way the reader should interpret the results. Unfortunately, identifying the purpose of a test is something that is sometimes overlooked by readers and test designers alike. A test designed for a given purpose may be irrelevant for another. The answer to a question is not necessarily useful for another question.

3.3. What does 'detection' means?

This seems like a stupid question, but when evaluating the performance of a detector, defining precisely when the operator using the detector has actually detected something is important. If an operator looks for anti-personnel mines with a metal detector and finds an alarm two meters away from the only mine present, the alarm is clearly a false alarm. But how close to the mine should the alarm indication be in order to consider that this is a true detection and nor a false alarm?

A second reason to be careful is when the operator has to make a decision based on a signal, whether it is audio, visual or otherwise. It is important to make the difference between the performance of the detector and the ability of the operator to locate precisely a target. In addition to defining properly what detection is, defining what an alarm is—whether it turns out to be true detection or a false alarm—must also be done carefully.

The guidelines to evaluate the performance of metal detectors described in T&E Protocol 14747-1 address the two problems [8]. For instance an operator should sweep the detector over the target five times "*to determine whether detection is consistently indicated*". They also propose a method to decide if an alarm should be considered as detection or a false alarm. Detection is defined when the distance between the indication location and the actual target is shorter than a given distance value. The detection halo of a target is defined as the circle whose radius is this given distance. According to Ref. [8], B6, page 59 "*The radius of the detection halo shall be half of the maximum horizontal extent of the metal components in the target plus 100 mm*".

This definition was designed for metal detector. It may have to be adapted for new detectors. Experience shows that other definitions of halos have been used. For instance in a report from DGA, 'precise detection' is defined as a detection within 20 cm of the target and 'close detection' when detection is within 40 cm.

In 2009 the Royal Military Academy was asked by the International Test and Evaluation Programme for Humanitarian Demining (ITEP) to send a scientist to invigilate a trial of metal detectors and dual sensors that combined metal detectors and ground-penetrating radars. Public results can be found in Ref. [9]. But an unpublished early result proved to be quite interesting. The protocol of use of the dual sensor was as follows: the operator would first use the metal detector to determine a suspicious location, mark it with a plastic chip of a certain colour and then use the ground-penetrating radar to confirm detection. In case of confirmation the operator will place a plastic of a different colour at the location. Sometimes as illustrated in **Scheme 1** results showed that the alarm indications of the metal detector were outside the defined halo but the alarm indications of the ground-penetrating radar were within.

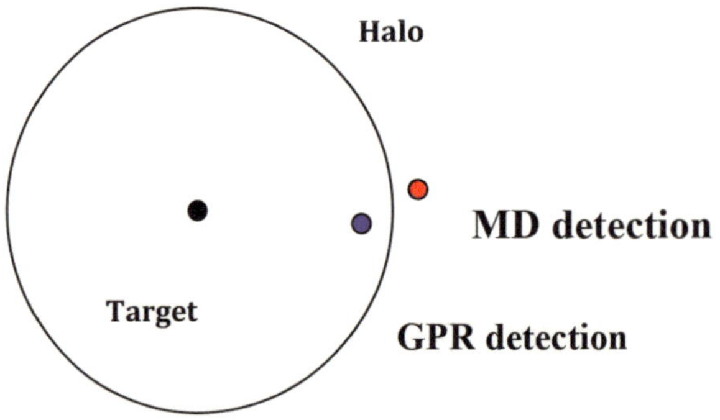

Scheme 1. Example of a test result where the alarm indication of a metal detector (in red) is outside the halo of the target (in black) and the alarm indication of the ground-penetrating radar (in blue) used as a confirmation detector is inside.

Since the footprint of the ground-penetrating radar is smaller than the metal detector footprint —making it easier to pinpoint with the radar than with the metal detector—and some targets have metal parts at their borders and not at their centres, it was indeed possible that an operator

http://dx.doi.org/10.5772/65780

using the dual sensor puts some metal detector chips just outside some halos and the corresponding radar chips just inside the same halos. By a strict definition of a hit the conclusion may be that the probability of detection of the radar, although used as a confirmation sensor, may be higher than the probability of detection of the metal detector.

This is counterintuitive. How to solve this problem?

One possibility would be to consider that there is a dual sensor hit if there is both a metal detector hit and a radar hit. This solution may lose the advantage of the radar, including its pinpointing ability.

Another possibility would be to consider that, if there is a radar hit, we should consider that there is a metal detector hit, even if the metal detector chip is a little outside the halo. Changing the definition on a metal detector hit in this manner may make it difficult to compare the dual sensor metal detector and the corresponding stand-alone metal detector.

Defining clearly a dual sensor hit is important. During the tests, some operators even wondered whether they should put the chips where they had a signal or where they thought the target was.

As a side note, during this test it appeared that one operator of the dual sensors was found out that he was colour-blind!

We could also consider that if the metal detector chip is just outside the halo and the radar chip is just inside, then the problem is that the halo is too small for the metal detector. The metal detector chip may have been outside the halo ('miss') but it was close enough to the target to allow the operator to find the target with the radar. It can be argued that the target was actually detected with the metal detector but the pinpointing was not accurate enough. Should the halo be increased then? If so, how to define the new halo?

If we want to increase the halo, then we may want to increase it also for metal detector trials. This may lead to a modification of Ref. [8].

In the report to the ITEP Executive Committee in Oberjettengberg, Germany, on 14 November 2009, a graph was presented showing what happens if the halo is doubled. With the original halo the ratio of detection was between 0.5 (for laterite soils) and 0.9 (with humus). For most detectors during the tests, multiplying the halo size by two led to a ratio of detection above 0.9.

It means that 90% of the targets had chips around them. Does that mean that the detection was correct but the pinpointing was inaccurate, or that there were so many chips on the lanes that when the halos are sufficiently increased most halos end up containing chips?

Increasing halos have an impact on both the probability of detection and the false alarm rate. When the halos are increased, more chips tend to be inside halos and therefore the probability of detection tends to increase. The false alarm rate is defined as the ratio of the number of false alarms to the area of the lane outside any halo. When the halos increase, both numbers tend to decrease. It is therefore difficult to predict how the ratio will change. If the halos increase too much they may intersect with each other. This may create problems for both the probability of detection and the false alarm rate. How should a chip inside two halos be considered? As one hit? And if so, for which target? Or as two hits? Intersecting halos have impact on the false alarm rate too. When halos intersect computing the area outside halos is a more difficult than if halos do not intersect. It is therefore easier to have halos that do not intersect with each other. How far can the halos be increased without having them intersecting? It depends on the smallest distance between targets.

When designing a new test site, using larger halo sizes may require increasing the smallest distance between targets.

Theoretically, the ratio of detection depends on many factors: the detectors, the targets, the soil, the operators, etc. But there is no reason why it should also depends on the halo size, which is only a tool to analyse the data. Halo sizes, however, have an impact on the number of what is defined as hits. A clear difference should therefore be made between probability of detection and the ratio of hits to the number of targets, which we can call the hit ratio or the ration of detection.

The probability of detection is what we want to estimate by the trials. Its value is unknown. During the trials we measure the hit ratio as exactly as possible. From this hit ratio and other information such as the number of targets we can estimate the probability of detection and some confidence intervals. Hit ratio is known exactly. There is no confidence interval for it. Probability of detection is estimated. So is its confidence interval.

This difference is important because the halo size, as a tool to analyse data, has no influence on the real performance of a detector in a given situation, but it has one on the hit ratio that is measured.

Figure 1 shows the hit ratio as a function of the false alarm rate for a halo size ranging from the definition in Ref. [8] to the double for three detectors that were tested. Note the scale used for the hit ratio.

When halo sizes increase, the hit ratio increases, as expected.

Sometimes the increase of hit ratio is large; sometimes it is small. There are two reasons, at least, why it could be small.

One reason is when the hit ratio is already high. With few missed targets that could become detected when halos increase, the hit ratio can only increase a little.

The other reason is when the false alarm rate is small. With few false alarms that could become hits when halos increase, the hit ratio increases only a little.

On the other hand, the case where the hit ratio could benefit the most from an increase of halo sizes is when the hit ratio is low to begin with and the false alarm rate is high.

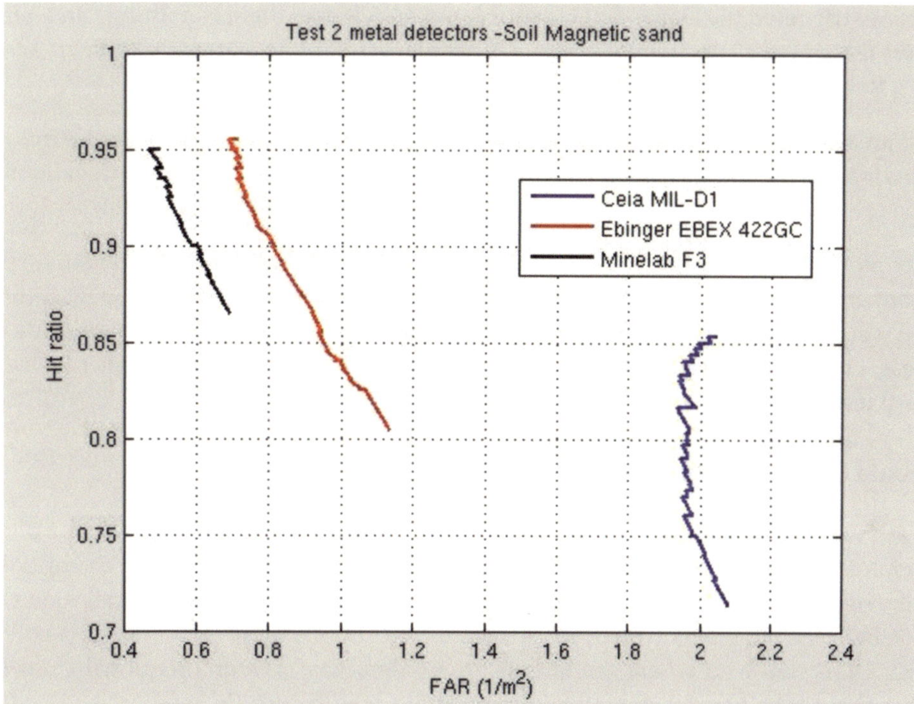

Figure 1. ROC curves as functions of the halo size.

3.4. What does 'false alarm' means?

This question is of course strongly related to the definition of detection.

When testing a detector it is not only important to estimate its capability to detect relevant target, namely its sensitivity, but also whether what it detects is a legitimate target, that is, its specificity. The false alarm rate is often used to quantify this. It can be expressed in number of false alarms per square metre, although the ratio of number of false alarms to the number of detection is sometimes used. The question then rises about the exact nature of a false alarm.

When testing a metal detector, should the presence of a small amount of metal instead of a mine be considered as a source of false alarms? For operators during clearance operations, it is a false alarm because their objective is to find mines, not scraps of metal. On the other hand, manufacturer would consider that metal detector should detect metal and then consider the detection of small metal pieces as a positive detection.

In order to solve these difficulties, [8] 8.1.2, recommends that *"the presence of metal objects around, above and under the test areas shall be avoided as far as possible"*.

With other technologies a work around is not so easy. Ground-penetrating radars, for example, can detect soil disturbances. Such disturbances can happen when a test target is buried in the test areas. Therefore, in case of detection it is unclear whether the target itself or the soil disturbance triggered the alarm. But creating a test area where there is nothing *"around, above and under the test areas"* that would create a radar alarm might be extremely difficult and may lead to a test site that is not representative to reality.

If a test area contains a small number of locations that create unwanted radar alarms, these locations may be mapped. Then test targets should not be buried close to these locations. And after the trial, radar alarms close to these locations should be removed from the analysis.

Another method would be not to use local soil but soil with known characteristics, such as sand-containing grains of a given diameter. But the radar waves that would propagate through this soil would reflect at least partially at the interface between the soil and the outside of the test areas, creating false signal that might affect detection. This method might require large and deep test areas.

3.5. Should the test be blind? Double blind?

We all know that we tend to see what we want to see. For this reason, when testing a detector, the operator must not know where the targets are hidden. A test where the operator does not know the right answer to the test is said to be a blind test. Moreover, a clear separation should be made between people who analyse the results and anyone involved in the data collection of a test. Observers who where the targets are hidden may give unintentional clues to the operator about whether the operator is performing well or not. In medical study doctors checking the health of a volunteer for a test should not know whether the volunteer has received the medicine under test or a placebo. When a test is designed in such a way that anyone involved in the data collection have no knowledge that may influence their behaviour the test is said to be double blind. Double-blind tests are considered much better than blind tests.

3.6. How to deal with human factors?

Carrying out double-blind trials is a way to avoid some human factors. Having several operators for the same detector is also a possibility, although it increases the number of tests.

Human factors are all the more important when the detectors require some interpretation from operators. It could be because the sensors output an audio signal or provide complex display, such as images.

An extreme case is ADE 651, a sensor that has been proved to be totally ineffective. It is composed of a horizontal antenna that can rotate freely around a handle. When used properly it was said to point in the direction of a source of explosive, narcotics or any other sub-stance the system was calibrated to detect. During a trial supervised by the Royal Military Academy, a bag was located in a test area. The operator did not know whether the bag contained the sub-stance to detect or something else. The operator would move around the bag along a square-like trajectory. He would first walk along the first side of the square, along a few metres. Either the antenna would point to the bag or not. Then he would move along the second side of the square, as seen in **Scheme 2**. The antenna would then either confirm the first result or not. In case of confirmation, either positive or negative, the operator would usually stop and make his claim. Then we would test a second bag. If during the second side of square the antenna would give a different indication than with the first one, the operator would try again with the third side; he would also try again the second or first side or both. After some time, he would make his claim. Almost every time his final claim was consistent with the indication of the antenna during the first side of the square. Was it coincidence or did the operator made up his mind after the first side, even if unconsciously and then try, again unintentionally, to confirm his first impression?

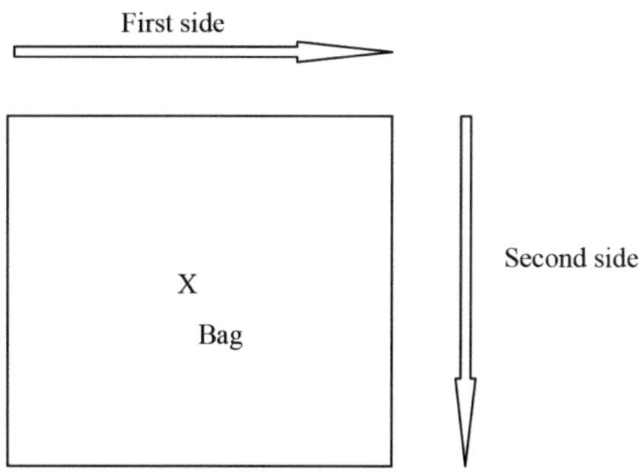

Scheme 2. Design of the test area for ADE 651.

3.7. Can a test be both realistic and statistically relevant?

There are many parameters that have an influence on the evaluation of the performance of equipment. For a detector they include, but are not limited to, the types of targets used during the test, the environmental conditions and the operator. If you want to study the influence of these factors you may want to make tests where they vary. You may want to use different kinds of targets, locate them in different situations, use several operators, etc. When factors may vary the test designed is said to be factorial and it leads to a number of runs that can become very large. In order to get statistically relevant results, the same tests might have to be carried out several times in similar conditions in order to reduce the impact of unwanted factors. But then such a test may require a long time and huge resources, and ultimately cost a lot of money. For these reasons a statistically relevant test may be carried out on a smaller set of conditions. But then the operational relevance is questionable. On the other hand, a small scale test with few redundant data collection may be fast, cheap and give clear results but it might be difficult to guarantee the relevance of the results. A trade-off must therefore be found between realism and statistical relevance.

Several ways to tackle this problem exists. One is to clearly distinguish between factors that have a random effect, and for which several replicable tests may be required and conditions that are more deterministic and for which only a few experiments are enough. Another way is not to use all runs of a factorial design, that is, not use all combinations of all factors, but only a fraction. This is called fractional factorial design. Several options exist to choose that runs to use.

The fractional factorial design is based on a delicately balanced data set. What happens if some data are not collected because of some external reason, such as lack of time or bad weather?

An example of fractional factorial design is given in Ref. [10] and is reproduced below.

Detectors are identified by two numbers: one is the model and the second is the specimen. For instance detector 2–3 is model 2, specimen 3. Operators are denoted by letters. Operators A, B, C and D are trained to operate detector 1. Operators E, F, G and H are trained for detector 2 and operators I, J, K and L for detector 3. **Tables 1** and **2** show the two rounds of six consecutive 'starts' in six different lanes. The document states that: *"After both rounds are executed, each detector will have been operated in each lane by all four persons trained for that detector"'*.

Round 1	Test #1	Test #2	Test #3	Test #4	Test #5	Test #6
Lane 1	A 1-1	G 2-2	J 3-1	D 1-2	E 2-1	K 3-2
Lane 2	E 2-1	K 3-2	B 1-1	H 2-2	I 3-1	C 1-2
Lane 3	I 3-1	C 1-2	F 2-1	L 3-2	A 1-1	G 2-2
Lane 4	B 1-2	H 2-1	I 3-2	C 1-1	F 2-2	L 3-1
Lane 5	F 2-2	L 3-1	A 1-2	G 2-1	J 3-2	D 1-1
Lane 6	J 3-2	D 1-1	E 2-2	K 3-1	B 1-2	H 2-1

Table 1. Example of a fractional factorial design (round 1).

Round 2	Test #1	Test #2	Test #3	Test #4	Test #5	Test #6
Lane 1	B 1-1	H 2-2	I 3-1	C 1-2	F 2-1	L 3-2
Lane 2	F 2-1	L 3-2	A 1-1	G 2-2	J 3-1	D 1-2
Lane 3	J 3-1	D 1-2	E 2-1	K 3-2	B 1-1	H 2-2
Lane 4	A 1-2	G 2-1	J 3-2	D 1-1	E 2-2	K 3-1
Lane 5	E 2-2	K 3-1	B 1-2	H 2-1	I 3-2	C 1-1
Lane 6	I 3-2	C 1-1	F 2-2	L 3-1	A 1-2	G 2-1

Table 2. Example of a fractional factorial design (round 2).

What would happen if one data set was missing?

Suppose Round 2, Test #6, lane 6 is missing (operator G with detector 2–1, in red in the table). On lane 6, detector 2 is no longer used by all four operators trained for it.

In order to have detector 2 used by the same number of operators on all lanes, one possibility is to remove all data sets with operator G (in yellow in the table). Then detector 2 will have been operated in each lane by three operators trained for that detector.

But then detector 2 is operated by three operators and the other detectors by four operators. In order to have the same number of operators for each detector, it may be necessary to remove all data set for one operator trained for detector 1 (say, operator A, in green in the table) and one operator for detector 3 (say, operator I, in blue in the table). Then after both rounds are executed, each detector will have been operated in each lane by three operators trained for that detector. And this reduced data set is as balanced as the original.

Then for one missing data set, 17 others (more than 20% of the available data) should be deleted.

3.8. Should the test be in controlled or realistic conditions environment?

There is a test and evaluation protocol to test mechanical equipment for mine action: T&E Protocol 15044 [11]. These guidelines explain how to create test lanes. Test centres exist with premises that permit testing machines based on this document.

The advantage of following such guidelines is that tests can be reproduced. They form reference to compare machines.

But the conditions of these tests might be different from conditions encountered in real mines areas. The type of soils, the vegetation, the slope, etc. may differ in reality from the test conditions. It has been argued that some machines could succeed in test sites and fail in reality in some situation. Therefore, some people prefer trials to take place in a mined country, close to real mined areas and close to where the machines will be actually used.

These field tests have also some drawbacks. The environmental conditions may vary from one place to the next, and the same equipment may have to be tested several times in different locations of the same country.

3.9. Which data to show and how to display them?

Ideally, all data should be reported and raw data should be displayed. But if there are a lot of data or if the data are noisy, some filtering may be used. Usually, models are used to filter data and fit a line, or any other curve, to them. These mathematical tools must be used carefully because they may alter how results are perceived.

Here is an example based on a real case. The test is designed to evaluate the performance of a mine detector combining a metal detector used as primary detector and ground-penetrating radar used as a confirmation tool. The metal detector is first used and when there is an alarm the ground-penetrating radar is switched on. That way, the ground-penetrating radar can only reduce the number of false alarms but cannot improve the number of detections.

Let us consider an experiment where the detector must detect targets buried at 0, 5, 10, 15 and 20 cm and where 10 targets are buried at each depth. Now let us assume that the results of the test were as described in **Table 3** and illustrated in **Scheme 3** where the ground-penetrating radar confirms all alarms from the metal detector but one at 5 cm depth.

Depth (cm)	0	5	10	15	20
Nb of targets detected by the metal detector	10	9	7	3	2
Nb of targets confirmed by the GPR	10	8	7	3	2

Table 3. Example of detection results of a metal detector and a GPR.

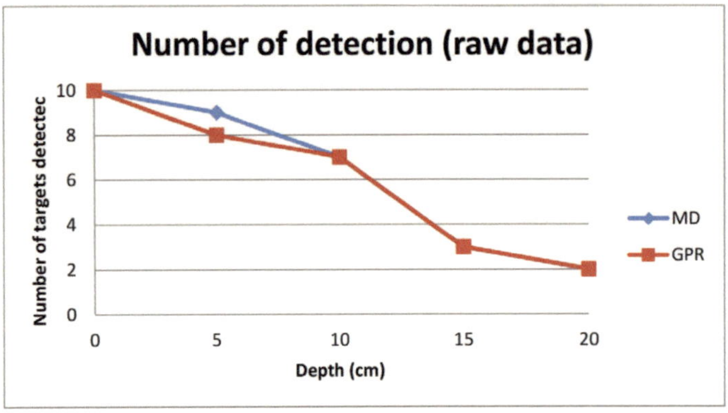

Scheme 3. Example of raw data.

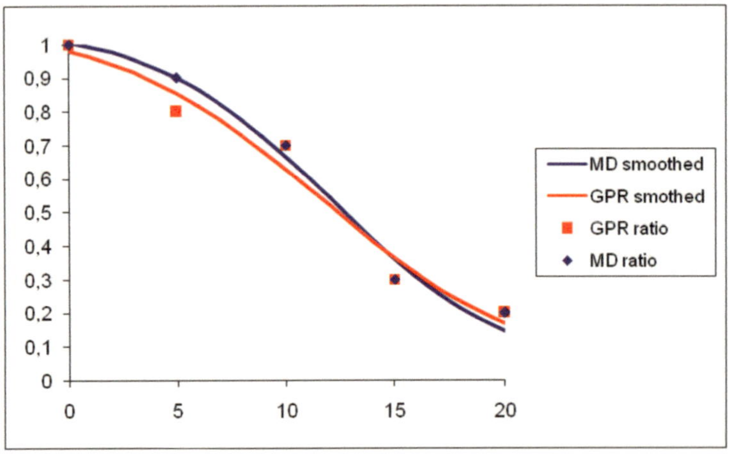

Figure 2. Example of discrete data wrongly displayed as continuous curves.

In order to display the results, let us smooth the data to fit the following curve to them:

$$y = \frac{a}{1 + e^{bx/c}}. \tag{1}$$

where a, b and c are parameters to estimate by least-mean square.

http://dx.doi.org/10.5772/65780

The results can be seen in **Figure 2**. The GPR curve goes above the metal detector curve for depths larger than 15 cm. The raw data are correct, but the graph is not.

Each point on the curve depends on the number of targets used. If confidence intervals for the probabilities of detection were displayed on all points of the curve it would make it clear that the difference between the two curves were within the confidence intervals.

Figure 3 assumes a number of targets per depth and displays the confidence interval as a function of the hit ratio. The dashed line is for five targets and the solid line for 10 targets.

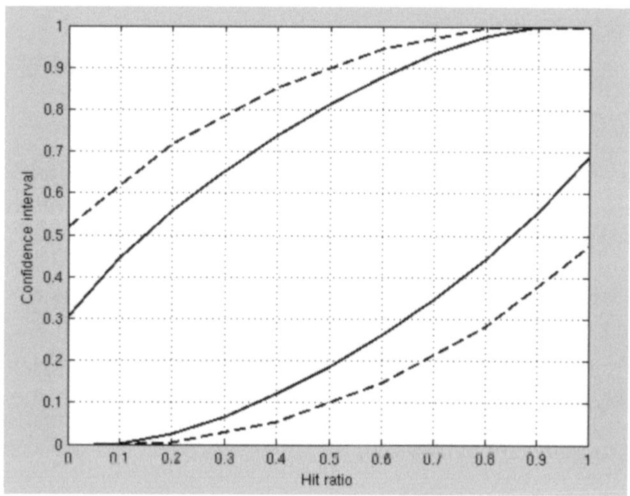

Figure 3. Confidence intervals as a function of the hit ratio for 5 targets (dashed line) and 10 targets (solid line).

For instance, if 20% of the targets are detected (hit ratio of 0.2) the confidence interval is (0.01, 0.72) for five targets and (0.03, 0.56) for 10 targets. The small difference between the results of the metal detector and the ground-penetrating radar for the depth of 20 cm (both with a hit ratio around 0.2 for 10 targets) in the previous graph is clearly not significant compare to the (0.03, 0.56) confidence interval.

Another solution would be to assume a model for the probability of detection as a function of depth and fit it to the data.

4. The case of ADE 651

4.1. Introduction

On 23 April 2013, Jim McCormick was convicted on fraud in the United Kingdom. He had sold a device called ADE 651 that, he said, could detect tiny amount of explosive from a great distance through any barrier or obstacle, but was absolutely inoperative [6].

None of the explanation given by Jim McCormick's company, ATSC or any vendor, made any sense and all tests of the device proved that it did not work better than random chance.

In 2010 the Royal Military Academy had the opportunity to evaluate the performance of a version of ADE 651 designed, according to their producers, to detect both narcotics and explosives. This test, which proved the inefficiency of the device, led to a testimony during Jim McCormick's trial in 2013. The protocol that was used could be seen as an example of an entry test [12].

The test protocol was based on some basic principles.

4.2. Testing in optimal conditions

Since there were legitimate concerns about the very principle of ADE 651 the objective of the trial was to evaluate the performance of the device in optimal conditions.

The substance to detect would be present in sufficient quantity for detection, as agreed upon with an experienced or trained operator. Prior to the trial the operator could check whether the quantity could be detected.

The location of the trial was chosen so that there were no unwanted alarms prior to the trial. No obstacles would be placed between the substance and the operator except a screen to hide the presence or absence of substance. For instance the substance to test could be hidden in an opaque bag and case.

The device should be calibrated, if needed, to be able to detect the sub-stance.

There were no time limits for detection.

4.3. Double-blind evaluation

The test protocol requires at least three persons. First, a trained operator will operate the device under test. The samples containing either the substance to detect or something that cannot be detected would be prepared by a second person, called the sample provider, who will have no contact whatsoever with the operator. A third person, called the handler, will take the samples one at a time without knowing if they contain the substance of not, take them to the agreed-upon location and after the operator has made a determination will tell it to the sample provider. Only the sample provider will know if any given sample contains the substance to detect or not and will not have any direct contact with the operator. Moreover, the operator will have contact only with the handler, who will not know the contents of the sample.

4.4. Statistical relevance

The content of each sample, the substance to detect or not, is selected trough a random process. The sample provider prepares the sample so that neither the handler not the operator can guess whether it contains the substance or not. Then the handler brings the sample to a specific location. The operator then uses the device to determine whether the sample contains the substance. This process is repeated with different randomly prepared samples.

http://dx.doi.org/10.5772/65780

When testing the ADE 651 in 2010 the Royal Military Academy used a quantity of cannabis of 7–10 g. The trials used 66 samples chosen so that each sample had a uniform probability of 0.5 to contain the substance. This random selection led to 26 samples containing cannabis and 40 containing small pieces of paper.

The results of the tests are given in **Table 4**.

		Reality		
		Cannabis	**No cannabis**	
Claims	**Drugs**	10	14	**Positive predictive value 42%**
	No drugs	16	26	**Negative predictive value 62%**
		Sensitivity 38%	**Specificity 65%**	

Table 4. Results of a double-blind test of ADE 651.

We can draw the following conclusions:

- When cannabis was present it was successfully detected 10 times out of 26 (detection rate, also called true positive rate or sensitivity, of 38%)
- When cannabis presence was claimed it was true 10 times out of 24 (positive predictive value of 42%)

Other ways to interpret the results are the following:

- False positive rate or type I error is 35%
- False negative rate or type II error is 62%

The poor sensitivity and positive predictive value obtained in optimal conditions showed no raison to consider this device for further testing. A more complete statistical analysis of these results can be found in Ref. [12].

5. Evaluating external effects on performance

The effect of external conditions on the performance of a device can be important. It is therefore paramount to take that into account during a test, either by considering several different conditions or by selecting a set of conditions and evaluating the equipment during these conditions.

An alternative way to deal with this problem is to evaluate the effect of some conditions on the performance. An easy way could be to measure proxies to this effect. The Royal Military Academy worked in this direction when it co-ordinated the work of drafting a test protocol to evaluate the effects of soils on the performance of metal detectors and ground-penetrating radars.

This work was performed during a CEN Workshop chaired by the Royal Military Academy [13]. This Workshop produced a document, T&E Protocol 14747-2, which provides some information about how soil may affect the performance of metal detectors and ground-penetrating radars and guidelines to estimate these effects and the soil characteristics that cause them [14].

It first gives guidelines that cover measurements that can be done during field operations. These are easy to follow and give a rough indication of the effect of soils on performance.

It also gives a list of characteristics to measure and methods to measure them. This includes magnetic soil properties such as susceptibility, effective relative electric permittivity, effective electrical conductivity, attenuation coefficient, characteristic impedance, the surface roughness or soil water content, characteristic of targets such as its electric size, how to document weather conditions, soil texture, how to describe vegetation, roots, rocks, surface cracks, etc.

6. Existing test protocols for mine action

The European Committee for Standardisation (CEN) has published several CEN Workshop Agreements containing test protocols for mine action. They are now called T&E Protocols and are available together with IMAS.

T&E Protocol 14747-1:2003, formerly known as CWA 14747-1:2003, provides protocols to test and evaluate metal detectors [8].

T&E Protocol 14747-2:2008, formerly known as CWA 14747-2:2008, focuses on characterising the effect of soil on the performance of metal detector and ground-penetrating radars [14].

T&E Protocol 15044:2009, formerly known as CWA 15044:2009, provides protocols to test demining machines [11].

In addition, many reports of test campaigns carried out under the ITEP programme are publicly available on the GICHD website.

7. Conclusion

Is a new technology better than current technologies? How do we define 'better' equipment? Should I purchase a given tool or another? What are the performance and limitations of a given solution? Trials are a method to provide answers to such questions. It is therefore important to clarify the questions and to design a test protocol that will permit to gather evidence to provide the answers. The importance of test protocols therefore should not be underestimated. In case poor protocols are used no reliable conclusions might be drawn and the trials might be a waste of time and resources and their conclusions disputed.

http://dx.doi.org/10.5772/65780

Test protocols are an integrant part of a test report. A test report cannot be correctly analysed and no valid conclusions may be drawn from it without an understanding of the protocol that was used.

Designing a good test protocol is not easy. This chapter gives a list of things to keep in mind, based on the long experience of the Royal Military Academy.

Fortunately many protocols, test reports and guidelines have already been published. These publications may be good starting points when designing your own protocol.

Moreover, published test results may already give you the answers you are looking for without having to carry out trials yourself.

Acknowledgements

The author of this chapter would like to thank the Belgian Ministry of Defence and the European Union.

The trial of ADE 651 was carried out with Pascal Druyts (RMA) and the support of the Police Zone of Geel-Laakdal-Meerhout.

Author details

Yann Yvinec

Address all correspondence to: yann.yvinec@rma.ac.be

Royal Military Academy, Department Communication, Information, Systems & Sensors, Brussels, Belgium

References

[1] Enright J.T., "Testing dowsing: the failure of the munich experiments", Skeptical Inquirer 23(1), January/February 1999. http://www.csicop.org/si/show/testing_dowsing_the_failure_of_the_munich_experiments, accessed on 8 September 2016

[2] Ioannidis J.P.A., "Why most published research findings are false", PLoS Med 2(8), e124, doi:10.1371/journal.pmed.00201242005, Published 30 August 2005

[3] Button K.S., Ioannidis J.P.A., Mokrysz C., Nosek B.A., Flint J., Robinson E.S.J. and Munafo M.R., "Power failure: why small sample size undermines the reliability of

neuroscience", Nature Reviews Neuroscience 14, 364–376, May 2013, doi:10.10.1038/nrn3475

[4] Johnson V.E., "Revised standards for statistical evidence", Proceedings of the National Academy of Sciences of the United States of America, 19313–19317, doi:10.1073/pnas.1313476110, https://www.researchgate.net/publication/258446201_Revised_Standards_for_Statistical_Evidence, accessed 8 September 2016

[5] Australian Research Centre for Health of Women and Babies, "Effectiveness of Homeopathy for Clinical Conditions: Evaluation of the Evidence", Overview Report Prepared for the NHMRC Homeopathy Working Committee by Optum October 2013. https://www.hri-research.org/wp-content/uploads/2016/02/Literature_review_of_public_submissions.pdf, accessed on 8 September 2016

[6] BBC, "Fake bomb detector seller James McCormick jailed", 2 May 2013, http://www.bbc.com/news/uk-22380368, accessed on 14 January 2016.

[7] Gülle D., Gaal M., Bertovic M., Müller C., Scharmach M., Pavlovic M., "South-East Europe Interim Report Field Trial Croatia—Continuation of the ITEP-Project Systematic Test and Evaluation of Metal Detectors (STEMD)" available at www.gichd.org/fileadmin/pdf/LIMA/STEMETAL DETECTOR_Interim_Croatia_final.pdf, accessed on 31 January 2013.

[8] "T&E Protocol 14747-1:2003", formerly known as CWA 14747-1:2003, available at www.mineactionstandards.org

[9] Takahashi K. and Gülle D., "ITEP evaluation of metal detectors and dual-sensor detectors", Journal of ERW and Mine Action, 14(3), pages 76-79, Centre of for International Stabilization and Recovery at James Madison University, Fall 2010

[10] "Guidelines for Reliability Tests of Dual Sensors in Humanitarian Demining", 21 June 2010, http://www.gichd.org/fileadmin/pdf/LIMA/DSTESTguidelines2010.pdf, accessed on 15 January 2016, pages 76 to 79

[11] "T&E Protocol 15044:2009", formerly known as CWA 15044:2009, available at www.mineactionstandards.org

[12] Yvinec Y. and Druyts P., "A Simple protocol for a double-blind test on an explosives/drugs long-range detector" in International Conference dedicated to Hazardous Materials: Issues of Detection and Disposal, Koscierzyna, Poland, May 2010.

[13] Druyts P., Yvinec Y., Acheroy M., "Relating soil properties to performance of metal detectors and ground penetrating radars", in Using Robots in Hazardous Environments, Elsevier, 2011, Copyright © 2011 Woodhead Publishing Limited. All rights reserved.

[14] "T&E Protocol 14747-2:2008", formerly known as CWA 14747-2:2008, available at www.mineactionstandards.org

Chapter 11

The Special Case of Sea Mines

Olga Lucia Lopera Tellez, Alexander Borghgraef and Eric Mersch

Additional information is available at the end of the chapter

http://dx.doi.org/10.5772/66994

Abstract

In this chapter, work carried out at the Royal Military Academy regarding sea mines and mine countermeasures is summarized. Three sensors used for the detection and identification of sea mines are studied here: sonar, gradiometer and infrared camera. These sensors can be applied to detect different types of sea mines. Some signal and image processing techniques developed to extract relevant information for the detection of underwater objects are presented in this chapter. These techniques are validated using data collected in the frame of different European and NATO projects.

Keywords: sea mines, mine countermeasures, sonar, gradiometer, infrared camera

1. Introduction

Land mines and sea mines have more in common than one could think. Generally speaking, land/sea mines are intended to prevent the use of or passage through an area in land or sea. Considering that maritime areas provide a vital dimension of the world's economy (90% of global trade and about half of the world's oil are transported by sea), sea mines are a complex hazard of international concern.

Defensively, military forces deploy sea mines to discourage undesired entrance into territorial waters. Offensively, sea mines are deployed to hobble a target's naval assets or channel an enemy through a designed route. Mine warfare is an old military technique and has been applied in many conflicts since the nineteenth century. Like land mines, sea mines can still be active after many years of being deployed. More than 550,000 sea mines were laid during World War II (WW-II); some of them were removed after the conflict but part of them were missed [1, 2]. The legacy of recent conflicts in the north of Africa means that even today, sea mines can be laid in European waters (and in any critical maritime space where vital resources are transported) and therefore remain a very real threat to shipping, sailors and any maritime exploitation of ocean and coastal resources.

To address this problem, governments around the world have traditionally relied on a standing mine countermeasures (MCM) force, which involves both passive and active tactics. Passive countermeasures entail changing the specific target vessel characteristics or signatures, since these trigger sea mines (see next section for more information about sea mines). These might include building vessels with fibreglass or wood instead of steel or even attempting to alter a steel vessel's magnetic field through degaussing. Alternatively, active countermeasures aim to discover mines using specially designed ships or platforms for the purpose of either avoiding or destroying them. Mine sweeping and mine hunting are two of these measures. Mine sweeping uses either a contact sweep, a wire dragged through the water by one or two ships to cut the mooring wire of floating mines, or a distance sweep that mimics a ship to detonate the mines. Mine hunting is very different from sweeping. It requires searching for all the mines in an area before disposing them. The latter is generally decomposed in four stages:

1. Detection: finding targets or contacts from signals of different sensors (e.g. acoustic, magnetic).
2. Classification: determining if the contact is mine-like or friendly object.
3. Identification: using additional information (e.g. from a diver or autonomous underwater vehicle (AUV)-based camera) to validate the classification result.
4. Disposal: neutralizing the mine.

As mines have become smarter, MCM have become more complex and sophisticated. Modern navies employ an array of different mine warfare countermeasures. The trend in mine hunting operations goes towards keeping the human operators out of the minefield. In order to achieve this, modern MCM forces use (semi)autonomous platforms (e.g., AUV, remotely operating vehicles [ROV], unmanned surface/underwater vehicles [USV/UUV]), equipped with high-resolution (HR) sensors (e.g. sonars, magnetometers, optical cameras) and employing computer-aided detection (CAD) and classification (CAC) algorithms as well as automated target recognition (ATR) processes.

Not only high-tech hardware and software are developed for mine hunting operations, but biotechnology is also applied. Just as dogs and rodents can be trained to detect land mines, maritime mammals such as dolphins are trained by the United States and Soviet navies to detect sea mines while keeping the deminer out of the minefield. Dolphins are trained to detect and mark the location of tethered mines floating off the bottom, mines on the sea floor or mines buried in sediment. They learn to use their natural echolocation bio-sonar to find a target and to report back to their handler, giving particular responses to communicate whether a target is detected. The handler then sends the dolphin to mark the location of the object by releasing a buoy. The US Navy reported that mine-clearance dolphins were deployed in the Persian Gulf during the Iraq war in 2003. The Navy said these dolphins were effective in helping to detect more than 100 sea mines [3].

This chapter reviews research carried out at the Royal Military Academy addressing the applicability of acoustic, magnetic and optical sensors used in mine hunting operations and introduce some of the challenges these techniques encounter in order to detect and classify sea mines.

http://dx.doi.org/10.5772/66994

2. Types of sea mines

Sea mines typically are explosive devices designed in different forms whose dimensions can vary from around 50 cm up to almost 3 m and can contain from 50 kg to over 1000 kg of high explosives. Aside from their dimensions and explosive charges, mines can be further categorized by their method of deployment, position in the water column and method of actuation (or detonation) [2].

Sea mines can be laid in many different ways, which depend on the size of the mine itself. Certain conventional surface vessels, e.g. war vessels and patrol vessels, can have sea mines deploying capabilities. Some smaller, unusual surface vehicles (e.g. smaller fishing boat) could drop sea mines too. They can also be deployed from air using either fixed or rotary wing aircraft or deployed underwater using the torpedo tubes of submarines.

Depending on the type, sea mines can be positioned at different altitudes in the water column. The most common type of sea mines deployed during both World War I (WW-I) and WW-II was drifting mines. They float freely on the water surface following the direction of water currents or winds. The latter have been banned by international law for over 70 years; however, some countries still use them in their defensive/offensive strategies in conflicts since 1980 [2]. Highly versatile drifting mines can be deployed by small boats, warships or aircraft. Drifting sea mines can be activated remotely or by contact.

As their name suggests, bottom sea mines stay on the seafloor utilizing negative buoyancy or are partially/totally buried at the bottom of the sea. Most of these sea mines are influence mines; a few ones can also be activated remotely or by contact. Since they rest on the bottom of the sea, bigger explosive charges (around 1000 kg) can easily fit inside them. To be able to sense vessels on the surface and reach the ship's hull floating near the surface with their explosive energy, bottom sea mines tend to be used in relatively shallow water (less than 50 m). Bottom mines are generally laid by aircraft or submarine.

In contrast with bottom or drifting sea mines, moored sea mines must maintain buoyancy to float at certain water depths. In order to achieve this, part of their structure must be designated to maintaining their buoyancy and therefore the weight and size are key factors in their design. Moored sea mines are filled with smaller amounts of explosives than bottom sea mines. Additionally, they can be tethered to allow for various altitudes in the water column. The latter are by far the most sophisticated and expensive types of sea mines. Deployed most commonly by conventional surface ships, moored mines can be contact, influence or remote-controlled mines.

The most easy-to-use mines, contact sea mines, actuate when a vessel physically comes into contact with (or very close to) the mine in the water. Contact sea mines have been fitted with many ingenious anti-sweep devices including explosive charges to cut sweeping wires and ratchet devices that enable a sweep wire to pass through the mooring cable without cutting it [2]. Contact mines are available at costs between US$2000 and US$15,000 [1].

Influence sea mines are becoming more sophisticated, complex and technically advanced. These categories of sea mines do not need physical pressure to detonate. Instead, these mines hold different kinds of sensors that can respond to different vessel signatures. The latter sense magnetic, acoustic, pressure or underwater electrical potential signals. Originally, magnetic sea mines used the vertical component of the vessel's magnetic field to activate themselves when a given field density was reached. During the 1930s, magnetic mines worked off the horizontal component of the ship's magnetic field [2]. This apparently insignificant change made it possible to design a mine fuse which responded to the rate of change of field strength rather than absolute field strength. This made defence against magnetic sea mines by degaussing and magnetic sweeping procedures far less effective. Magnetic mines are available at costs between US$10,000 and US$25,000 [1].

Due to the vessel's machinery, the design of the hull and the propellers (and many other factors), all ships and submarines have a specific acoustic signature. This principle is used in the fabrication of acoustic sea mines. Delay clocks can be included to leave the sea mine inert (and thus unsweepable) for some days after it has been deployed, while the incorporation of counters means that the sea mine will only be detonated after a certain number of impulses have been received. This means a sweeper would have to make a large number of passes over a suspected field before it could be sure that all the mines within the field had been exploded. The cost of acoustic sea mines differs dramatically depending on the capability of its system (bandwidth). They are available at costs between US$50,000 and US$150,000 [1].

Pressure mines are invariably bottom mines since they measure the absolute drop in pressure associated with the difference between the known pressure due to water depth and the depth of water under the hull of a passing ship. This differential is directly related to the depth at which the mine is situated. Pressure mines are inevitably unsweepable, are shallow-water inshore weapons and are specifically intended to destroy amphibious craft. They are located in such shallow water (even surf zone) that even minute pressure signatures would detonate them. Pressure mines have been adopted for an entire generation of small anti-invasion mines [2]. They are available at costs between US$25,000 and US$50,000 [1].

In this chapter, focus will be made on bottom and drifting mines. **Table 1** lists some of them with their characteristics and dimensions.

Mine name	Shape	Weight (in air, kg)	Dimensions (mm)	
Murena	Cylindrical	780	Length	2096
			Diameter	533
Manta	Truncated cone	220	Height	440
			Diameter	980 (max)
Rockan	Wedge	190	Length	1020
			Height:	385
			Width	80

Table 1. Some bottom type mines and their characteristics.

http://dx.doi.org/10.5772/66994

3. Sonar applied to mine hunting

3.1. Underwater sonar principles and limitations

SONAR is an acronym for SOund Navigation And Ranging, and its basic principle is to locate objects by using acoustic (or sound) waves. Sound is a series of pressure perturbations that travel as a wave and exhibit phenomena like reflection, diffraction and interference. These acoustic vibrations can be characterized by their sound velocity c, frequency f and wavelength $\lambda = c / f$ [4].

The propagation velocity of an acoustic wave is determined by the local propagation medium characteristics: the density ρ and the elasticity E,

$$c = \sqrt{\frac{E}{\rho}}. \tag{1}$$

In sea water, c varies between 1450 and 1500 m/s, depending on salinity, temperature, pressure and pH (as a remainder, acoustic wave propagation velocity in air is approximatively 340 m/s).

The frequency usable for a particular underwater sonar application is constrained by the sound wave attenuation, which increases very rapidly with frequency and limits the reachable distance (or range). In mine hunting, the frequencies of underwater sonars vary between 0.1 and 1 MHz (and their range between 1 and 0.1 km, respectively).

Underwater sonar is composed of hydrophones, which can transmit and receive acoustic waves or only receive echoes from the underwater. The first case is referred as active sonar and the latter as passive sonar. For the purpose of sea mine detection, active sonar is applied. **Figure 1** shows a scheme of a generic active sonar system. When an hydrophone transmits an

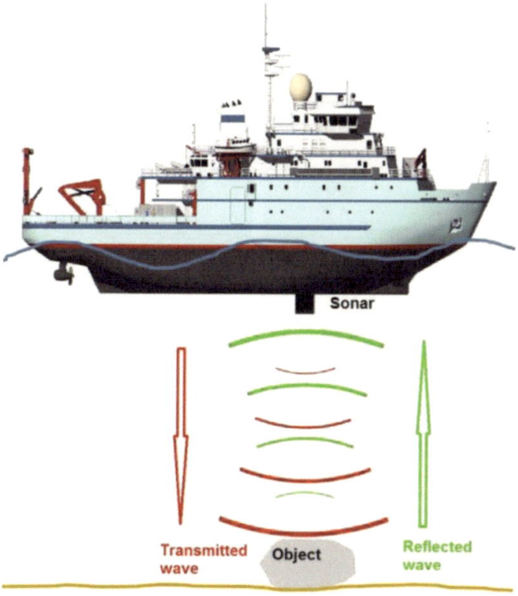

Figure 1. Schematic representation of a hull-mounted active sonar system.

acoustic wave, this propagates underwater and is reflected by any interface with a different characteristic impedance $Z = \rho c$, where ρ is the density of the interface and its unit is Rayl or kg/sm^2. The reflected wave is then recorded by an hydrophone, and the time delay τ between transmission and reception determines the distance R at which the interface is located, $R = c\tau / 2$. **Table 2** lists characteristic impedances of some materials.

The underwater propagation medium is limited by two well-defined interfaces: the sea surface and the sea bottom. Considering the reflection coefficient

$$V = \frac{Z_1 - Z_0}{Z_1 + Z_0} \tag{2}$$

and replacing the values by the impedance of sea water $Z_0 = 1.54 \times 10^6$ and of air $Z_1 = 415$ from **Table 2**, the reflection coefficient is then $V \approx -1$. Therefore, when an acoustic wave hits the sea surface (assumed flat) from below, it is reflected with no change in amplitude and a phase shift of π. A series of multiple paths are generated by unwanted reflections at these two interfaces, and the performance of a sonar system can be highly degraded by these parasite signals [4].

When acoustics waves propagate, the most evident effect is their loss of intensity because of the *geometrical spreading* and *absorption* of acoustic energy by the propagation medium itself. As the transmitted acoustic wave travels over a larger and larger surface, its intensity decreases proportionally to the inverse of the surface (geometrical spreading loss). Additionally, the acoustic pressure decreases exponentially with distance as part of the transmitted wave energy is absorbed by sea water through viscosity or chemical reactions [4].

There are other aspects associated with the propagation of acoustic waves in sea water that are not introduced here, since it is not the scope of this chapter. For more information, the interested reader should refer to Ref. [4].

3.2. Imaging the sea bottom using synthetic aperture sonar

Acoustic imaging is an inverse problem where the goal is to construct an image which represents the locations of discrete targets. This image results from the estimation of the reflectivity (or backscattered acoustic energy) for all ranges and in all directions. In sonar, it is common

Material	Characteristic acoustic impedance (in Rayl)
Air	415
Sea water	1.54×10^6
Clay	5.3×10^6
Sand	5.5×10^6
Sandstone	7.7×10^6
Granite	16×10^6
Steel	47×10^6

Table 2. Some materials and their characteristic acoustic impedance.

to have a horizontal array of receiver hydrophones. The azimuth (or along-track) resolution is given by the array length measured in wavelengths. The range resolution is given by the bandwidth of the system. The along-track resolution can be increased either by increasing the array length or by increasing the frequency. As very large arrays are impractical, the along-track resolution is usually improved by increasing the signal frequency. However, high-frequency signals suffer from attenuation and will therefore limit the application to shorter ranges. This reduces the area coverage rate and makes the sonar impractical for surveying large areas. A solution (adapted from radar) is to use synthetic aperture processing, where successive sound pulses (or pings) are coherently combined to synthesize a longer array. An excellent introduction to synthetic aperture sonar (SAS) can be found in Ref. [5].

Figure 2 shows the acoustic image (in grey scale) of an area of the Belgian continental shelf surveyed using the ROV SHADOWS. SHADOWS is a SAS which works in the range of 100–300 kHz, depending on the mode of operation, and it can achieve a resolution of 5 × 5 cm^2. It has two horizontal arrays (lateral sonars) of 36 receiver hydrophones each, one located at the port side and the other at the starboard (see **Figure 3**). The length of the arrays is 192.5 cm and that of the SHADOWS platform is 220.6 cm. In order to gather data, SHADOWS is towed at the back of a vessel (in this case, at the back of the oceanographic research vessel BELGICA, or RV BELGICA [6]). Data in **Figure 2** were collected by moving the platform following several lines (or tracks) from south-west to north-east and back at

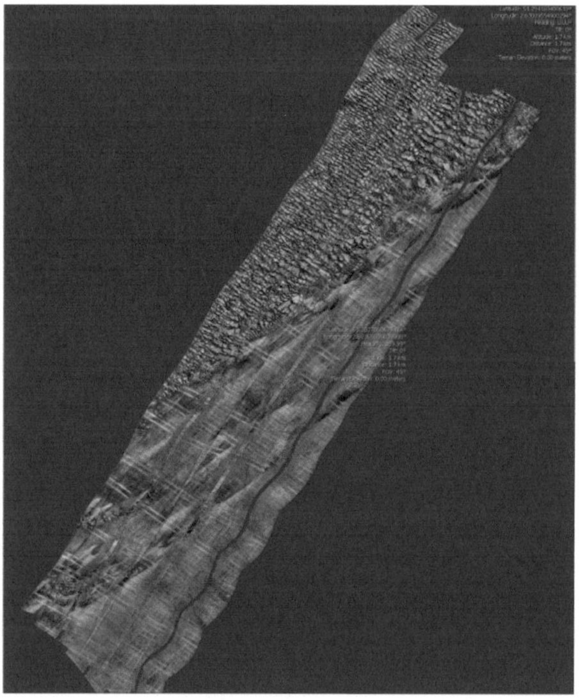

Figure 2. Image of a surveyed area at the Belgian Continental Shelf using the ROV SHADOWS SAS.

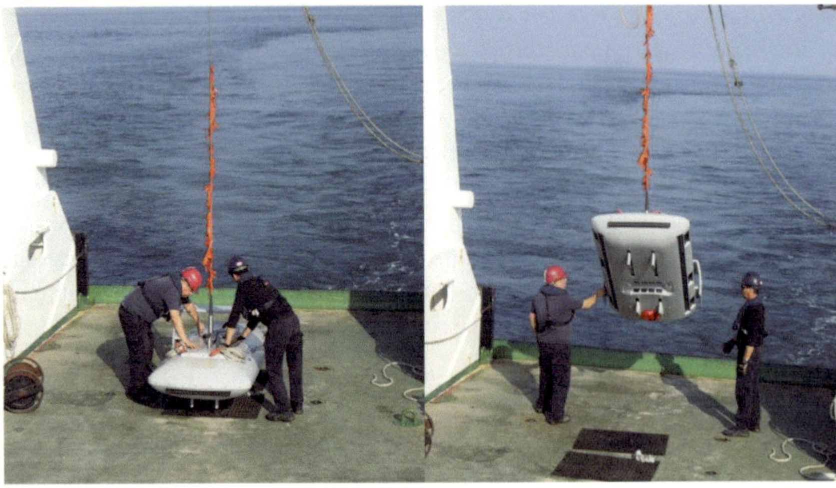

Figure 3. The ROV SHADOWS SAS and its deployment from the RV BELGICA.

an average ground speed of 2 m/s and at an altitude of about 18 m (water column height is 28–30 m). The total covered area is about 2600 × 700 m^2. In the acoustic image, two main zones with different sea bottom characteristics can be observed. In one part of the image, a rocky and rippled area appears, and it is well differentiated from a second, flatter area. The reflections of some bottom trawling lines across track are also visible. These man-made structures are created by a fishing method which involves a towed bottom fish net or trawl.

3.3. Detection and classification of targets using SAS images

Generally speaking, CAD/CAC and ATR schemas are based on the calculation and extraction of different types of image characteristics, which can mainly be grouped in three categories [7]:

1. Texture-based features, e.g. patterns and local variations of the image intensity.
2. Spectral (or radiometric) features, e.g. colour, energy.
3. Shape-based (or geometrical) features, e.g. length, area.

CAD/CAC and ATR systems developed to classify objects from real aperture or lateral sonar images are generally built on a combination of different features extracted from the spectral highlight response produced by the target and the geometrical characteristics of its shadow on the seafloor. In SAS, since the resolution is improved compared to a real aperture or lateral sonar, more complete geometrical information from the object shadow and highlight (or echo) is available, which could be exploited to improve classification results. Therefore, the approach presented here is based on geometrical features obtained from the increased image accuracy available in both object echo and shadow acoustic responses.

3.3.1. Detection

Before extracting geometrical features, two pre-processing steps are applied: (i) a speckle reduction filter is applied to improve the quality of the image and (ii) a segmentation algorithm is applied to detect echo and shadow pixels from the target. Multiplicative speckle noise is always present in SAS images, which is due to the coherent nature of scattering phenomena from rough interfaces. In Ref. [8], some speckle reduction techniques from synthetic aperture radar images are proposed and developed or adapted to be applied to SAS HR images. Those filters include anisotropic diffusion filter, adaptive neighbourhood and mean adaptive and are compared with other linear and non-linear filters, such as the Lee filter, Kuan filter and a wavelet-based methodology. A performance analysis is carried out considering image quality and better speckle suppression index. Results of the study concluded that the anisotropic diffusion filter was the best choice for this application. Some results after applying anisotropic diffusion and comparison with other filters are presented in **Figure 4**. Data were collected using the MUSCLE AUV from Centre of Maritime Research and Experimentation, NATO Science and Technology Organization (CMRE-STO) during the COLOSSUS measurement campaign along the Elba Island (Italy). MUSCLE AUV is equipped with a 300-kHz interferometric SAS, and it can achieve a resolution of 2.5 cm along track and 1.5 cm across track.

To detect the echo and shadow pixels from the despeckled image, some segmentation techniques are presented and evaluated in Ref. [8]. A rudimentary methodology for segmenting

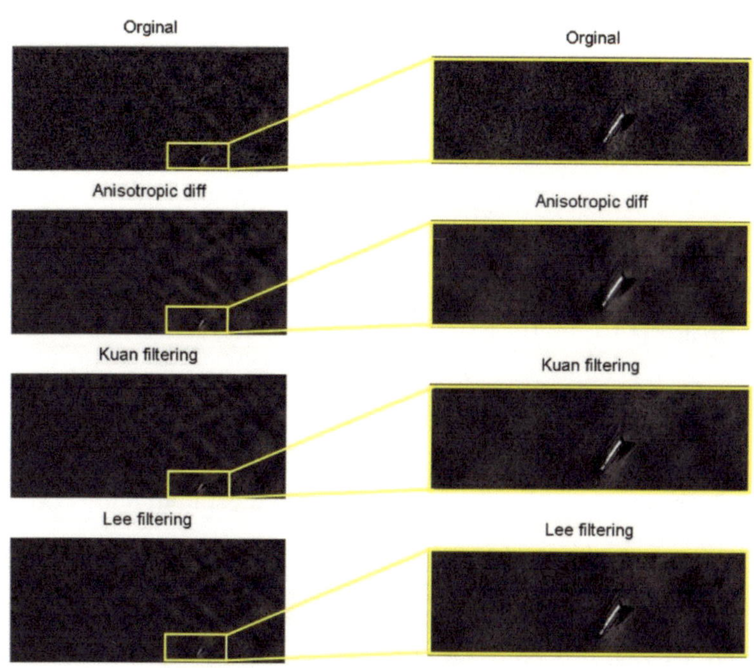

Figure 4. Results after applying different spackle reduction filters to a snapshot of MUSCLE data.

the image is to set an intensity such that all pixels below a level α become part of a mask or class (further referred to as *shadow* and *echo*), while pixels under that level do not. This basic approach has major downsides, notably (i) the selected threshold is frequently not obvious; and (ii) fixing a hard threshold will usually end in resulting masks that are imprecise; first, by the lack of continuity in the algorithm, that is, pixels at level α are part of the class, but that pixels at level $\alpha + 1$ are not, and second, that by not considering the pixel's local area, an important feature is ignored which could perhaps help in the description of the mask itself. Other techniques include Markov random fields to approach the differing grey level zones and the pixel correlations within the zones, and statistical snakes, which uses a closed curve that segments the image into an object area and the surroundings. Although these two methodologies give a good division procedure, they require iterative, computationally expensive calculations, especially when processing large datasets as it is the case here. Therefore, a segmentation method that creates a soft threshold based on fuzzy sets and takes into account the pixel's neighbourhood intensity level presented in Refs. [8, 9]. In fuzzy sets, the degree of membership μ_A to a segment A is distributed into the unit interval [0, 1], with values close to 1 implying a higher degree of membership. In this case, the degree of membership of a given pixel to one of the classes taken into account here (shadow or echo) is described by using a fuzzy set whose membership function is calculated using the pixel's intensity value i and the pixel connectivity value c. In the case of the shadow class, an adequate degree of membership is presented in Ref. [9] as

$$\mu_i = \begin{cases} 1 & \forall i \leq \alpha \\ e^{-(i-\alpha)^2/2\sigma_i^2} & \text{otherwise}' \end{cases} \tag{3}$$

where α is an original threshold, and σ_i is the standard deviation of the distribution. The variable σ relates the 'speed' at which the membership decays for $i > \alpha$. Higher values of σ show a tendency to include more pixels as part of the shadow, and lower values show a tendency to refuse more.

This membership function μ_i is integrated following the Boolean logic function AND with a membership function based on the pixel connectivity. Connectivity is computed by convolving the image describing the current shadow class (an image where pixels belonging to the current shadow having a value of 1 and 0 elsewhere) with a kernel $n \times n$. One example of this kernel can be an ones-3 × 3 matrix [1 1 1; 1 1 1; 1 1 1]. The membership function is then defined as

$$\mu_c = \begin{cases} 1 & \forall c \geq \beta \\ e^{-(c-\beta)^2/2\sigma_c^2} & \text{otherwise} \end{cases} \tag{4}$$

where c is the number of pixels in its immediate neighbourhood which are shadow pixels, σ_c is the standard deviation of the distribution and β is the maximum number of convolutions determined by the kernel.

The next processing level uses morphological filtering. The latter is described by two functions, opening and closing, which are based on two tasks, erosion and dilatation. Imaging a structure S and a kernel k, the erosion of S by k can be expressed as the group of pixels approached by the centre of k when k moves inside S. The dilatation of S by k can be defined by the group of pixels contained in k when the centre of k moves inside S. The opening of S by

http://dx.doi.org/10.5772/66994

k is calculated by the erosion of S by k followed by dilatation of the remaining form by k. The closing of S by k is calculated by the dilatation of S by k followed by erosion of the remaining form by k.

For classification purposes, synthetic data are needed in order to train the algorithms. SIGMAS (Synthetic Image Generator for Modelling Active Sonar) simulator is used to generate the training data. SIGMAS has been developed by CMRE-STO and allows the generation of SAS test sets for arbitrary target models, including mine-like objects (MLO) and friendly objects (FO). SIGMAS also includes bottom topography effects such as ripples, sea bottom slope variations and partial burial of targets [7]. Results for the segmentation approach applied to both synthetic and real data can be seen in **Figures 5** and **6**, respectively.

3.3.2. *Classification*

Three regions are considered for the feature extraction. Two of them (shadow and echo) have been already calculated following the *fuzzymorpho*-technique described before. The third selected region is the area in between the two previous classes, which corresponds to a low-backscatter area, i.e. the area in the sonar image between the shadow and the echo classes. The features are then calculated as function of some characteristics of the segmented region, such as along and across track region length, the area of the region, and the ellipsoid fitted to the region. It is important to mention that in order to benefit from these three areas, high-resolution images are needed. For images where the echo and low-backscatter areas are just a few pixels large, these characteristics could not be pertinent. If this is the case, characteristics from the shadow region are more likely to be used.

Since there is a large set of features for this application, an appropriate methodology to classify the detected target signatures could be the Markov Chain Monte Carlo approach (MCMC).

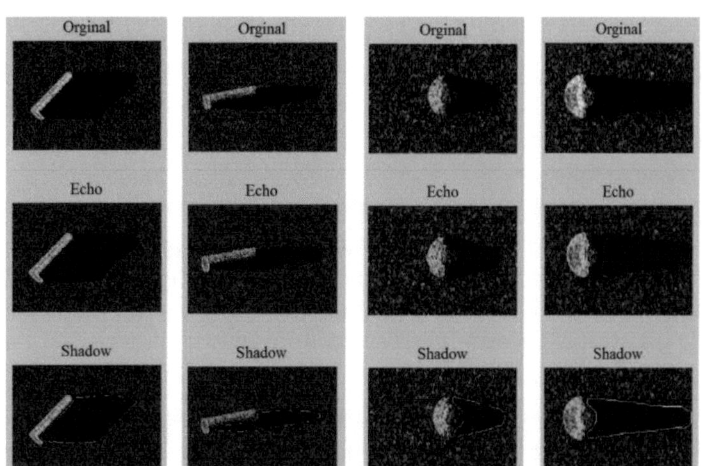

Figure 5. Segmentation of echo and shadow pixels performed on synthetic data for a cylindrical (left) and truncated cone (right) MLOs. Two angles at the same range are shown for the cylindrical MLO and two ranges are shown for the truncated cone.

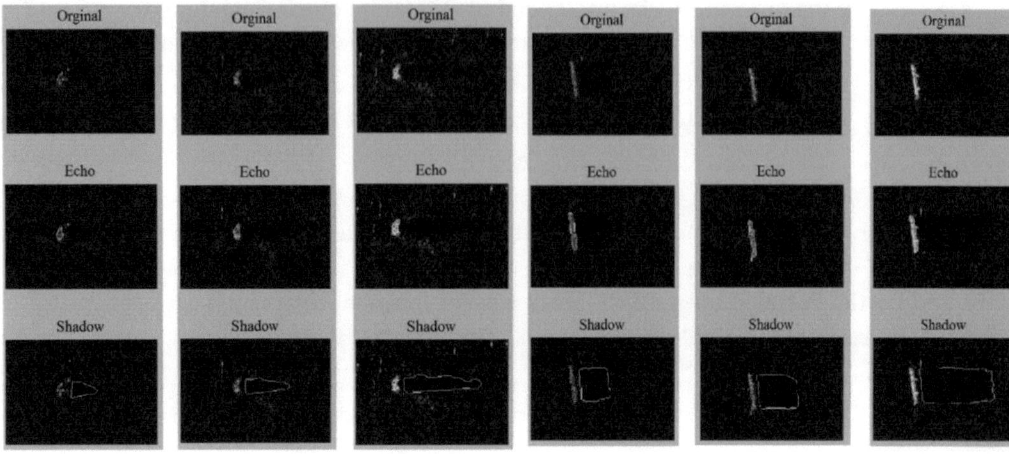

Figure 6. Segmentation of echo and shadow pixels performed on real data for a truncated cone (left) and a cylindrical (right) MLOs at different ranges.

An MCMC simulation is applied to list the parameters from the most indicative ones down to the less revealing ones, according to their discriminative power across distinct classes of data. This is realized by establishing as an importance criterion, the accuracy, obtained using the specific parameters as inputs in the classifier model.

Considering a training set T which consists of a number N of objects (regions of interest, ROIs) $O_{i,j}$, $i = \{1,\dots, N\}$ of class j, $j = \{1,\dots, M\}$. Each object is parameterized by a feature vector of size K, that is, $O_{i,j} = (f_{i,j,1}, f_{i,j,2},\dots, f_{i,j,K})$. These characteristics would be used to select suitable ways to differentiate between distinct classes. The feature vector is assumed as a succession of variables such that the next value of the succession is only determined by the previous one. Let us examine successive characteristics $fi,j,k+1$ and fi,j,k. The characteristic $k + 1$ is distributed according to $P(fi,j,k+1 \mid fi,j,k)$. Then, the observed probability vector connected to object i of class j is $pj = (pj[1],\dots, pj[K])$. These are the 'true' values for the obtained probabilities attached to objects of class j. The methodology for listing the most indicative parameters down to the less revealing ones is as follows:

1. Sample a large number, F, of observations from each observed probability vector resulting from class j. (Samples consisting of distinct characteristics, carefully chosen in accordance with their component probabilities.) Under the assumptions that (i) the feature vectors for different objects are mutually independent, and (ii) the observed feature vectors resulting from each class are well-parametrized by characteristics, the probability distribution of the sample sizes obtained from this sampling are straightforwardly integrated. The symbol $\mathrm{num}_j[k]$ is used to indicate the number of times feature fk, $k = \{1,\dots, K\}$, which is selected by the sample for class j. The probability expressed in terms of the p-parameters (probability parameters) of selecting these samples is:

$$P(\mathrm{num}_j|p) \propto \prod_{k=1,\dots,K} p_j[k]^{\mathrm{num}_j[k]}, j = 1,\dots,M \qquad (5)$$

2. The p parameters are updated (once the number of features of each size for each sample is obtained). This can be done by noting that the random variable $pj[k]$ is distributed as

$$\gamma_j[k]/\sum_{l=1,\ldots,K}\gamma_j[l], \quad (6)$$

with $\gamma_j[1],\ldots, \gamma_j[K]$ mutually independent gamma random variables with respective shape parameters, $\mathrm{num}_j[k]+\alpha_j[k]$. Stated another way, we take advantage of the fact that the Dirichlet distribution is similar to that of proportions of gamma random variables whose shape parameters match the above exponents.

3. The MCMC simulation consists in reiterate processing levels 1 and 2 explained before until convergence.

We index the features by assessing an amount of variation Var[k] for the p's at each feature k

$$\mathrm{Var}[k] = \sum_{j=1}^{M}\frac{(p_j[k]-\overline{p}[k])^2}{\mathrm{SD}^2};\ \overline{p}[k] = \frac{\sum_{j=1}^{M}p_j[k]}{M}, \quad (7)$$

where SD is a (MCMC-based) measure of the 'noise' in the posterior distribution of the p's.

In order to measure the dissimilarity of objects, a suitable distance function can be used. Although the common Euclidean distance function is well explained for a set of parameters of subjective dimension, it reveals major limitations with respect to similarity measurement. This is especially the case if the individual parameters of the feature vector are treated as independent from of each other. In some situations (mostly for spatial data), the components of the feature vector may be correlated. Considering this correlation, methodologies that apply a weighted Euclidean distance have been analyzed. It is relevant to mention that the ordinary Euclidean distance may be studied as being a special case of the weighted Euclidean distance, where the similarity matrix is the identity matrix. There is a large number of choices for these similarity matrices containing correlation coefficients between attributes.

An alternative technique based on the MCMC methodology introduced before is proposed in this section. Listing the parameters from the most indicative ones down to the less revealing ones in each MCMC simulation from until convergence results in a discriminative power weight that can be designated to each characteristic. Assuming that the MCMC simulation is reiterated *sim* times and features were indexed from the most revealing to the less ones by the number of times r_k they were chosen, the equivalent weight of feature k is $w_k = r_k/sim$. Each attribute weight thus ranges between [0…1], assigning to a feature a discriminative power. The proposed weighted Euclidean distance is:

$$\mathrm{wD}_{i,j} = \sqrt{\left(\sum_{l=1}^{k} w_l\left(p_i[l]-p_j[l]\right)^2\right)}, \quad (8)$$

and can be used as a distance measurement between the observed probability vectors p_i and p_j representing the initial objects O_i and O_j.

Figure 7 shows an example of the classification approach proposed here. We select a certain synthetic image (object) as a query one, and we retrieved the most similar objects from the snapshots selected out of the real data set. The similarity was defined with respect to the weighted Euclidean distance of the initial normalized feature vectors. In this figure, a query

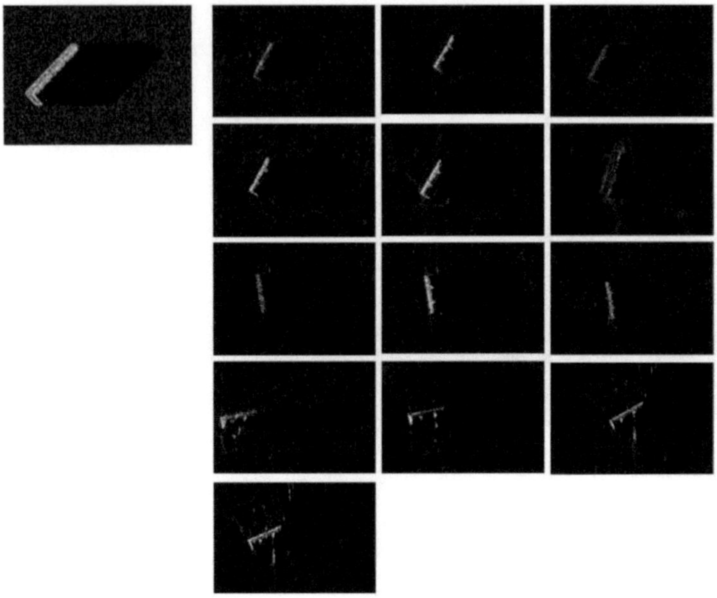

Figure 7. A query object (up, left) and the corresponding retrieved ones using the MCMC-based proposed approach.

cylindrical object is shown in the upper left corner followed by the corresponding retrieved ones from a COLOSSUS data subset. Note that the recovering system (metallic bars attached to the body of the dummy mine) can be easily recognized in the sonar images (four last cropped images in **Figure 7**). This means that the reflected signature (echo) is distinguished as well, having an influence on the classification results. Despite the fact that such objects are designated as cylinder-like, their similarity values are rather lower than the values calculated for the other targets. There is still some analysis to do regarding this effect.

It is important to note that the approach presented here does not transform the initial feature domain into a complicated set of characteristics, and therefore, it is very useful for the mine-hunting application, where both classification accuracy and indicative feature dependencies are important.

4. Magnetic gradiometry applied to mine hunting

Bottom mines can be challenging to detect since they can be buried in the sediment or in the sand. Since sea mines usually contain magnetic materials, a possible technique to detect them is to use a magnetic detector. Several kinds of magnetic detectors can be used to detect sea mines. Total field magnetometers measure the magnetic field magnitude; they can be used to detect dipole targets [10]. Three axis magnetometers measure the three magnetic field vector components and can be applied to mine hunting [11]. The same is true for tensor gradiometers which measure the gradient of the magnetic field [12]. We here consider a total-field gradiometer. It is composed of three total field magnetometers which measure the magnitude of the magnetic field with a sensitivity of 0.01 nT. The gradiometer is towed behind a ship, and its measurements are used to calculate an estimation of the gradient of the magnetic field magnitude.

A mine can be modelled as a dipolar magnetic source. In this section, a new technique is presented to localize and characterize bottom mines based on a dipolar model. Experimental results achieved in the bay of Gdansk in Poland are also presented.

4.1. The magnetic gradiometer

A sketch of the gradiometer is presented in **Figure 8**. The considered gradiometer is composed of three magnetometers: starboard, port and down, separated by distances $L_{trans} = 1.5$ m and $L_{vert} = 0.75$ m. They measure the magnitude of the magnetic field $\left|\overrightarrow{B_s^i}\right|$, $\left|\overrightarrow{B_p^i}\right|$ and $\left|\overrightarrow{B_d^i}\right|$ at N different locations $\vec{r}^i = (x^i, y^i, z^i)$, $i \in \{1, \dots N\}$. From these measurements, it is possible to compute an estimation of the gradient of the magnetic field magnitude:

$$\begin{pmatrix} G^i_{\text{trans}} \\ G^i_{\text{long}} \\ G^i_{\text{vert}} \end{pmatrix} = \begin{pmatrix} \dfrac{\left|\overrightarrow{B_s^i}\right| - \left|\overrightarrow{B_p^i}\right|}{L_{\text{trans}}} \\ \dfrac{\left|\overrightarrow{B_s^i}\right| + \left|\overrightarrow{B_p^i}\right| - \left|\overrightarrow{B_s^{i-1}}\right| - \left|\overrightarrow{B_p^{i-1}}\right|}{2\left|\vec{r}^i - \vec{r}^{i-1}\right|} \\ \dfrac{\left|\overrightarrow{B_s^i}\right| + \left|\overrightarrow{B_p^i}\right| - 2\left|\overrightarrow{B_d^i}\right|}{2\,L_{\text{vert}}} \end{pmatrix} \tag{9}$$

This is only an approximation since the components of the gradient are calculated by finite differences. This vector can be expressed in an absolute reference frame by multiplying by a rotation matrix:

$$\overrightarrow{G^i} = \begin{pmatrix} G^i_x \\ G^i_y \\ G^i_z \end{pmatrix} = \Omega^i \begin{pmatrix} G^i_{\text{trans}} \\ G^i_{\text{long}} \\ G^i_{\text{vert}} \end{pmatrix} \tag{10}$$

The aim of the inversion method is to estimate the magnetic moment $\vec{M}$ and the position $\vec{r}_t$ of the target given the measurements $\left(\vec{r}^i, \Omega^i, \left|\overrightarrow{B_s^i}\right|, \left|\overrightarrow{B_p^i}\right|, \left|\overrightarrow{B_d^i}\right|\right)$.

4.2. The inversion method

Traditionally, magnetic targets are localized by looking at the maxima of the magnitude of the gradient [10]. This is achieved by producing a map of this quantity. This method is not optimal because a single target can exhibit several maxima at low latitude. Some interpolation artefacts are also often present, leading to ambiguous interpretation. Moreover, it does not provide precise information about the burial depth or the magnetic moment, which can be used to reduce the false alarm rate [14]. We propose an alternative method for the localization of the target. Instead of producing a map, we directly use the raw data collected by the gradiometer and inverse the dipole model of the field. This allows for a real-time detection. Our method also provides the magnetic moment of the target and its depth. It works regardless of the latitude.

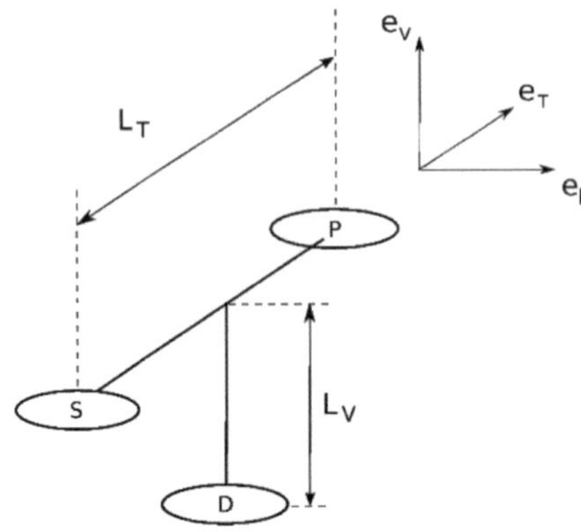

Figure 8. The considered gradiometer is composed of three magnetometers.

4.2.1. *Large distance approximation*

The gradiometer measures a finite difference approximation of the gradient. If the distance between the gradiometer and the target is large compared to the size of the gradiometer, this approximation will be satisfying. This allows one to estimate the position of the target by solving a linear system of equations. This is called the Euler method.

We assume that the magnetic field measured by the magnetometers has two contributions: the earth's magnetic field $\vec{B}_e$ and the field due to a dipole target $\vec{B}_t^i$:

$$\overrightarrow{|B^i|} = \left|\vec{B}_e + \vec{B}_t^i\right|, \tag{11}$$

where

$$\vec{B}_t^i = \frac{\mu_0}{4\pi} \frac{3\vec{R}^i\left(\vec{M}\cdot\vec{R}^i\right) - \left|\vec{R}^i\right|^2 \vec{M}}{\left|\vec{R}^i\right|^5}, \tag{12}$$

and

$$\vec{R}^i = \vec{r}^i - \vec{r}_t. \tag{13}$$

The anomaly field is defined as $A^i = \left|\vec{B}^i\right| - \left|\vec{B}_e\right|$. If the earth's magnetic field is large compared to the field of the target, it can be shown that the anomaly is given by the projection of the field due to the target on the direction of the earth's magnetic field:

$$A^i \approx \vec{B}_t^i \cdot \frac{\vec{B}_e}{\left|\vec{B}_e\right|}. \tag{14}$$

A function $f(r)$ is said to be homogeneous of degree k if $f(\lambda r) = \lambda k f(r), \forall \lambda > 0$.

A homogeneous function of degree k follows the Euler's homogeneity equation:

$$\vec{r} \cdot \vec{\nabla} f(r) = k f(r). \tag{15}$$

One can easily see that if $\left|\vec{B}_t\right| \ll \left|\vec{B}_e\right|$, the anomaly field is a homogeneous function of degree −3 of $\vec{R}^i$. The Euler's homogeneity equation takes the form

$$\vec{R}^i \cdot \vec{G}^i = -3\left(\left|\vec{B}^i\right| - \left|\vec{B}_e\right| \right). \tag{16}$$

This forms a first linear system of equations that allows to estimate the position of the source. This system can be solved by the linear least mean square method which does not require any initial parameters.

Once the position is known, one can estimate the magnetic moment by solving another linear system. Indeed, Eqs. (12) and (13) show that the anomaly is a linear function of the magnetic moment, and the magnetic gradients are also linear in M. Eq. (10) forms a second linear system of equations that can be solved by the linear least mean square method.

In practice, the hypothesis that the gradiometer measures the true gradient can be a bad approximation at small distance from the target. Estimating the parameters with the linear method leads to inaccurate results, even without noise. Therefore, a more general method is developed and presented here.

4.2.2. General case

Eqs. (11) and (12) can be injected in Eq. (10) to give a system of equations that are nonlinear in terms of $\vec{M}$ and $\vec{r}_t$.

This nonlinear system of equations can be solved by the Levenberg-Marquardt algorithm that finds the solution of the least mean square problem. This method is iterative and requires good initial parameters to avoid problems of local minima. We take the output of a linear method as initial values.

Figure 9 shows a comparison of the different localization techniques in the absence of noise on the basis of simulations. It allows one to compare the localization accuracy of the map

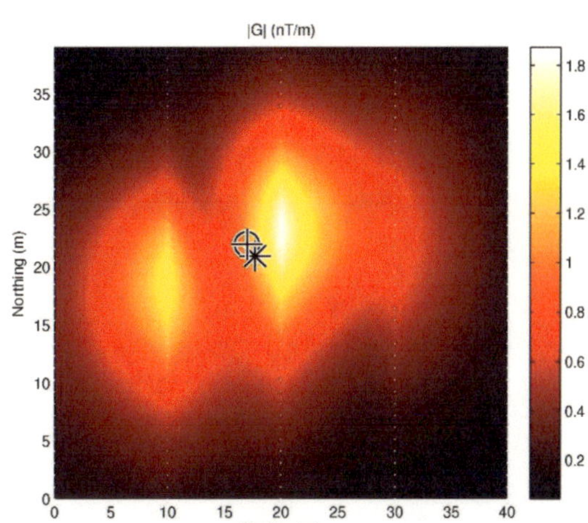

Figure 9. Comparison of the different localization methods. The map represents of the simulated magnitude of the gradient due to a dipole target. The map is produced on the basis of synthetic data taken on five tracks separated by a distance of 10 m. The circle is the true position of the dipole. The star is the position found with the linear method and the cross is the position found with the nonlinear method. The location is best estimated with the nonlinear method.

method, the Euler method and the nonlinear method. The nonlinear method provides the best results.

The effect of the noise on the precision of the estimated parameters is discussed in Ref. [15].

4.3. Experimental results

Some experiments were carried out in the bay of Gdansk, Poland, in September 2015. Two dummy bombs (T_1, T_2) and four reference targets (T_3, T_4, T_5, T_6) containing magnets of 67 Am^2 were placed at different burial depth as well a magnet of 30 Am^2 (T_7). **Figure 10** is a picture of the system during recovery from sea trials. **Figure 11** shows the signal due to the target T_4. **Figure 12** shows the positions of the targets measured acoustically after their positioning compared to the positions estimated from the gradiometer measurements together with the map of the gradient magnitude. The root mean square errors are $\Delta_x = 4.3$ m, $\Delta_y = 3.1$ m and $\Delta_z = 2.3$ m. The main limitation to the target localization accuracy is the accuracy of the gradiometer localization. The reference targets are detected several times, allowing for an estimation of their magnetic moment. The mean estimated magnetic moment magnitude of targets T_3, T_4, T_5 and T_6 is of 100 Am^2 to be compared to 67 Am^2. The smallest detectable magnet was of 30 Am^2. The main limitation to the detection of smaller targets is the noise level. Those preliminary results have to be refined but are already promising. Better gradiometer localization accuracy and smaller noise level should be reachable, and thus better target localization accuracy and detection performances should be reachable.

4.4. Partial conclusions

The original inversion method developed to retrieve the position and the magnetic moment of a dipole target from magnetic gradiometer measurements works regardless of the latitude, and it can be performed in real time. The knowledge of the horizontal position is important for the localization of the target and the knowledge of the burial depth and the magnetic moment can be used to reduce the false alarm rate. The method is tested on the basis of real

Figure 10. The gradiometer was attached to a platform with remotely controlled depth/altitude. The system was towed behind a boat. This picture was taken by TNO.

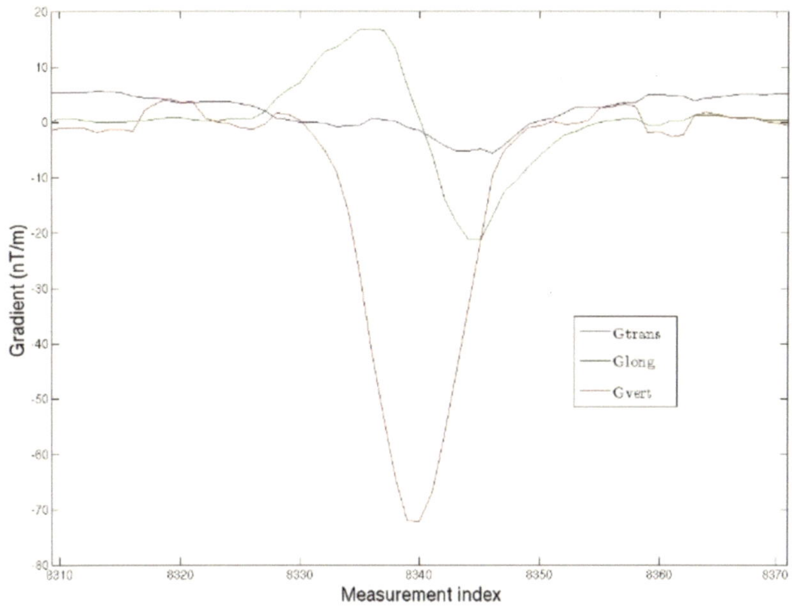

Figure 11. Signal collected by the gradiometer at the vicinity of the target T4.

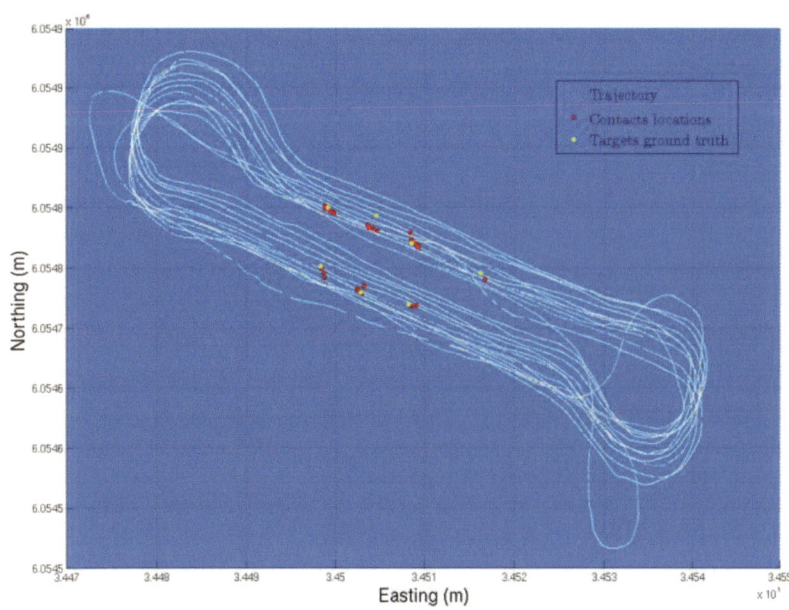

Figure 12. Targets positions (lighter dots) and position estimated on the basis of the gradiometer measurements (darker dots) together with the platform trajectory.

measurements and shows promising results. Further refinements need to be done and an analysis of the noise should allow for the computation of the target presence probability distribution and for data fusion with other sensors.

5. Behaviour-based detection of drifting mines

The threat of untethered drifting sea mines became apparent during the Gulf War of 1990–1991 between Iraq and the international coalition lead by the US. International regulations impose that floating sea mines have to be moored in place, and supplied with a self-disarming mechanism in case anchoring fails. Most nations, even those against international regulations, are inclined to follow these laws, since a buoyant mine carried by random currents is a very inefficient weapon, and could promptly leave the battle area.

Nevertheless, when the hostilities take place in more limited areas with several crafts navigating through, the damage probability rises significantly and floating sea mines can be transformed into a very effective anti-ship weapon. An example of such a region is the Persian Gulf, where a large number of sea mines drifting freely were found by the coalition forces during the Gulf War.

The efficacy of this military tactic was evident when US Navy's amphibious assault carrier Tripoli was brought to a standstill by a sea mine. After this incident, floating sea mines were classified as a principal menace to US Navy aircraft carriers. As a result, warfare was considerably decelerated and risk-reluctant shipping companies heavily diminished their number of times they transit in the conflict area.

Since then, floating sea mines have continued to show up, as recently as during the NATO intervention in Libya in 2011. A similar concern has arisen from the recent terrorist threat: floating IEDs released within a commercial harbour could cause tremendous economic damage, though no such cases have emerged until now.

5.1. Detection methods

Conventional subsurface mines are mainly detected through sonar techniques. This proved inadequate for spotting floating mines; on the one hand, due to the practical matter of downward sensor inclination and on the other hand, because of the cluttered background caused by the active sea surface. The same applies above water: the target is small and partially submerged, and the sea surface acts as a dynamic background hiding the target. In practice, the only reliable detection method so far proved to be a human visual detection and identification. Therefore, an automated detection system would provide a huge benefit in military campaigns, in clean-up operations and even for a number of civilian applications such as collision avoidance, salvaging and search and rescue.

The conditions required here are for a ship-based system to be able to operate 24 hours a day and in different weather conditions. Although the diameter of the expected targets is rather limited, it has been demonstrated that it is possible for a human operator to visually detect the target. Therefore, an infrared video imager is selected for this application.

Automated detection of the floating mine is hard, because waves on the sea surface are similar in size, shape and thermal emissivity as a partially submerged floating mine in thermal equilibrium with the water. This makes even advanced background subtraction methods such as

http://dx.doi.org/10.5772/66994

EGMM[1] and ViBE[2] [13], which are capable of dealing with significant amounts of noise in the image, unsuitable for detecting objects on the dynamic scene presented by the sea surface [14].

A better feature to discriminate between object and background is the motion behaviour of the dynamic elements in the scene. Waves propagate in a linear fashion across the scene along the wind direction, whereas the drifting mine shows a vertical oscillation and a slower drift with the subsurface current (not necessarily along the wind's vector). Both may change shape and appear/disappear, complicating matters.

Different target-based tracking techniques, e.g. mean-shift tracking [15] and particle filters [16], could be applied to create a background motion model. The latter could be useful when classifying object tracks as foreground or background. Essentially, the main disadvantages of these approaches are dealing with the large quantity and high variability of the dynamic background targets, and with the restrained target information delivered by infrared image arrays [17]. Optical flow [18] and block-matching [19] techniques have been applied with promising results. Besides, real-time performance is needed in order to make this application useful, and therefore, algorithms operating at the pixel level are preferred here.

5.2. Behaviour subtraction

The method we describe here is the behaviour subtraction method [14] first described in Ref. [20]. This is a pixel-based method in which behaviour signatures over a time window are evaluated and compared to a learned background model. It is fairly simple, computationally quick and easily extended to use multiple features. We will continue to describe this method in the following sections.

The behaviour subtraction method has at its core the idea that one should look at events in a video sequence, not at objects in individual frames. Then, the features of these events can be extracted, analysed, used to construct a background model and to detect outliers to that background model.

For this, activity in the scene needs to be detected first. The background subtraction methods mentioned earlier provide us with the necessary tools for this: they detect change in the image on a per-pixel level. Now, while state-of-the-art background subtractions such as EGMM or ViBE try to minimize unnecessary detections, we actually prefer having lots of events. This, and computational simplicity, is why typically a very simple running mean background subtraction is used.

Background subtraction converts the video stream into a stream of binary activity images, where pixels contain the value 1 if change was detected (and hence an event is occurring) or 0 if not. When we look at a set time-window's worth of these, each pixel will be characterized by a behaviour signature, a sequence of 0's and 1's describing the activity in this scene. From this, a number of simple activity features can be extracted [21, 22]:

[1]Extended Gaussian mixture model.

[2]Visual background extractor.

- The total activity in the time window, summing all of the activity bits in it and giving a general measure of the event's 'business' in that pixel.
- The number of transitions between activity and nonactivity, giving us some form of frequency information.

These two measures are ideally suited for the drifting mine problem, as both generic movement behaviour and frequency are deemed characteristic features distinguishing between the mine and the surface waves.

The next step in behaviour subtraction is the construction of a background model and a method for detecting outlier behaviour. A probabilistic approach is possible [23], but a more pragmatic and computationally simpler method is possible. For this, behaviour features are calculated for a set of training data. If we just look at one behaviour feature, say the total activity in a time window of set length, for each pixel, we can find the maximum value of the said feature. The resulting image will be called the behaviour image representing the background model.

Outlier detection then simply becomes calculating the online behaviour image at time *t*, subtracting the trained background model from it and thresholding the result, as can be seen in **Figure 13**; hence, the name behaviour subtraction.

5.3. Validation

In order to validate the performance of the algorithm on actual measurements, we used the Elbit MicroCompass MWIR camera system mounted on the P901 Castor patrol vessel of the Belgian Navy (see **Figures 14–16**). Since policing belongs to the tasks of the ship, recording capability is provided, allowing us to obtain video footage from the infrared camera.

Measurements with this system were done in the harbour where the Castor was at anchor. A small practice mine was released in the dock at a distance of approximately 250 m, and the camera was zoomed out to cover the entire width of the mouth of the dock. This resulted in the target being observed with an object size of about 6 pixels, making detection a hard (but not impossible) task for both a human observer and our detection algorithm.

The mine was released in the middle of the mouth of the dock by a RHIB and let free to be carried by the current. Sea state was approximately 1–2, so unfortunately the amount of wave activity was limited and not representative of tests at sea. Still, the tests provide us with a best-case scenario using an MWIR sensor against which future tests may be compared.

Figure 13. From left to right: IR image, activity image, behaviour subtraction.

http://dx.doi.org/10.5772/66994

Figure 14. P901 Castor patrol vessel.

Figure 15. MicroCompass sensor.

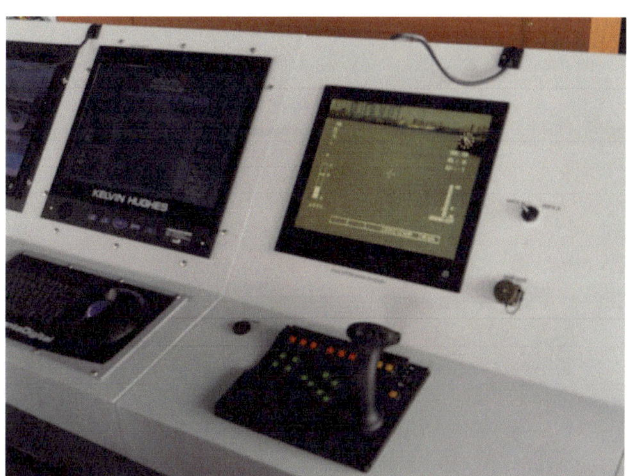

Figure 16. Operator station for the MicroCompass camera on P901 Castor.

Four sequences of video footage of the drifting mine were recorded during these tests, along with a number of sequences of swimmer and small boat activity. **Figure 17** shows a snapshot of one of those videos. Ground truth annotation was performed on the mine sequences for performance evaluation of the detection algorithm.

Video files amount to 10 minutes of footage and are encoded to MPEG-4 format using the H264 codec with a resolution of 720 × 576 pixels at 25 fps. Compression settings were not accessible, and neither were temperature ranges, but no problematic compression artefacts were observed in the data.

We applied the behaviour subtraction algorithm to these video sequences varying training sets, behaviour features, time window lengths and thresholds, and compared the detection results to the annotated ground truth position of the drifting mine, in order to obtain sets of true positive ratio (TPR) and false positive ratio (FPR) values. These values are plotted and the resulting convex hull provides us with an ROC curve for each parameter set describing the algorithm's performance. The calculated ROC curves can be seen in **Figures 18–20**.

Behaviour subtraction accumulates per-pixel activity features over a sliding time window in order to build up a 'behaviour image' of the scene, which is compared to a prior learned behaviour model in order to detect behavioural anomalies. The two main behaviour features are the activity sum, adding all the pixels in which a background subtraction has indicated that an activity is taking place, and the transition count, in which all of the changes between active and non-active during the time window W are counted.

The size or length of the time window W is one of the defining parameters of the algorithm. We chose to evaluate values of 5, 25, 50 and 100 frames, which given the 25fps frame rate of MicroCompass corresponds to lengths of 1/5, 1, 2 and 4 s. We evaluate the algorithm's performance by plotting the TPR (true positive ratio) and FPR (false positive ratio) pairs for multiple runs of the algorithm with varying thresholds in a graph, and subsequently plot the convex hull enveloping these points. This convex hull is known as the ROC (receiver operating characteristic). When we evaluate the ROC curves for behaviour subtraction using the activity sum feature, we see that the 1- and 2-second time windows outperform the other values. We also immediately see that for these values we obtain a very good ROC curve, approaching ideal performance. In a way,

Figure 17. Drifting mine in the Zeebrugge harbour as captured by the MicroCompass MWIR sensor on the P901 Castor.

http://dx.doi.org/10.5772/66994

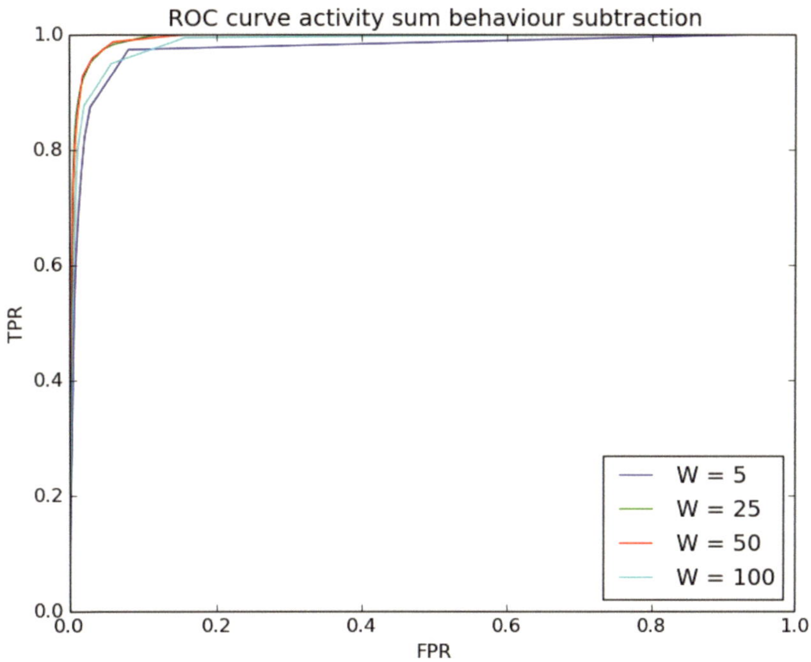

Figure 18. Comparing activity sum performance for different time window sizes.

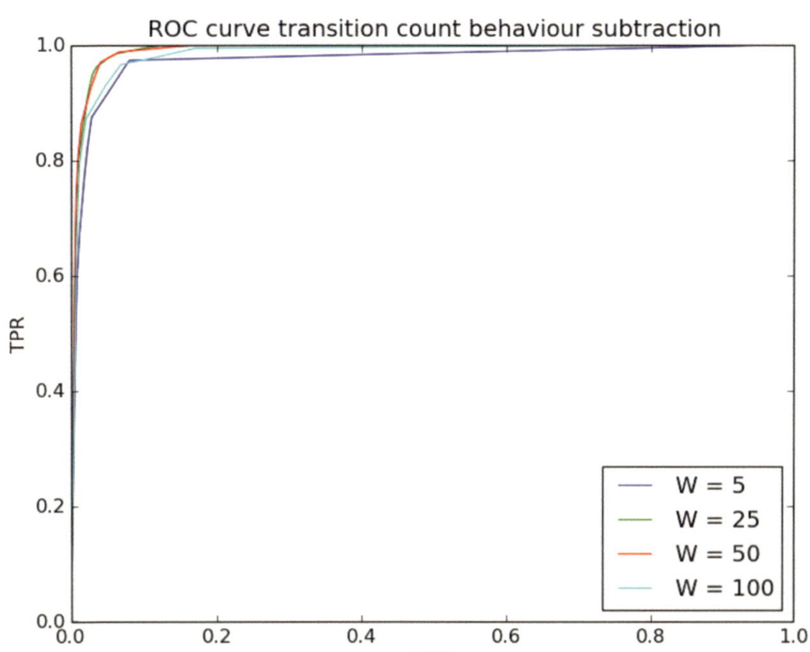

Figure 19. Comparing transition count performance for different time window sizes.

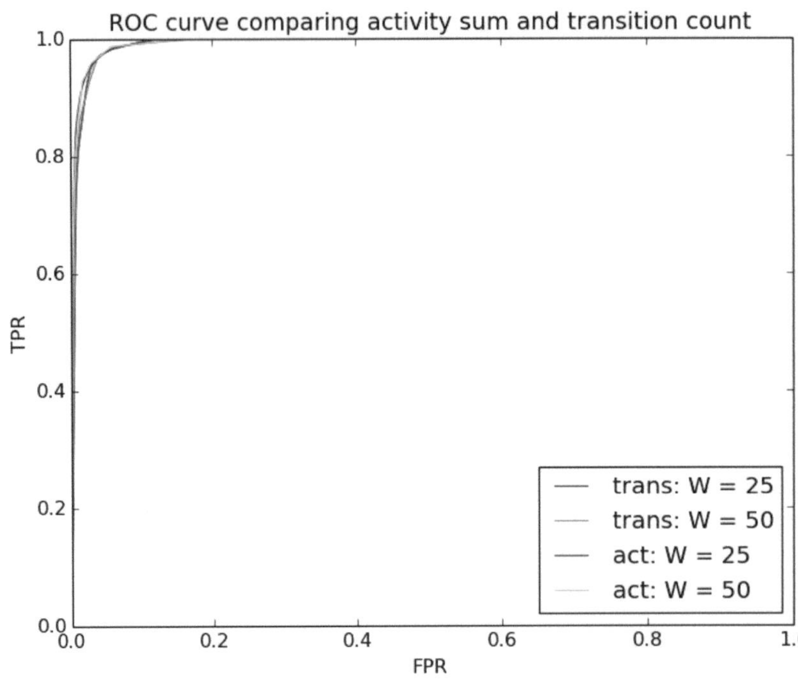

Figure 20. Comparing activity sum and transition count performances.

this was to be expected for the near-ideal conditions for a drifting mine on a sea with sea state 1–2, yet in practice, the mine was not always that easy to spot for a human observer, meaning that the algorithm remains promising.

When we evaluate the transition count feature for the same time window sizes, we obtain very similar results. Both behaviour features perform equally well, something we also see when comparing both in the same plot for the optimal window sizes of 25 and 50 frames.

These results do not allow us to make a choice as to whether the 25- or 50-frame time windows are better suited to the problem. We assume that in higher sea state conditions, the 50-frame window will perform better, but this will have to be determined with additional data.

These results demonstrate that the detection of floating mines (and similar objects) on the sea surface using passive sensors and behaviour analysis is possible. The behaviour subtraction algorithm seems to be an ideal method for the initial detection of floating targets. Further classification procedures can build upon this, with or without human operator involvement.

To implement this as an operational collision avoidance system on a moving ship, we would recommend a scanning setup, in which a stabilized camera system (thermal or optical) scans staggered sections of a designated safety zone for periods of 5–10 seconds.

In future work, we would like to see to what extent behaviour subtraction can be improved by using different or multiple activity features. Furthermore, we would like to validate the method on data obtained at greater distances and higher sea states.

http://dx.doi.org/10.5772/66994

6. Conclusion

Some techniques developed at RMA in order to detect and/or classify sea mines are summarized in this chapter. These techniques include different sensors and apply different advanced signal processing approaches. Due to the complexity of the ocean environment, these techniques have their advantages and drawbacks. Each of the methodologies presented here is focused in a particular type of sea mine, and therefore, they are complementary. Further research should be focused in the integration of these techniques in a MCM system that accomplishes an end-to-end mission from search to disposal using autonomous vehicles. Processing and decision-making capabilities should be integrated on-board of the sensor's vehicles. In order to improve mine hunting capabilities in conditions that are difficult for conventional systems, new sensing modalities should also be investigated.

Acknowledgements

The work related to the sonar is carried out at the Royal Military Academy of Belgium for the Belgian Navy Component, in the scope of research projects MRN09 and MRN17, both funded by the Belgian Ministry of Defence. Ship time on board of the RV Belgica was provided by BELSPO and RBINS–OD Nature. OD-NATURE and the RV Belgica are gratefully acknowledged for outstanding support in collecting SHADOWS data. CMRE is gratefully acknowledged for sharing MUSCLE data through the Joint Research Program MIAMS.

The work related to the gradiometer is carried out at the Royal Military Academy of Belgium for the Belgian Navy Component, in the scope of research projects MRN10 and MRN16, both funded by the Belgian Ministry of Defence. The experiments were done within the EDA UMS BURMIN project, in collaboration with TUS, ATLAS, TNO, CTM, IPHT Jena, Fraunhofer Institute and ECA.

The work related to drifting mines detection is carried out at the Royal Military Academy of Belgium for the Belgian Navy Component, in the scope of research project MRN13, funded by the Belgian Ministry of Defence. The P901 Castor patrol vessel of the Belgian Navy is gratefully acknowledged for outstanding support during the experiments.

Finally, the authors would like to thank Adm Yves Dupont for his help in improving this manuscript.

Author details

Olga Lucia Lopera Tellez*, Alexander Borghgraef and Eric Mersch

*Address all correspondence to: olopera@elec.rma.ac.be

CISS Department, Royal Military Academy, Brussels, Belgium

References

[1] Morris DA. The Mine Warfare Cycle: History, Indications, and Future. Globalsecurity.org; 1997. Available from: http://www.globalsecurity.org/military/library/report/1997/Morris.htm [Accessed: September, 2015]

[2] Rios JJ. Naval mines in the 21st century: Can NATO navies meet the challenge? [Master's thesis]. Naval Postgraduate School, 2005, pp. 12-16

[3] US Navy Maritime Mammal Program. Available from: http://www.public.navy.mil/spawar/Pacific/71500/Pages/default.aspx [Accessed: September, 2015]

[4] Lurton X. An Introduction to Underwater Acoustics–Principles and Applications. 2nd ed. Xavier Heidelberg: Lurton Springer; 2010

[5] Hansen R. In: Kolev N, editor. Introduction to Synthetic Aperture Sonar, Sonar Systems. InTech; 2011. DOI: 10.5772/23122. ISBN: 978-953-307-345-3. Available from: http://www.intechopen.com/books/sonar-systems/introduction-to-synthetic-aperture-sonar

[6] http://www.howbigisbelgica.be/en/index.php

[7] Vera J, Coiras E, Groen J, Evans B. Automatic target recognition in synthetic aperture sonar images based on geometrical feature extraction. EURASIP Journal on Advances in Signal Processing. 2009;**1**:1-10. DOI:10.1155/2009/109438

[8] Lopera O, Dupont Y. Combining despeckling and segmentation techniques to facilitate detection and identification of seafloor targets. In: IEEE eProceedings of Oceans, Santander, Spain. December, 2012. pp. 1-4

[9] Myers V. Sonar image segmentation using iteration and fuzzy logic. In Proceedings on CAD/CAC Conference, Halifax, Nova Scotia, Canada; 2001

[10] Sheinker A, Shkalim A, Salomonski N, Ginzburg B, Frumkis L, Kaplan BZ. Processing of a scalar magnetometer signal contaminated by 1/f α noise. Sensors and Actuators A: Physical. 2007;**138**(1):105-111

[11] Sheinker A, Lerner B, Salomonski N, Ginzburg B, Frumkis L, Kaplan BZ. Localization and magnetic moment estimation of a ferromagnetic target by simulated annealing. Measurement Science and Technology. 2007;**18**(11):3451

[12] Wynn WM. Detection, localization, and characterization of static magnetic dipole sources. Journal of Detection and identification of visually obscured targets. 1999;**1**:337-374

[13] Li X. Understanding 3D analytic signal amplitude. Geophysics. 2006;**71**(2):L13

[14] Yvinec Y, Druyts P, Dupont Y. Detection and classification of underwater targets by magnetic gradiometry. In: Proceedings of International Conference on Underwater Remote Sensing, Brest, France. I. Quidu et al. (ed) 2012

[15] Mersch E, Yvinec Y, Dupont Y, Neyt X, Druyts P. Underwater magnetic target localization and characterization using a three-axis gradiometer. In: Proceedings of OCEANS'14 conference, Taipei, Taiwan; IEEE; 2014

[16] Barnich O, Van Droogenbroeck M. ViBE: A powerful random technique to estimate the background in video sequences. In: IEEE International Conference on Acoustics, Speech and Signal Processing, 2009. ICASSP 2009. Taipei, Taiwan; IEEE; 2009. pp. 945-948

[17] Borghgraef A, Lapierre F, Barnich O, Vandroogenbroeck M, Philips W, Acheroy M. An evaluation of pixel-based methods for the detection of floating objects on the sea surface. EURASIP Journal on Advances in Signal Processing. 2010;**2010**:11

[18] Comaniciu D, Ramesh V, Meer P. Kernel-based object tracking. IEEE Transactions on Pattern Analysis and Machine Intelligence. 2003;**25**:564-577

[19] Khan Z, Balch T, Dellaert F. An MCMC-based particle filter for tracking multiple interacting targets. In: European Conference on Computer Vision; T. Padja and J. Matas (ed); 2004, vol. IV. pp. 279-290

[20] Cheng J, Yang J. Real-time infrared object tracking based on mean shift. Progress in Pattern Recognition, Image Analysis and Applications. 2004;**3287**:45-52

[21] Borghgraef A, Lapierre FD, Acheroy M. Motion segmentation for tracking small floating targets in IR Video. In: 3rd International Target and Background Modeling and Simulation (ITBMS) Workshop. 3rd ed., Toulouse, France; 2007

[22] Wei Z, Lee D, Jilk D, Schoenberger R. Motion projection for floating object detection. Advances in Visual Computing. 2007;**II**:152-161

[23] Jodoin P-M, Saligrama V, Konrad J. Behavior subtraction. IEEE Transactions on Image Processing: A Publication of the IEEE Signal Processing Society. 2012;**21**(9):4244-4255